Complete
Systems
Analysis

Complete Systems Analysis

James & Suzanne Robertson

foreword by Tom DeMarco

Dorset House Publishing
353 West 12th Street
New York, New York 10014

Library of Congress Cataloging in Publication Data

Robertson, James.
    Complete systems analysis : the workbook, the textbook, the
answers / by James Robertson, Suzanne Robertson.
      p. cm.
    Includes bibliographical references and index.
    1. System design--Programmed instruction. 2. System analysis-
-Programmed instruction. 3. Computer-aided software engineering-
-Programmed instruction.   I. Robertson, Suzanne. II. Title.
QA76.9.S88R633 1994   93-44616
004.2'1--dc20   CIP
ISBN 0-932633-25-0
ISBN 0-932633-50-1 (pbk.)

Cover and Interior Page Designs, Hardcover Edition: Ellen Moorcraft
(Designs Revised for the Softcover Edition by David McClintock)

Copyright © 1998, 1994 by James and Suzanne Robertson. Published by Dorset House
Publishing, 353 West 12th Street, New York, New York 10014.

Distributed in the English language in Singapore, the Philippines, and Southeast
Asia by Toppan Co., Ltd., Singapore; in the English language in India, Bangladesh,
Sri Lanka, Nepal, and Mauritius by Prism Books Pvt., Ltd., Bangalore, India; and
in the English language in Japan by Toppan Co., Ltd., Tokyo, Japan.

Printed in United States of America

Library of Congress Catalog Number 93-44616

ISBN: 0-932633-50-1   12 11 10 9 8 7 6 5 4 3 2

For Reginald and Helen

CONTENTS

ACKNOWLEDGMENTS

We have always disliked the acknowledgments sections of books. The authors present an incredibly humble face to the reader, which, we assume, is to make the reader think that the authors are humble, and therefore likable people. We also assume that this display of humility is intended to make the reader like the book all the more.

We now know better.

Authors are humble because writing a book is an incredibly humbling experience. Once written down, thoughts that are entirely clear and lucid appear awkward and ambiguous. Explanations that work in normal conversation become completely inappropriate when set in type. Diagrams that we are convinced are models of ingenuity and perfection are exposed as faulty and somewhat pedestrian.

What this all means is that without a lot of help, authors, or at least the authors of this book, would never publish anything readable. We therefore make these (humble) acknowledgments to the people who have helped us to bring you this book.

We acknowledge the inspiration over the years, and help with this book, given to us by our fellow principals at the Atlantic Systems Guild: Tom DeMarco, Tim Lister, Steve McMenamin, and John Palmer. They all have provided us with many of the insights we present in this book.

We should also mention the major influences of DeMarco's *Structured Analysis and System Specification,* McMenamin and Palmer's *Essential Systems Analysis,* and Matt Flavin's *Fundamental Concepts of Information Modeling.*

Drafts of this book have been reviewed and corrected by many people. We acknowledge the invaluable help given by the team working in Rich Cohen's CompuServe forum: Darrin Chandler, Thor Christensen, James Curran, Michael Diehl, Harry Holt, Jim Hudson, Bob Koss, Jim Langendoen, Lucy Lockwood, Rud Merriam, Tom Ochs, Sue Petersen (who suffered more than most to bring you this book), Russ Ranshaw, Jeff Schweiger, Bryan de Silva, Mark Washick, and Mark Weisz. Special thanks to Rich Cohen for his Herculean efforts in keeping the group focused.

We also acknowledge contributions from our colleagues Truitt Allen, Gary Austin, Therese Lange, Gary Schuldt, Andy Smith, and Nina and Lee Snett. Their suggestions have made this a better book.

This book has seen the insides of upwards of a dozen Macintosh® computers, starting with the first drafts on early Macintosh Plus machines, and ending with the typesetting on a Quadra. The text was written using Microsoft Word™ and most diagrams were drawn with Aldus Freehand™. The Piccadilly Project models were built using TurboCASE™. We gratefully acknowledge the help given us by Shang-cheng Chyou of StructSoft. His words of advice saved us many hours of work.

The original project at Associated Television provided us with the wonderful experience of working with some very talented people. We particularly want to acknowledge the work of Kenda Harris and Wendy Wakley (who is even better at her job than her Stepney Green counterpart in this book).

The design of this book and the cover is the work of Ellen Moorcraft. We think her work speaks for itself. Paula Gair helped with the preparation of the manuscript by organizing and checking its hundreds of components. Meanwhile Ian Gair kept our consulting business running so smoothly we didn't mind taking the necessary time off to write this book. We wish to thank many of our clients for their encouragement during the book's long gestation.

Finally, the publisher. No book would ever see the light of day without some amazing work by the publisher. David McClintock, Janice Wormington, and Tony Yip of Dorset House patiently guided us through the process, made our English readable, and, most importantly, provided encouragement and guidance during the dark hours of despair when it seemed that we would never finish.

To all of the above, a very grateful, and humble, thank you.

FOREWORD

The past decade has been one of striking change for the discipline of systems analysis. As recently as the mid-1980s, analysts everywhere were still inclined to dictate to their users what the new system would be like. They typically passed down their pronouncements in the form of a written specification. Today, that approach will no longer fly anywhere. Today, we analysts find ourselves serving not so much as inventors of the new system but as catalysts for the invention process. Instead of text, we rely on models of all kinds: function models, data models, object schemas, state models, prototypes, GUI frameworks, and the like.

At the key interface between analyst and user, the dialogue has changed, too. Before, we tended to say, "Here's what you're getting." Now, we show one of the models and ask, "How about something like this?" Once a model is on the table, the process comes alive. The users get their hands on the model and start to reshape and remold. "Close," they say. "If only we could change it, though, to something like this . . ." When that begins to happen, I know the project is on track.

When we first show users a model, we know it is, at best, close to what is required. We show it specifically to elicit change. As such, the model isn't supposed to be an exact replica of the system, but rather an example of the kind of system we'll probably be building. Modern analysis is, accordingly, a "by-example" discipline. We show examples at each and every stage of the process.

In *Complete Systems Analysis,* Suzanne and James Robertson have hit upon the charming idea of guiding you through this by-example discipline with a by-example presentation. From the very first page, an extraordinary and wonderful page, you will know this book is fundamentally different from any other analysis texts you may encounter. It doesn't lecture at you, it doesn't take up your time telling you anything you already knew. It's a book that you don't exactly read at all; instead, you sort of ski through it, along a path of your own choice. (Well, they'll explain all about that.)

When my own analysis book was still in manuscript, the publisher sent it out for evaluation to a number of referees. Among the comments that came back to me from this process was the following one: "I guess you can't really like this book unless you're willing to like the author." Since that particular referee had never shown herself to be "willing to like the author," I knew the comment was supposed to be a criticism. But I felt just the opposite. I felt it was the nicest possible compliment. After all, I wasn't trying to hide behind the page, to make myself anonymous. The books I had most admired in our field (Fred Brooks's *The Mythical Man-Month,* for example, or virtually any of Jerry Weinberg's works) were personal and personally revealing and more meaningful because of it. That was the kind of book I had been trying to write.

Complete Systems Analysis is similarly personal. You'll come to know its authors, Suzanne and James Robertson, as you read through their work. I predict you'll find them (like the book itself) to be honest and on-target and funny and inventive and curmudgeonly and wise . . . but, most of all, honest. There is something of their own lives and professional experience on every page. The work they share with you is a real one, based on their extensive involvement in automation within the British television industry, as well as numerous other organizations. The lessons they drew from that project at the time and redraw for you now are not simplistic—there are no easy answers in systems analysis—but they are useful lessons. They help you to discover some of what is the most useful, more or less the way they discovered it themselves. You will probably make a few of the same mistakes they made in the discovery process, and if you can profit from making those mistakes in the safe environment of these pages instead of in your own work, the benefits should be obvious.

October 1993
Camden, Maine

Tom DeMarco

SECTION 1

The Project

YOUR PROJECT STARTS HERE **1.1**

Your Client

Nestled amongst the soft green hills of the English Midlands is the market town of Nuffield-on-the-Moor. A good place to start exploring the town is from the river. Cross Upminster Bridge and walk through Stonebridge Park until you come to the Elephant and Castle public house. Turn left here and you'll find yourself in a large cobbled square. This is the market square, and, today being Wednesday, it is crowded with farmers, artisans, housewives, and children from the surrounding districts. Today is market day.

Figure 1.1.1: The market town of Nuffield-on-the-Moor.

Walk around the market and sample some of the regional products. See the homemade, unpasteurized green cheese, taste the fresh pepper pickles, buy a dozen smoked quail eggs, try a pint of freshly brewed malt ale, or buy a hand-thrown Nuffield pot as a souvenir. When you have eaten and drunk your fill, look around the charming Norman church in the southeast corner of the square. The church warden will lend you the key to the bell tower. Now climb the one hundred and forty-two steps to the top of the tower. You may get out of breath, but it's worth the effort. Spread below you are some of the richest farmlands and prettiest villages in England.

However, green fields are not all you can see. In the distance, the factory chimneys, cooling towers, motorways, and all the other clutter of industrial complexes and large towns bring you back to the twentieth century. Now turn your attention to the west. Perched at the top of Nuffield hill, you can see the Piccadilly Television transmission tower.

Piccadilly Television holds the franchise for this part of the Midlands of England. Nuffield-on-the-Moor is located at the geographic center of Piccadilly's franchise area, so by locating the transmission tower here, Piccadilly ensures that all the households in the area get good television reception.

A franchise entitles the holder to be the sole transmitter of television programmes and commercials within a defined geographical area. While a franchise holder has a monopoly in one region, most products are advertised nationally. A commercial television company is therefore competing with companies in other regions for a share of an advertiser's national budget. This is a very competitive business, and that is the reason you're here in Nuffield-on-the-Moor: Piccadilly is about to launch a project with the objective of building a new computer system to help get more of the advertisers' money. The new system must be the best in the industry to give it an edge on the other commercial television franchise holders. Piccadilly management has decided that the best way to take maximum advantage of the latest technology is to have the project team study most of Piccadilly's operations. The final decision on what is to be computerized will be made when the analysis is complete.

You are the chief systems analyst on the project. To help you get started, Piccadilly has provided some background material on how the British television industry works. Read through it, then we'll discuss ways to tackle the project.

Introducing the British Television Industry

The Broadcasting Board has the authority to issue an eight-year franchise to a commercial television company. As the franchise holder is the only commercial broadcaster, the Board imposes strict conditions on programming standards. There is a defined balance between drama, comedy, children's programmes, sports, and other types of entertainment. There are also rules about what sort of programmes can be transmitted at certain times, and more rules about the content of programmes and commercials. These rules are taken very seriously. Any franchise holder who does not abide by the rules is in danger of losing the franchise and hence the whole of the business. This may sound pretty tough, until you know what a franchise does for the holder.

A franchise means that a television company is the only supplier of broadcast commercial airtime within its area. If an advertising agency wants to reach an audience in the Midlands, Piccadilly is the only source of supply. There are cable and satellite stations active in the same area, but broadcast television commands the lion's share of the audience. Commercial airtime is expensive, and while having a franchise is often likened to a license to print money, the franchise holder's rates must be competitive to attract its share of the national advertising budgets.

The success of a commercial television company depends upon its ability to sell advertising. Before spending money with Piccadilly, advertisers must believe that people will watch Piccadilly Television's programmes and the commercials broadcast along with them. Selling advertising, then, is all about convincing advertisers that enough of the people likely to buy their product will watch Piccadilly's programmes.

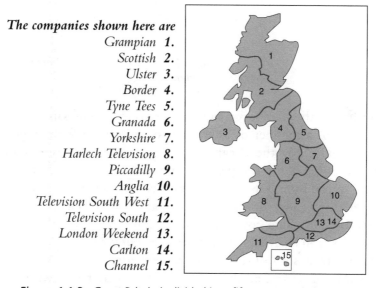

The companies shown here are

Grampian **1.**
Scottish **2.**
Ulster **3.**
Border **4.**
Tyne Tees **5.**
Granada **6.**
Yorkshire **7.**
Harlech Television **8.**
Piccadilly **9.**
Anglia **10.**
Television South West **11.**
Television South **12.**
London Weekend **13.**
Carlton **14.**
Channel **15.**

Figure 1.1.2: Great Britain is divided into fifteen transmission areas. The franchise for each area is held by one commercial television franchise holder.

Here's how it works. Let's say that an advertising agency is running a television campaign for a product targeted at homemakers. Remember that each of the fifteen television companies is restricted to a specified transmission area. If the advertising agency wants to reach householders in the south of England, the agency must spread the budget over the five television companies operating in that part of England. These companies are competing for a share of the agency's advertising budget. The one that can offer the largest numbers of householders watching its programmes and that can sell time at the most attractive rates will get the biggest share of the budget.

Audience measurement bureaus track the number of viewers of each of the television channels, with a combination of questionnaires, surveys, and electronic monitoring equipment. These audience numbers, or ratings, are analyzed by programme type, audience type, time of day, television company, and any other breakdown that makes the ratings salable to the television companies. Every week, the bureaus provide the ratings to the television sales executives who use the information as ammunition for selling airtime to the advertising agencies.

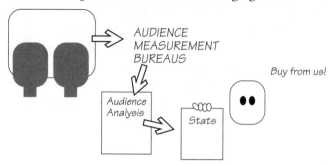

Figure 1.1.3: Audience measurement statistics are an important aid in selling airtime.

But just having the numbers is not all there is to it. The same advertising time slot can be sold for a number of different rates and, naturally enough, the advertising agency wants to pay the cheapest one. However, it doesn't always pay to be a cheapskate. Some of the cheapest rates are sold on the basis that another buyer who is willing to pay more for the time can preempt the first buyer. This results in the first buyer's losing advertising time that might be a key element in a campaign. The rate structure for selling commercial television time is complex, and discovering all its intricacies is an analysis treat that lies ahead of you.

There are all sorts of rules about when certain advertisements may or may not be shown. For instance, alcohol advertisements may be shown only after 9 p.m. If an actor is in a programme, a commercial containing the same actor may not appear within the forty-five minutes preceding or following the programme. If an advertisement for floor cleaner is broadcast, then no other floor cleaner advertisements may appear within the same commercial break. As you work on the project, you will come across other rules like this. Keep in mind that the Broadcasting Board can, and probably will, change any of these rules at any time.

How to Do Your Project

You are about to start an exciting project for Piccadilly Television. Your task is to analyze the requirements for a new system, whose principal activity is selling commercial television airtime.

This project is based on a real analysis that we did for one of the British television companies. (For that reason, we are using the British spelling of "programme" to refer to anything relating to Piccadilly Television programmes.) We have condensed the most interesting bits of our project into the case study, so that you can get the maximum practice in a reasonable amount of time. In the original analysis, we used process models and data models that we built from both the physical and essential viewpoints. Don't worry if you don't know these terms: As you work through this project, you'll learn about these models, and you'll use them to build the specification for Piccadilly's airtime sales system. Also, don't be concerned if you don't know anything about the television industry. It will be progressively introduced to you as you work through the project.

You are here to learn systems analysis and/or to improve your analysis skills. Once you finish the Piccadilly Project and the practice exercises along the way, you will have enough hands-on experience to be able to apply these analysis techniques to your own projects.

How You and the Project Come Together

In the next few paragraphs, we'll explain how you'll do the systems analysis for Piccadilly Television. For now, ignore all unfamiliar terms, and keep on reading. We'll give you a complete explanation of them before long.

The Project Section and the Textbook Section of this book teach the modeling techniques you will be using. We will guide you on a trail between the Project and Textbook chapters as the need arises. Later we'll tell you more about the structure of the book and how the trails work, but now let's concentrate on how you will do the Piccadilly Project.

At first, since you will be unfamiliar with the Piccadilly organization, you will build models that will be a faithful reproduction of the current business system. We refer to these models as having a physical viewpoint, and you will take the appropriate trail through the book to read about viewpoints and physical models before you have to build one for Piccadilly. You'll start the analysis by defining the boundaries, and by developing a context diagram.

Analyzing the stored data of a system helps you get a better understanding of the system. That's why early in the Project, you'll build a data model. As before, if you are unfamiliar with this type of model, we shall guide you through the data modeling chapter in the Textbook Section, where you can work on some practice exercises before tackling the Piccadilly data model.

After the data model, you'll need to begin the data dictionary, and then expand your context diagram by building some lower-level physical models. Once you have done these, naturally with the help of the Textbook (if you want it), the Project shifts up a gear and you will start to look at the essential requirements.

The essential requirements are critical to your analysis. The Textbook provides chapters on the essential viewpoint and on event-response models, which are used in the Piccadilly Project to determine the essential, or real, requirements for the television company.

Your next assignment will be to define the essential processes using mini specifications. You will write some for Piccadilly. If you are unfamiliar with developing and using this type of specification, a Textbook chapter will tell you how.

The Project then enters a stage where you will consolidate all of the work you have done to date, and flesh out the analysis by building more event-response models. By now, you will have a good enough understanding of the analysis process to proceed without help from the Textbook—that is, until you come to defining the new requirements.

At this point, you may need help from the Textbook before modeling the additional functions and data that Piccadilly needs to complete the new system. Once the requirements are complete, you will move on to look at how they might be implemented. Here you will use the new physical viewpoint to model your proposals for the computers and human organizations that can successfully carry out the requirements you have gathered during the systems analysis.

This is a long adventure in systems analysis, but we know that by the time you reach the end of the adventure, you will have a complete and practical knowledge of the art and craft of systems analysis.

Now let's see how you can get the best value out of this book.

How to Make This Book Work for You

This book is a self-discovery learning tool. It contains a complete analysis project and a state-of-the-art textbook. You can make use of either, or both. Here's how.

The book is divided into four sections. Each section is relatively self-contained in that it deals with a separate aspect of learning systems analysis. The sections are not intended to be read sequentially. You will read and work through each section in the order that is appropriate for your level of knowledge and skill. We will provide you with guidance and an appropriate trail to follow.

Section 1 contains the analysis project that you will work through. Each of the eighteen chapters in this section adds to your knowledge about the business to be analyzed and asks you to build various types of requirements models, to make some strategic decisions, and to raise questions about the business. In other words, the Project Section simulates the task of systems analysis.

We don't know your exact level of systems analysis experience, so you will want to consult the Textbook in Section 2 as you need to while you go through the Project chapters. Rather than intermingling the text and the case study, we've presented the Project and the Textbook in separate sections to let you decide how much, and when, you want to make use of each of them.

The Textbook is an up-to-date treatise on systems analysis. Even after you have finished the Piccadilly Project, you will want to refer to the Textbook from time to time. Having it as a separate section makes it more convenient for ad hoc referencing and reading.

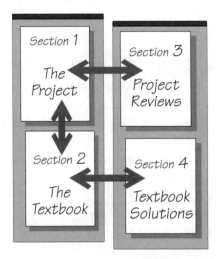

Figure 1.1.4: *This book has four sections. Section 1 contains a complete analysis project. You will learn systems analysis by doing the project. Naturally, we give you sample answers to the project exercises we assign; you'll find these in Section 3, along with a discussion of our solutions. Along the way, you will use Section 2, the Textbook, which provides coaching on how to build the models and viewpoints used in the project. The Textbook contains exercises to practice your skills, and the answers are provided in Section 4.*

For example, before you can build the requirements models for the Piccadilly Project, you have to know the modeling language. The Textbook contains tutorial material on how to build data flow diagrams, entity-relationship diagrams, event-response models, the data dictionary, and all the other analysis models. Before you can build these models, however, you will need something more. Analysis models do not show the entire system, but rather focus on one particular aspect at a time. We call this focus a viewpoint. You will use viewpoints to emphasize the information that is necessary at the time, which makes the analysis of complex systems much easier. So the Textbook includes a discourse on effective viewpoints, describes ways of modeling them, and discusses when each viewpoint is useful.

We strongly believe that when a book assigns exercises, or asks questions, it should provide the answers. As you work through the Project exercises, you will need to refer to Section 3, where we provide a suggested solution to each problem, along with a discussion of how we came up with our answer. We think you'll find the discussion almost as educational as doing the Project.

The Textbook section also introduces short exercises to build proficiency using a model. We suggest you complete these exercises before using each model in the Project. Naturally, there are answers and discussions for the exercises. You will find them in Section 4.

This arrangement means that you will be jumping back and forth between sections: from reading about the Project to the Textbook; reading the Textbook and doing the exercises; jumping to the Textbook Solutions, back to the Project; doing the Project assignment; studying the Project Reviews; turning to the Project and back to the Textbook again. The precise route you'll take depends on your level of experience, and what use you want to make of this book. Navigation through the chapters may seem difficult at first, so we have introduced Trail Guides to assist you in finding your way. The Trail Guides are explained below.

How to Work Your Way Through This Book

Millions of people in the world today enjoy skiing. Almost all of them can ski in complete safety because the ski resorts mark their trails with symbols to indicate the degree of difficulty of the terrain. Thus, skiers can ski on trails most appropriate to their skiing ability. Alternatively, adventurous skiers can find more excitement by selecting more difficult trails. The trail guides used at ski resorts look something like this:

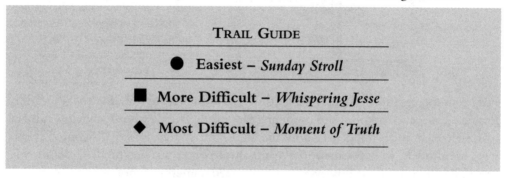

Figure 1.1.5: These ski trail symbols are used to advise skiers of the difficulty of alternative trails. The skier uses the trail guide to select the most suitable way down the mountain. You will use the same kind of trail guide as you work through the Project.

Ski trail guides work so well that we borrowed the idea. At the end of each chapter, you'll find a Trail Guide that points you to various chapters. There are four different trails through this book. Select the one that suits both the way you want to work and your current level of experience. Read through these descriptions to find the right trail for you.

● Easiest Trail

Follow the Easiest Trail if you have little or no experience with systems analysis or if your exposure to analysis has given you only a fragmentary knowledge of the subject. This trail guides you through all the chapters. Before attempting each Project chapter, you will be directed to the appropriate Textbook chapter to learn how to use that model, and to do some exercises. The exercises are designed to give you some practice before using the model on the Project case study. You will also visit the viewpoints chapters to learn how to look at a system in different ways and to build a model that displays the most useful characteristics of the system for each situation. This trail is marked by the ● symbol.

If you are in any doubt as to which trail to follow, pick the Easiest Trail. It covers all the analysis material. You can, at any time, switch to something harder if you find you are covering familiar ground.

■ More Difficult Trail

The More Difficult Trail will suit you if you've had some exposure to analysis, but are not confident that your skills are as up to date as they could be. For example, you may already have some knowledge of how to draw data flow diagrams and how they are supported by the data dictionary and mini specifications. Similarly, you may have a reading acquaintanceship with data models. While you have used these models, it may have been some time ago, and you would like to refresh your memory.

The More Difficult Trail takes you through the Project chapters, as well as the viewpoints chapters of the Textbook. The Trail Guides will steer you past the Textbook chapters that explain the analysis modeling language. If you choose this trail, we assume that you want to use the Project assignments to sharpen your modeling skills, and that you need the minimum exposure to modeling theory. This trail is marked by the ■ symbol.

◆ Most Difficult Trail

This is the black diamond run. If you are confident that you already know how to build analysis models and that you understand the viewpoints, the Most Difficult Trail is for you. This trail will lead you through the Project Section without any help from the Textbook Section. At the end of each Project chapter, you will be directed to sample models and a discussion of the answer in Section 3.

If you choose this Most Difficult Trail, we assume that you already have all of the necessary analysis modeling skills, and are using this book because you want a demanding case study to give you some real experience in applying those skills. Along with the case study experience, this trail will also teach you about choosing the best analysis strategy for a given project.

Don't be intimidated by this trail. If at any stage you find it too tough, the Trail Guides provide an escape by showing you an easier path through the book. This trail is marked by the ◆ symbol.

❄ Promenade Trail

The Promenade Trail is intended for managers, project leaders, and supervisors of a team of systems analysts. We also recommend it for those of you who aren't interested in doing the Project work, but who want to know what systems analysis is all about. It gets its name from the many people who ride chair lifts to the top of mountains, but who have no interest in skiing. For them, to stroll around and enjoy the spectacular scenery is what matters. They have no interest in "guts or glory."

If you take this trail, your reading order is somewhat different from the other trails. You will not be asked to do any Project work, but we will lead you through the Textbook and, toward the end, some of the case study. This trail is marked by the ❄ symbol.

Choosing Any Trail

If you pick a trail that turns out to be too easy, switch to a more challenging trail. Alternatively, if your trail is turning out to be too tough, you can always jump onto an easier trail. As you become familiar with the progression of the chapters, you can switch back and forth between trails, or select chapters to suit your own purposes.

Whichever trail you take, as you work through the case study, you will be guided to Section 3 for the answers to each Project chapter. This section gives more than just answers. Analysis is a human skill, and different analysts produce different answers to the same question. *The real skill of analysis is raising all the questions.*

Section 3 gives you alternative answers, and a discussion of why and how our answers were formulated. There is also a discussion of how to conduct the analysis, and the problems that you are likely to have when doing analysis under the circumstances posed with the problem. Section 3 is valuable. Don't miss it.

At the beginning of each chapter, you will find a short list of the chapters that you should have passed through to reach that point. If you need a particular skill to complete part of the Project, that skill will be mentioned. If you need to understand a particular viewpoint, that, too, will be listed. The list has another benefit: If you wander off your trail or forget where you are, the list will help you rejoin your trail. To help you keep track of where you are in this book, we have listed the chapters visited by each trail. You will find these trial guides listed inside the back cover of this book. Once you start following a trail, you may wish to chart your progress on one of these guides.

No matter which trail you take, if you get into trouble, there is always the Ski Patrol.

✚ Ski Patrol

As a skier, you can rely on the Ski Patrol to help you if you have a problem. For example, if you have a bad fall (is there such a thing as a *good* fall?), most ski areas have a Ski Patrol ready to pick you up within a few minutes and get you safely back on your trail. Similarly, if you have a (metaphorical, not physical) fall during the Project chapters, you can expect help from our Ski Patrol.

The Ski Patrol appears in the Project Reviews chapters in Section 3. Its purpose is to discuss problems that you may be having with the models. We have based these discussions on problems encountered over our many years in systems analysis. We cannot guarantee that we will always anticipate your precise problem, but we will come close. The Ski Patrol offers advice and suggests remedial actions aimed at overcoming any temporary difficulties you may be having. The Ski Patrol is marked by the ✚ symbol.

You Don't Need a CASE Tool

Computer-aided software engineering (CASE) tools were invented to run on personal computers or workstations and to help analysts with record keeping and with some rule and consistency checking. This means that a computer, rather than paper, is used to store and access the data flow diagrams, data models, data dictionary, and the other components of the requirements specification.

CASE tools can be very useful, and, no doubt with the very large projects that lie in your future, CASE is going to be a necessity. However, while the Project in this book is quite large, you are never asked to build such uncontrollably large models that they cannot be handled using pencil and paper.

If you are new to systems analysis, we recommend that you not use a CASE tool until you feel comfortable with the modeling techniques. Some tools have procedures for building models that will confuse you and that will get in the way of your learning about systems analysis.

But You Do Need ...

Lots of writing paper, pencils, and erasers (possibly more erasers than pencils) are definitely in order. A stack of index cards is useful for building your data dictionary. Some small Post-it™ brand notes are helpful for marking your place in each section. Analysts who are concerned about their ability to draw neat models can use a graphic designer's template. However, don't be overly worried about producing great art. Your main concern is with the idea of using models as a common language. Models help you to raise questions with the users, get answers to your questions (and probably raise more questions) to understand the system, and communicate that understanding to others.

Another important thing you'll need is time in a quiet place where you can think. We also suggest something or someone to help you celebrate when you finish the Project. Systems analysis is enjoyable because you take control of your own work. In that sense, at least, we're confident you'll enjoy this book.

What to Do
Your first Trail Guide appears below. Select the degree of difficulty that you wish to follow and begin your trail. As Americans say, "Have a good one!" Or as the British would have it, "Tally ho!" Or maybe you prefer the simple French, "Bon ski!"

Trail Guide
● Easiest: Go to Chapter 2.1 *Analysis Models* for an explanation of why systems analysis is best done by modeling the system.

■ More Difficult: You should already know about context diagrams. Go to Chapter 1.2 *Start with the Context* to begin the Piccadilly Project.

◆ Most Difficult: You plan to do the whole Project without any coaching from the Textbook. Start the Piccadilly Project in Chapter 1.2 *Start with the Context*.

✳ Promenade: We can spare you the work of doing the Piccadilly Project. In Chapter 2.1 *Analysis Models,* you will find the reasoning behind our approach to systems analysis.

START WITH THE CONTEXT 1.2

Before You Reached Here ...

If you are following a trail other than the ● Easiest Trail, review the description below to ensure that you can build the required models and can understand the viewpoint used in this chapter.

● Easiest: This stage of the Project asks you to build a context diagram to define the scope of the analysis project that you are to undertake for Piccadilly Television. To prepare you, this trail has led you through Chapters 1.1 *Your Project Starts Here*, 2.1 *Analysis Models*, and 2.2 *Data Flow Diagrams.*

■ More Difficult: Your trail assumes you already know how to build analysis models. This chapter asks you to build a context diagram for Piccadilly Television, based on the information you gained in Chapter 1.1 *Your Project Starts Here*. If you have any doubts about your modeling skills, take a small diversion through the appropriate Textbook chapters as cited for the ● Easiest Trail above. While the ✚ Ski Patrol will pick you up if you fall down along the way, you may save yourself some head–plants* by doing the basic lessons first.

◆ Most Difficult: You have elected to do the Project without benefit of the Textbook. Naturally, you may take a temporary diversion through an easier trail if you want to brush up on a particular skill. This trail's required reading to date is Chapter 1.1 *Your Project Starts Here.*

❋ Promenade: This is a Project chapter and therefore not on your selected trail. However, because the context diagram discussed here has special significance for managers, you might like to join the ● Easiest Trail for this part of the Project. If so, we suggest you read Chapters 2.1 *Analysis Models* and 2.2 *Data Flow Diagrams* before attempting this chapter. If you wish to rejoin the ❋ Promenade Trail, you can pick it up in Chapter 2.1 *Analysis Models.*

---

*A skiing term meaning to fall such that your head gets jammed into the snow.
This is sometimes painful, and always embarrassing.

The Story of Piccadilly Television

You have already learned a little about the British television industry in Chapter 1.1 *Your Project Starts Here.* That information, plus the following description of Piccadilly, will enable you to build a context diagram for the analysis project.

As you know, Piccadilly Television holds the commercial television franchise for the Midlands of England. The people who buy the commercial airtime are the advertising agencies, and most agencies have their offices in central London. To be nearer their customers, Piccadilly has a sales office in London.

The sales executives are delighted with their offices in the charming mansion at Victoria Square. The leafy garden in the center of the square lends a calm and peaceful atmosphere to this part of London, which is well supplied with grand restaurants, corner cafés, and everything in between. You'll get to know this area very well, as most of your time will be spent in London analyzing the requirements for a system to sell commercial airtime.

Piccadilly produces some of its own programmes, and buys others from a variety of programme suppliers both in England and overseas. These programme suppliers inform Piccadilly of their offerings, which include first-run films, sporting events, documentaries, talk shows, and old movies. Some of the programmes, such as the talk shows and documentaries, may be a series with a number of episodes. When the programme schedulers hear about a programme they want to buy, they send a programme purchase agreement to the supplier. Of course, they don't buy all the programmes being offered, but choose programmes with the best potential for attracting viewers, and the best fit into the overall plans for scheduling.

Piccadilly's programme schedulers have the complicated job of deciding the date that each programme should be transmitted, and where in the programme the commercial breaks should be placed. To make these decisions, the schedulers use the weekly ratings that are supplied by the audience measurement bureaus and that tell them how many people are watching which programmes. The schedulers must also follow the Broadcasting Board's rules for placement of programmes and for the number and placement of commercial breaks within those programmes. Four times a year, the schedulers set a new programme transmission schedule for the coming quarter.

This schedule is sent to the Broadcasting Board for its review, as well as to all of the advertising agencies so that they know what commercial breaks will be available for sale in the next quarter. A commercial break is usually two or three minutes long, and is composed of spots of varying lengths that agencies buy to air their commercials. Piccadilly tries to maximize its revenue by completely filling the available breaktime with commercial spots.

The advertising agencies buy commercial spots that make up campaigns to advertise the products they represent. Each agency sends its campaign requirements to the Piccadilly sales executive who deals with that agency. The executive then models the campaign by selecting commercial breaks for the spots to occupy that will be profitable to Piccadilly, and that will deliver the required ratings to the advertiser. When the executive is satisfied with his selections, the suggested campaign is communicated to the agency. The agency responds by selecting spots from the executive's suggestions and informing him of the choices. The executive finalizes the deal by sending the agency written confirmation of the agreed spots that make up the campaign.

One way that Piccadilly tries to attract advertisers is by making its rates as flexible as possible. Piccadilly researchers and management revise the rates quarterly, and ratecards for the period are sent to the agencies.

30-SECOND SPOT RATES

Rates effective from 1 January 1994

SEGMENT Monday-Friday	Fixed	Broad	ROD	ROW
Up to 16.30	£1500	£1000	£ 750	£ 500
16.30-18.00	£3600	£2400	£1800	£1200
18.00-22.40	£8250	£7000	£5250	£3500
22.40-23.40	£3600	£2400	£1800	£ 900
23.40-Close	£1200	£ 800	£ 600	£ 300
Saturday				
Up to 16.00	£2400	£1600	£ 800	£ 400
16.00-17.50	£3600	£2400	£1800	£1200
17.50-22.40	£8250	£7000	£5250	£3500
22.40-Close	£2400	£1600	£1200	£ 600
Sunday				
Up to 14.00	£1500	£1000	£ 750	£ 500
14.00-19.20	£3600	£2400	£1800	£1200
19.20-22.40	£8250	£7000	£5250	£3500
22.40-Close	£1200	£ 800	£ 600	£ 300

Figure 1.2.1: Part of Piccadilly's ratecard illustrates the variety of rates that can be paid for a thirty-second spot. Fixed spots are the highest priced and are not moveable. Agencies use fixed spots to guarantee transmission during a targeted programme. Broad spots are moveable within a segment. Run-of-day, or ROD, means the spot can be moved within comparably priced segments on a given day. Run-of-week, or ROW, means that Piccadilly can transmit the spot almost any time. There are similar ratecards for other spot durations.

The rate paid for a spot is made up of two dependent factors: price and the degree of moveability. Simply speaking, the lower the price, the more moveable the spot. If the agency pays the lowest price, the spot may be transmitted at almost any time during a given week, subject to Piccadilly's own scheduling requirements. The highest-priced spots are fixed in the breaks selected by the agency.

Some of the lower-priced rates have no guarantee that the spot will be transmitted at all. If the commercial breaks have all been filled, someone prepared to pay a higher rate can replace a previously sold lower-priced spot. A spot that is replaced by another one is *preempted*.

Agencies often buy spots at a low rate initially and upgrade if there is a danger of being preempted by a higher bidder. When such a danger threatens, the Piccadilly sales executive phones a preemption warning to the agency. In this case, the agency usually requests a spot upgrade, and the executive confirms the upgrade. Of course, the executives prefer to sell time at the higher rate at the start. It guarantees higher revenue, and saves administrative time in renegotiating the rates for spots.

Meanwhile, the advertising agencies design and write the copy for the campaign, and hire film production companies to make video recordings for each commercial and to supply them to the Piccadilly Programme Transmission Department. Some advertisers use several different commercials in a campaign, and the agency must send instructions on which copy is to be transmitted in each spot.

After their broadcast, the Piccadilly Programme Transmission Department sends details of each spot's actual transmission time to the Computer Department, which generates an invoice for the agency. The Computer Department also produces revenue reports that go to Piccadilly management and that are used to help set the instructions for sales targets.

Your Strategy

By now, you have a general idea of how the British television industry works and how Piccadilly runs its business. However, a general idea is not enough.

You are being asked to analyze Piccadilly's system with the goal of automating most of the airtime sales functions. However, Piccadilly management wants you to look at more than just the sales side of the business. The managers believe you may discover other areas where computerization may be valuable, although there are several areas they do not want disrupted. Therefore, you have to show them what you intend to study.

Your assignment, then, is to propose the scope of the project by drawing a context diagram. This diagram summarizes, in one big bubble, all the processes of the system you are studying. The context diagram defines the system boundaries by showing the connections between your context of study and the companies, organizations, individuals, and other external bodies that interact with the system.

Think big. Include as much as you think useful in your study. Once you have drafted your context diagram, Piccadilly management will negotiate a final version with you before you begin the detailed analysis. Although you were hired to do the analysis for selling the airtime, you may well find that other functions within the company are linked to the airtime sales, and so should be included in your study. At this stage, even though you may doubt the usefulness of some functionality, it's better to play it safe and include it than to later regret a missed opportunity.

What to Do

Use the description of Piccadilly Television given above, as well as any relevant information from Chapter 1.1 *Your Project Starts Here,* to draw a context diagram for the Piccadilly system you are about to study. If you have any doubts about what a context diagram looks like, refer to Chapter 2.2 *Data Flow Diagrams.*

When you have finished, compare your answer with the sample in Chapter 3.1 and read the discussion of the solution.

Trail Guide

If you came back here after reviewing Chapter 3.1, proceed as follows:

● Easiest: Go to Chapter 2.3 *A Variety of Viewpoints* for a discussion of the viewpoints used to build analysis models.

■ More Difficult: Go to Chapter 2.3 *A Variety of Viewpoints.* This trail assumes that you already know how to build models, so it focuses on the viewpoints that you use to build them.

◆ Most Difficult: Go to Chapter 1.3 *What About the Business Data?* for more work on the Piccadilly Project.

❋ Promenade: The Project chapters are not on your trail, but if you have found this interesting, you might consider switching trails. Otherwise, you can pick up your own trail in Chapter 2.1 *Analysis Models.*

1.3 WHAT ABOUT THE BUSINESS DATA?

Before You Reached Here ...

You either have learned or already know how to build data models. (You may know them by some other name—information models, entity-relationship diagrams, Chen diagrams, or some other variation.)

● Easiest: You have read Project Chapters 1.1 *Your Project Starts Here* and 1.2 *Start with the Context*. You'll use your context diagram as input to this part of the Piccadilly Project. From the Textbook, you have read Chapters 2.3 *A Variety of Viewpoints,* 2.4 *Data Viewpoint,* and 2.5 *Data Models.*

■ More Difficult: You are assumed to be able to build data models, but Chapter 2.4 *Data Viewpoint* has added to your understanding of this part of systems analysis. You also have read Chapters 1.1 *Your Project Starts Here* and 1.2 *Start with the Context.*

◆ Most Difficult: No Textbook for you! However, you should have read Chapters 1.1 *Your Project Starts Here* and 1.2 *Start with the Context.* (Note: Do be certain that you understand data models before attempting this Project chapter. Check that you understand the material given as requirements for the easier trails before plunging on.)

✳ Promenade: Project chapters are not part of your trail except the first one, of course. If you have already read Chapters 2.4 *Data Viewpoint* and 2.5 *Data Models,* by all means attempt this part of the Piccadilly Project. A sample answer and discussion appear in Chapter 3.2. (If you're here because the book happened to fall open to this page, we suggest you go to the end of Chapter 2.2 *Data Flow Diagrams.*)

Your Strategy

A data model looks at a system from the point of view of the data stored within the context. In Chapter 1.2 *Start with the Context,* you built the context diagram for Piccadilly Television. You used this diagram to focus on the scope of the system by identifying the flows of data around the boundaries and by answering the questions, "What data enter my system, and what data does my system produce?"

Let's consider the underlying strategy you are using here. Why start by building a context diagram rather than a data model? The answer hinges on the word "start." You start the context diagram but do not necessarily complete it before starting the data model. In our experience, you can rarely build an accurate context diagram on the first attempt. However, the better your knowledge of the boundary data flows, the easier it is to construct the data model. By at least making a first attempt at the context diagram, you make it easier to build a relevant first cut of the data model. Then, the experience of working with the data

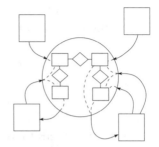

Figure 1.3.1: The context diagram, which defines the boundaries of the system. All of the data that flow into and out of the context are part of the data model.

model verifies your ideas for the context diagram. We therefore suggest you first attempt the context diagram, and then start the data model. The two models are completed in parallel, with the knowledge gained from one model helping you to build the other.

Now let's turn our attention to the data model. This model focuses on the business subject matter, and helps you to raise new questions by putting aside, for the moment, the system's processing. However, you can't complete the data model until you have thoroughly investigated all the processes. So the model you are about to build must remain a provisional model until you can confirm it by modeling the processes in the form of event-response process models. These will be discussed at a later stage of the Piccadilly Project.

To build the first cut of the data model for Piccadilly Television, start with a detailed description of the system's policy. "The Story of Piccadilly Television" in Chapter 1.2 *Start with the Context,* which provided the information for you to build your context diagram, is also the source of information for identifying entities and relationships. To help in this task, use the heuristics, or rules of thumb, presented in Chapter 2.5 *Data Models* to identify entities and relationships.

You are trying to convert a statement of policy into a data model, so there will be occasions when you are not sure whether you have discovered a legitimate entity or a relationship. The best approach is to put it in the model and mark it with a question mark for later resolution. One of the reasons you're building this model is it focuses your questions. Normally, you'd resolve your questions with the users. Because you don't have a user for this project, we'll try to anticipate your questions. We can't possibly anticipate all of them, but your questions won't remain unanswered. Later, when you do a detailed analysis of all the processes and data inside the context, this analysis will provide the answers and confirm all your entities and relationships.

The first-cut data model is not intended to be definitive, and, as we mentioned, there will be opportunities to confirm and correct it. This model is a statement of what you know about the business policy within the context. It should contain all of the entities and relationships that you believe are necessary for the system to remember.

"Necessary to remember" means that an entity must have a business purpose, and must be needed for later reference. For each relationship, ask if it has a business reason and if the user needs to know that connection between the entities. Add cardinality to the model by asking this question of the entities at each end of the relationship: "For one instance of this entity, how many of the other can participate in this relationship?"

What to Do

Use "The Story of Piccadilly Television" in Chapter 1.2 *Start with the Context* as your source of information about the business. Also, use the Piccadilly context diagram you have built in that chapter as a source of information about the data stored by this company.

Draw a first-cut data model for the Piccadilly Project. Your data model should contain an entity with a name something like ADVERTISING CAMPAIGN. List its attributes. Remember that this entity will contain some of the data elements remembered from the incoming flow CAMPAIGN REQUIREMENTS. Also keep in mind that some of the elements of CAMPAIGN REQUIREMENTS will be attributed to other entities.

Compare your answer with the sample answer in Chapter 3.2 and read the discussion.

Trail Guide

There is a Trail Guide at the end of Chapter 3.2, but if you came back here:

● Easiest: Go to Chapter 2.6 *More on Data Flow Diagrams*. You will leave data models for the moment to expand your knowledge of data flow models. You will rejoin the Piccadilly Project shortly.

■ More Difficult: Now is the time to consider an appropriate viewpoint for this stage of the Piccadilly Project. Go to Chapter 2.8 *Current Physical Viewpoint*.

◆ Most Difficult: Continue on with the Piccadilly Project, when you will now get some more background into the company. Go to Chapter 1.4 *The Piccadilly Organization*.

✳ Promenade: This wasn't intended to be part of your journey. However, if you have already been through the data modeling chapters (2.4 *Data Viewpoint* and 2.5 *Data Models*), look at the data model in Chapter 3.2 (Figure 3.2.8), then pick up your trail in Chapter 2.10 *Essential Viewpoint*.

THE PICCADILLY ORGANIZATION 1.4

Before You Reached Here ...

You should have learned how to build a current physical model of a system.

● Easiest: You'll be spending a lot more time with the Piccadilly Project. To do this part of the case study, you've read the Project Chapters 1.1 *Your Project Starts Here,* 1.2 *Start with the Context,* and 1.3 *What About the Business Data?* as well as the Textbook Chapters 2.6 *More on Data Flow Diagrams,* 2.7 *Leveled Data Flow Diagrams,* and 2.8 *Current Physical Viewpoint.*

■ More Difficult: You need to know leveled data flow diagrams, and to have read Chapter 2.8 *Current Physical Viewpoint.*

◆ Most Difficult: Before proceeding, check the requirements for the other trails and decide whether you want to make a side trip for review.

❋ Promenade: This chapter is not part of your trail. However, since you are here, you may want to read through the following description of "Piccadilly People" and look at the model in Chapter 3.3. To find out how this is done, refer to the chapters mentioned in the ● Easiest Trail.

Piccadilly People

In this part of the Project, we ask you to build a current physical model of the Piccadilly organization. This is to give you more background on how the company currently operates. Your future system may well improve on the current state of affairs, but first you need to understand what it is that you seek to change.

The following is a description of how the Piccadilly departments are currently organized. After you have read it, you will build a current physical model of this company. Your Project will affect these departments, so let's tour them all and find out what each of them does.

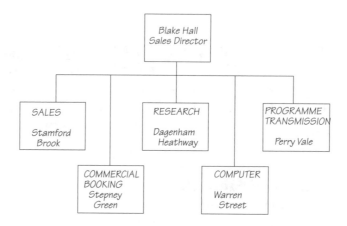

Figure 1.4.1: An organization chart showing the Piccadilly directors and high-level managers. Including all clerical and secretarial staff, Piccadilly employs approximately two hundred people.

Sales Department

Stamford Brook is the ambitious young manager of the Sales Department. Stamford would be the best-dressed member of the Piccadilly staff if he didn't wear such outlandish socks. Hosiery considerations aside, he is an excellent manager, and usually meets his sales targets.

Figure 1.4.2: Stamford Brook, the manager of the Sales Department. Stamford is an excellent manager and is well-respected by his staff.

Stamford is responsible for making sure that the airtime is sold at a rate that will meet the sales target. Every morning, Stamford gets the airtime analysis from the Commercial Booking Department. He uses this analysis when he meets with his thirty sales executives and reviews their progress against the quarterly sales target instructions set by Blake Hall, the sales director. Blake calculates the target based on the amount of money that he estimates will be spent on television advertising during the quarter.

Under Stamford's direction, the sales executives sell airtime to the advertising agencies. Some of the sales work is done on the telephone, and sometimes the executives meet their agency contacts face to face. This is a business in which personal contacts mean a lot. If you have lunch at *Le Pied de Cochon,* the small bistro facing onto Victoria Square, you will always find at least one of the sales executives entertaining a buyer from an agency. Meanwhile, in the public bar of the Royal

Oak, other executives are pushing the advantages of buying Piccadilly airtime to their agency contacts.

Let's review how airtime is sold. It all starts when the advertising agency informs the sales executive of the requirements for a new advertising campaign. The executive looks for advertising time on the breakchart where all the bookings and available time are recorded. The executive uses availability from the breakchart, together with the ratecard and the programme transmission schedule, to plan a campaign for the product. Naturally, the executive selects some higher-priced spots to help meet the sales target. When suitable spots have been selected, the executive sends his suggestions for the campaign to the agency.

Selecting spots for a campaign is a complex business and requires a great deal of experience. Later, you will talk to some of the sales executives to analyze their part of the business in more detail.

The agency reviews the suggested campaign, and tells the executive which spots the agency wants. Each spot has a number so that it can be uniquely identified. The executive finalizes the deal by sending the agency an agreed campaign document and, at the same time, tells the Commercial Booking Department people about the deal. They keep the breakchart up to date by recording the spots the executive has sold.

People from the Commercial Booking Department tell the sales executives if one of their spots has been preempted, or is likely to be. Whenever there is the likelihood of a spot being preempted, the executive warns the agency and recommends a higher rate that would protect the spot. Usually the agency agrees to upgrade the spot, and the executive confirms the upgrade with both the agency and the Commercial Booking Department.

Each advertising agency is assigned to one executive so that he can build a solid working relationship with the contact at the agency. This sort of service is vital in a business that relies so much on personal interaction. Whenever Piccadilly employs a new executive, the Personnel Department sends his qualifications to Sales so that a plan can be made for assigning agency responsibilities. Agencies are accredited, which means that they are obligated to pay for the airtime they buy. When a new agency presents its credentials to Piccadilly, the Sales Department sends a new agency form to the Computer Department so that the all-important computer records are kept up to date.

Commercial Booking Department

The manager of the Commercial Booking Department, Stepney Green, is admired for her ability to remain calm in the midst of frantic activity. Between 9 and 5, her department is very busy indeed. Sales executives rush in and out to find availability on the breakchart and to leave word of the airtime that they have sold. The spot administrators, who maintain the breakchart, have to move quickly to keep up with the rapid changes, particularly when transmission time nears. Overlaying all this activity is the almost constant ringing of the telephones.

Figure 1.4.3: Stepney Green heads the hub of airtime sales, the Commercial Booking Department.

The Commercial Booking Department first learns about the commercial breaks between programmes from the quarterly programme transmission schedule. When a new schedule is received, one of the spot administrators writes each break onto a form called a breaksheet. The breaksheets exist because selling activity on these breaks can be fairly slow until closer to transmission date, and there is not enough room to fit them onto the hanging breakchart files. A month before transmission date, the number of sales, moves, and upgrades increases greatly. To make the breaks more accessible to the spot administrators, the administrators transfer the breaksheet information to one of the hanging files on the breakchart. A copy of the programming rules set by the Broadcasting Board is pinned to the breakchart so that the spot administrators know the current restrictions on spot placements.

Figure 1.4.4: Piccadilly Television's breakchart, which consists of banks of hanging files. Each file can lie flat and show the pattern of the commercial breaks for a day. Each day has a certain number of advertising breaks, each of which contains a number of spots. When spots are sold, colored stickers are attached to the breakchart.

The job of the spot administrators is to record the sales of commercial airtime. They use stickers showing the duration, product name, moveability code, and unique number to record the spots on the breakchart. To make space for a new spot, the spot administrator often has to move one or more existing spots. A spot can be sold for a variety of different rates, and each rate has its own moveability rules. For instance, a spot sold at the lowest rate, known as run-of-week, is governed by moveability rules that allow it to be placed anywhere on the breakchart within a given week, whereas a spot sold at a middle rate, known as run-of-day, has its moveability restricted to any commercial break within a time segment on a specified day. Paying the highest rate, fixed, for a spot means that it cannot be moved from the break in which it was originally sold.

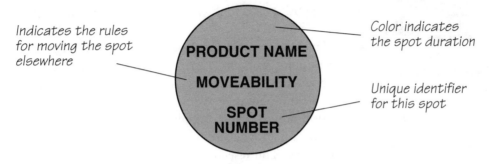

Indicates the rules for moving the spot elsewhere

Color indicates the spot duration

Unique identifier for this spot

PRODUCT NAME

MOVEABILITY

SPOT NUMBER

Figure 1.4.5: A booked commercial spot sticker.

When a sales executive confirms a sale, the administrator has to find a commercial break that conforms to the rate charged for the spot and satisfies the programming rules. This is called "slotting a spot." Whenever necessary, the administrator makes room by displacing a spot sold at a lower rate. Now a home must be found for the displaced spot, which may in turn displace another. Slotting continues until suitable homes are found for all the spots. Sometimes, there is no room left on the breakchart. In that case, the lowest-priced spot in the chain is preempted and loses its place on the breakchart. As soon as a spot is preempted, the appropriate sales executive is told about it so that he can break the news to the agency.

After each day's trading, the spot administrators analyze the breakchart. The analysis reveals any unsold airtime and any that has been sold at a rate considered too low for the commercial break. The sales manager uses this information to monitor the sales executives' work.

On the evening of the day before transmission, the planned break transmission schedule is sent to the Programme Transmission Department in Nuffield-on-the-

Moor. The transmission controllers in Nuffield use the schedule to identify the advertising in the commercial breaks.

Research Department

The main task for Dagenham Heathway and his twelve research assistants is to set the rates for the sale of airtime. The researchers are all mathematicians, and during their coffee breaks, they enjoy solving the tough puzzles or brain-teasers that paper the walls of the department. The work of setting the rates is another tough mathematical puzzle. The rates have to be low enough to allow the sales executives to be competitive, and high enough to meet the sales target set by management.

The sales target instructions come from management at the beginning of every quarter. The Research Department then calculates the new rates based on these instructions. However, the new rates are not just plucked out of the air. Television ratings reports supplied by the audience measurement bureaus provide information on who watches which programmes. The researchers use the ratings to identify the likely audience for each time period, or segment as they call it, in a day. The agencies want to buy into the segments with the biggest potential audience, and the researchers set a higher price for the spots making up the commercial breaks in these segments.

Figure 1.4.6: Dagenham Heathway has responsibility for setting the rates for the sale of commercial air-time.

The price for advertising reflects the spot's duration, moveability, and segment. The objective is to set rates such that if all the spots are sold at the price that the researchers anticipate, the sales target will be met.

After they have calculated the rates, the researchers publish a ratecard that is sent each quarter with the sales target instructions to the Sales Department. The ratecard is also sent to the Computer and Commercial Booking departments, as well as to all of the advertising agencies.

The ratecard specifies the price and the moveability of advertising spots. As you learned from reading about the Commercial Booking Department, a low rate means that a spot can be moved to a wide variety of places on the breakchart, whereas a higher rate guarantees that a spot is transmitted within a much narrower window of time. A spot bought at the highest rate has a guaranteed transmission within a specific commercial break.

Computer Department

Warren Street is the Computer Department manager. His staff of analysts and programmers are fully occupied maintaining the ten-year-old computer system. Along with revenue reports for Piccadilly management, the system produces invoices for the agencies. The invoices show the exact details for the spots that have been transmitted, and are generated from the transmission times sent from the Programme Transmission Department.

The users would like the computer to give them more facilities, but it just isn't possible given the current circumstances. Programming is done in an outmoded, low-level language so that the simplest modification takes weeks; the department's Tardis 2 computer is already stretched to the limit.

Programme Transmission Department

Figure 1.4.7: The quarterly programme transmission schedule is set by Perry Vale, the department's director in Nuffield-on-the-Moor.

Perry Vale runs the Programme Transmission Department in Nuffield-on-the-Moor. Perry's main task is to produce the quarterly programme transmission schedule. Here's how it's done: Programme suppliers from all over the world offer Piccadilly new programmes. Whenever Perry thinks that a programme will suit Piccadilly's requirements, he sends a programme purchase agreement to the supplier.

Now, Perry has to decide when the programme will be transmitted. He has to show it at a time of day that will attract the maximum audience for that kind of programme. To consider the competition for the audience at that time, he gets copies of the opposition's schedules to keep him current on which programmes the opposition's channels are planning to show. Although Piccadilly is the only commercial channel in the Midlands, there are two noncommercial channels run by the British Broadcasting Corporation and funded by television viewer license fees and government subsidies. These channels, BBC1 and BBC2, are well-known for the high quality of their programmes and provide Piccadilly with serious competition.

Perry must also keep the Broadcasting Board happy by presenting a balanced schedule. The Board will not approve a schedule dominated by game shows, soap operas, and football, but likes to see a mix of programmes that both educate and entertain. The Board's rules even specify the percentage of different types of programmes that must be shown during a three-month period. The point is to provide the public with what the Board considers to be varied and entertaining television viewing.

When Perry decides where he would like to place the new programmes, he sends his preliminary schedule to the Research Department. The researchers predict the rating for each programme in its proposed time slot. If Perry decides that the predicted rating is satisfactory, he adds the programme to the programming plan. This plan, which is in the form of a giant chart on the wall of the scheduling room, displays the schedule for the coming six months.

Every quarter, finalized programme transmission schedules are sent to the Sales, Commercial Booking, and Research departments, as well as to the Broadcasting Board for its review. The version of the schedule that Perry sends to all of the agencies highlights the new high-rating programmes in the hope that it will encourage them to book their spots early.

Perry's department also is responsible for the commercial copy, which is recorded on a type of videocassette. These automatic cassette recordings (ACRs) are sent to Nuffield-on-the-Moor by the film production company. The Piccadilly copy librarians store the ACRs in the library near the transmission room.

When the break transmission schedule arrives from London, the copy librarians know which products will occupy which commercial breaks. The copy library typically holds several different ACRs for the same product, so the agency must send copy transmission instructions to specify when each cassette should be used. The librarians refer to the transmission schedule and the copy transmission instructions when they load the ACRs on the trolleys ready for the transmission room.

It would be dangerous to keep outdated copy in the library, as it might be transmitted by mistake. When the agency decides that an ACR is outdated, it sends disposal instructions. The library staff destroys the recording and updates the copy register. At the end of each day's transmission, the actual transmission times for each spot are sent to the Computer Department in London.

Your Strategy

So far, you've built a first-cut context diagram and a data model for Piccadilly. These give you a fairly precise idea of the scope of the Project. However, they cannot yet be considered perfect.

You have just read a more detailed account of how Piccadilly Television currently does business. We now ask you to draw a data flow diagram of the organization to reflect your new level of knowledge. Is it appropriate to build such a model at this stage of the Project? There are several factors that indicate this data flow model is relevant:

1. Because you are now operating at a more detailed level, you will probably find some discrepancies between your context diagram and your new model.

Don't worry. This happens in all projects, and you simply bring the context up to date by correcting the balancing between your new model and the context diagram.

2. The context must be verified. When you model the organization as a whole, it gives you the chance to see how much of it can be changed by a new system. If everything that you model can be changed, your original context is not large enough. Keep in mind that the context of analysis is always larger than the extent of any new system.

3. Since you're probably not familiar with this business, it makes sense to spend some more time modeling the current physical view of the system. You need to become acquainted with the users and to build a working relationship with them. This data flow model, and other models like it, give your users a chance to become familiar with your analysis style.

4. If Piccadilly management is asking for a new system, the current one most likely has some problems. An organization-wide model helps to highlight the problems and possibly to suggest corrective action.

This data flow model will be your biggest yet in this book. Be prepared: It will take longer than the Textbook exercises. You will also find that there is too much information here in this chapter, and you'll need to summarize it in a meaningful way. The lessons from Chapter 2.7 *Leveled Data Flow Diagrams* should be useful, particularly the guideline about how much information you can reasonably have on one page.

The partitioning of this model should be understandable to upper management, which is appropriate when you are modeling almost all of the organization. Later, you will interview some of the people working in the departments and build more detailed models of their operations.

You will probably discover new entities and relationships. If you do, update the data model you built in Chapter 1.3 *What About the Business Data?*

Finally, as you build the model, keep a list of any questions that you'd like to ask the users. (Of course, since you can't ask the fictional users, we'll try to anticipate your questions.) As always, there will be fragments of the business policy that the text doesn't make completely clear. You analyze systems—you don't invent policy, nor do you guess it. When something is unclear, the best action is to draw what you think is the correct answer, and highlight it with a large question mark. We'll discuss possible questions when we examine the sample answer in Chapter 3.3.

What to Do

You've read about the Piccadilly departments in this chapter. Use this information to draw a current physical data flow diagram. This is the first level of decomposition and is called Diagram 0.

Check your balancing against the context diagram and make any necessary corrections. Add any new entities and relationships to your data model. When you are finished, compare your answer with Chapter 3.3's.

Trail Guide

If you returned here after Chapter 3.3, find your next destination.

● Easiest: The next step in the Piccadilly Project is to define the data flows and stores. To find out how to do this, go to Chapter 2.9 *Data Dictionary.*

■ More Difficult: If you already know about data dictionaries, proceed to Chapter 1.5 *Building the Data Dictionary.* If you need to brush up on some rusty skills, may we suggest a quick detour through Chapter 2.9 *Data Dictionary.*

◆ Most Difficult: Put your skill with data dictionaries into practice by going to Chapter 1.5 *Building the Data Dictionary.*

❊ Promenade: This chapter is not required reading for you. Turn to Chapter 2.8 *Current Physical Viewpoint,* where you can pick up your trail.

BUILDING THE DATA DICTIONARY 1.5

Before You Reached Here ...

Any trail except Promenade has taken you through all the previous Project chapters. This chapter builds on the Project information that you have already discovered.

● Easiest: You need to have worked through Chapter 2.9 *Data Dictionary*. This chapter asks you to write some dictionary definitions. You will also use the information from Chapters 2.4 *Data Viewpoint* and 2.5 *Data Models* to update the definition of some entities.

■ More Difficult: You should already know about data dictionaries. In this chapter, you're asked to define some of the data from the model you built in the previous Chapter 1.4 *The Piccadilly Organization*. If you want to review the subject before you start, we recommend Chapter 2.9 *Data Dictionary*.

◆ Most Difficult: You are in the right place. Proceed with the Project.

❊ Promenade: This chapter is not on your trail. If you have come here from Chapter 2.3 *A Variety of Viewpoints,* by all means take a look at the problem posed by this chapter and its answer in Chapter 3.4. The most likely place to rejoin your trail is in Chapter 2.8 *Current Physical Viewpoint.*

Your Strategy

You don't build analysis models with one attempt—you build them progressively as your knowledge of the system develops. The Piccadilly models that you have built so far are evidence of your growing understanding of a complex system.

The models provide convenient places to record the system's information as it becomes available. For example, at the beginning of the Project in Chapter 1.2

Start with the Context, you learned that a programme supplier tells Piccadilly about new programmes. You added the data flow NEW PROGRAMME to the context diagram to show this communication between your system and the outside world. Because the new programme information has to be remembered, you also added the entities PROGRAMME and PROGRAMME SUPPLIER to the data model.

As soon as you learn the contents of the NEW PROGRAMME data flow, you define it in the data dictionary. Whenever you define a new data flow, you often discover new entities and relationships, so it's back to your data model to record them. As you understand the attributes of the entities, you complete the cycle by recording them, too, in the data dictionary.

Every time you discover something new about Piccadilly, you continue to add to the appropriate models. Each time you add something, the model reveals some new facet of the system, or indicates other areas that are fertile ground to investigate. Eventually, the whole of the system will be discovered and recorded.

Index cards make a convenient repository for your data dictionary entries. Make one entry per card and file them alphabetically in a box. Alphabetically tabbed cards make it easier to find your entries. Or, if you prefer, write your entries on regular paper—it's just a bit harder to retrieve your entries later.

Using a word processor or database to store your entries makes your dictionary easier to manage. However, please don't spend too much time setting up elaborate schemas to store your dictionary. Remember that index cards are perfectly adequate for the number of entries you'll need in this book. If you are using a CASE tool to record your models, use the built-in dictionary.

We discourage you from using automated data dictionaries for this case study for the simple reason that you don't have to write enough entries to justify using automation (we will provide most of the dictionary for you). However, when you move on to analyzing your own projects at work, you'll have to write hundreds, perhaps thousands, of dictionary entries. The tasks of building and maintaining the data dictionary for large projects both need and benefit from automated help.

Read the following description of part of the activities of the Programme Transmission Department. Think about the data this department uses, because you are going to write some data definitions.

More About the Programme Transmission Department

When a programme supplier wants to sell a new programme, he tells Piccadilly's Perry Vale in the Programme Transmission Department about it. This appears as the data flow NEW PROGRAMME on the diagrams you've built.

Figure 1.5.1: The data flow NEW PROGRAMME must contain enough information for Perry Vale and his staff to decide whether to buy the new programme being offered.

Let's look more closely at these data. The programme always has a name, which gives some idea of its content. Programmes are categorized by type. For example, it may be a first-run film, sporting event, documentary, talk show, old movie, or one of the many other types of programmes. When the supplier describes the programme's contents, he sometimes provides a list of all the performers appearing in the programme. Some programmes include the names of the producer and director.

Another factor in deciding the suitability of the programme is its duration. For instance, Abel Gance's epic film *Napoleon* runs for five hours. There is no point in buying it unless you have a five-and-a-half-hour transmission slot available (Perry has to allow for the commercial breaks).

The supplier always tells Piccadilly the programme price. As Piccadilly is a British company, the price is always stated in pounds sterling.

What to Do

Record what you know about these data by writing a data dictionary definition for NEW PROGRAMME and for each of the data elements it contains. Before proceeding, you may wish to refer to Chapter 1.4 *The Piccadilly Organization* if you need background information.

When you have written your definitions, there are samples and a discussion in Chapter 3.4. You may wish to look at them before proceeding with this next part.

Now Study the Ratecard

Figure 1.5.2 shows a page from the Piccadilly ratecard. Study it. We will ask you to define it in your data dictionary.

COMPLETE SYSTEMS ANALYSIS

30-SECOND SPOT RATES

Rates effective from 1 January 1994

SEGMENT Monday-Friday	Fixed	Broad	ROD	ROW
Up to 16.30	£1500	£1000	£ 750	£ 500
16.30-18.00	£3600	£2400	£1800	£1200
18.00-22.40	£8250	£7000	£5250	£3500
22.40-23.40	£3600	£2400	£1800	£ 900
23.40-Close	£1200	£ 800	£ 600	£ 300
Saturday				
Up to 16.00	£2400	£1600	£ 800	£ 400
16.00-17.50	£3600	£2400	£1800	£1200
17.50-22.40	£8250	£7000	£5250	£3500
22.40-Close	£2400	£1600	£1200	£ 600
Sunday				
Up to 14.00	£1500	£1000	£ 750	£ 500
14.00-19.20	£3600	£2400	£1800	£1200
19.20-22.40	£8250	£7000	£5250	£3500
22.40-Close	£1200	£ 800	£ 600	£ 300

Figure 1.5.2: These are the various rates for 30-second spots. There are similar pages for 10-, 20-, 40-, 50 -, and 60-second spots.

A brief explanation is needed here to get you started. First, note that the day of transmission is important, and the rates vary depending on whether the spot is transmitted on a weekday, a Saturday, or a Sunday. The days are broken into segments that match the times when people watch television.

The headings—"Fixed," "Broad," "ROD," and "ROW"—specify the degree of moveability of the spot. A fixed spot will be transmitted in a designated break, on a designated date. (Note that "date" and "day" have different meanings in this context.) Advertisers buy fixed spots if they wish to take advantage of the audience of particular programmes. For example, to have an advertisement shown in a break during *Coronation Street* (Britain's highest-rated soap), you would have to buy a fixed spot.

A broad spot can be moved within its segment, but must be transmitted on the specified date. For example, a broad spot sold into the prime-time segment on a weekday can be transmitted in any break between 18.00 and 22.40 (6 p.m. and 10:40 p.m. in the U.S.). Broad spots are moved into breaks with lower ratings if another advertiser is willing to buy a higher-priced fixed spot.

Run-of-day, or ROD, means that the spot can be transmitted in any segment with the same price, on the designated day. For example, an ROD spot sold at £1800, could be shown between 16.30 and 18.00, or between 22.40 and 23.40, on the designated day. Run-of-week, or ROW, allows the spot to be moved to any same-priced segment within a designated week. If you bought an ROW spot for £1200, it will be shown sometime in the afternoon segment on any one of seven days.

Piccadilly naturally prefers to sell at the highest price, whereas the advertisers want to stretch their budgets by buying the cheapest spots they can. So the advertisers buy at the lowest price they think will neither be preempted nor moved to the poorest allowable break.

What to Do

Be sure to read the explanation of our solution when you compare your answers with those in Chapter 3.4.

Use the sample page of the ratecard in Figure 1.5.2 to write the data dictionary entry for RATECARD. Next, define each of its parts until you have defined all the data elements. If you are unsure of any data items, add a question mark to the appropriate entry. If you are not sure of the composition of a flow, simply write a comment and leave the body of the definition blank. If you are unsure of a component in a definition, highlight it with question marks.

Now look at your data model. Are there any new entities or relationships that are revealed by your data dictionary entries for RATECARD or NEW PROGRAMME? As you do this, remember that data flows are not entities, but carry data elements that may, if they are stored by the system, become attributes of entities. Look through the data flows to see if any of their elements can be meaningfully grouped into new or existing entities. If so, add them to your data model and define the entities and their attributes in the data dictionary.

Trail Guide

If you arrived here after Chapter 3.4, find your next destination:

● Easiest, ■ More Difficult, and ◆ Most Difficult: Go to Chapter 1.6 *Selling the Airtime.* There you will expand your model of the Piccadilly system.

❊ Promenade: The most appropriate destination for you is to rejoin your trail in Chapter 2.8 *Current Physical Viewpoint.*

SELLING THE AIRTIME 1.6

Before You Reached Here ...

The ● ■ and ◆ Trails have taken you through all the previous Project chapters.

● Easiest: This chapter gives you more practice drawing data flow diagrams. The relevant chapters for you are Chapters 2.6 *More on Data Flow Diagrams*, 2.7 *Leveled Data Flow Diagrams*, 2.8 *Current Physical Viewpoint*, and 2.9 *Data Dictionary*.

■ More Difficult: This chapter uses the information from Chapter 2.8 *Current Physical Viewpoint*.

◆ Most Difficult: Keep right on reading.

❋ Promenade: This chapter continues the work started in Chapter 1.5 *Building the Data Dictionary*. It is therefore of little interest to you. We suggest you rejoin your trail in Chapter 2.8 *Current Physical Viewpoint*.

Your Strategy

It is still early in the Piccadilly Project. As yet, you probably don't have a detailed enough picture of the business to go on to the next stage. Consequently, we suggest that you build more of the current physical viewpoint. This chapter gives you some more detailed information about the way Piccadilly Television sells commercial airtime. Use the information in this chapter to produce a data flow diagram that is a lower level or child of the diagram you produced in Chapter 1.4 *The Piccadilly Organization*.

You may notice some repetition and even some contradiction between the information in this chapter and the information you received earlier. This is normal in any project where you get information from a number of different sources. You, the analyst, should make your most reasonable interpretation, and mark it as a question to be raised with the users.

This next option is directed to followers of the ■ More Difficult and ◆ Most Difficult Trails. If you feel that building more of the current physical viewpoint is unnecessary, and if you already have a good idea of where you want the analysis to take you, go directly to Chapter 1.7 *Strategy: Focusing on the Essentials.* You can save some time by not building a physical model. However, there is the risk you'll misunderstand the current system. You can simulate the risk right now: There is a complete current physical model in Chapter 3.6. Try to complete the Project while minimizing your references to the current physical model. Whenever you find yourself needing to refer to it, return to this chapter and build just the fragment of the model that you really need, and check it against the one there. This approach will give you some experience with the strategy of physical modeling on demand.

If you are new to systems analysis or if you have chosen to give yourself maximum practice by following the ● Easiest Trail, then press on with the current physical model.

Interview with Stamford Brook, Sales Manager

Read this statement by Stamford Brook. Then, build a model of the Sales Department.

"The easiest way for you to understand how this business works is for me to give you an example of how we sell our airtime. The director I report to is Blake Hall. At the start of each quarter, Blake gives me a sales target that he and the Board of Directors have agreed to. The target tells me how much revenue I have to generate by selling airtime during the coming three months.

"I use the sales target to set the sales policy. Let me explain that. First, I know you've already had a look at our ratecard, so you know there are several possible rates for each commercial break.

Monday-Friday	Fixed	Broad	ROD	ROW
18.00-22.40	£8250	£7000	£5250	£3500

I know how much money we need to bill, so I figure out which of the rates will be the minimum for each break. I know that some breaks will be very popular. For instance, we could sell any break during a first-run movie many times over. It's silly to sell this time at ROD or ROW rates; it only results in preemptions later on. Instead, I set the breaks in a first-run movie at fixed or broad. Why don't I set all breaks at fixed and maximize the revenue? Because there are limits on how much advertisers will spend in a quarter. The sales target is a realistic estimate of Piccadilly's share of the total spend. Also, no advertiser will pay fixed rates for the commercial break at the end of the broadcast day when anybody in front of a television set is asleep. That's how I set my sales policy: by putting a recommended minimum rate against each commercial break.

"Every morning, I get an airtime analysis report from the Commercial Booking Department. This report tells me the rates and revenue for all the time that has been sold that week. The report helps me to keep track of our progress toward the sales target. Sometimes, depending on the demand for time, I adjust the minimum rates on the breaks.

"Our clients are the advertising agencies, which are hired by companies that want to run advertising campaigns for their products. When an agency tells us about a campaign, they quote the campaign budget and the budget amount that they intend to spend with Piccadilly. The campaign budget is not all spent with Piccadilly—it's a national budget that is shared among a number of regional television companies. Our aim is to persuade the agencies to spend more of their budget with us. We try to get a larger share by convincing the agencies that they will get better value for their money if they spend it with us.

Buy this!

"For instance, yesterday one of the agencies phoned with campaign requirements for a new type of liqueur chocolate bar. They told Dollis Hill, the sales executive, that of the £1,000,000 in their total budget, they intend to spend £250,000 with Piccadilly. The remainder will be spent with the other television companies.

"The campaign will run from December to the end of February. (I suppose the cold weather is likely to make people feel more like eating sweets.) As the chocolate is filled with alcohol, the target audience is adults. The agency also told us how many television rating points they want us to achieve. Their commercials will all be thirty seconds long. Dollis recorded the requirements in her campaign file, and then spent the morning planning the campaign.

"First, Dollis considered the spots that would be suitable for the campaign. She checked the breakchart for what times were available, and she looked up the predicted ratings stored in the programme schedules file to see which programmes are most watched by adults. She priced the time using my sales policy minimum rates for the breaks and the prices on the ratecard.

"Finally, she matched the priced time against the campaign requirements. She selected a mixture of spots that satisfied the requirements. In other words, the spots she chose would deliver the required audience, and the total came to a little over the agency's budget. (We always try to get a little more than our share.) Dollis recorded her choices in the campaign file and then phoned the agency with her suggestions for the campaign. Planning these campaigns takes a lot of time; we hope the new system will give us some computerized help.

"The agency phoned back in the afternoon and told Dollis which spots from her suggested campaign they wanted. She sent a notice to the agency that signified our agreement to the deal. She also sent a copy to the Commercial Booking Department so that they could record the selected spots on the breakchart. Dollis keeps details of the campaign in her file, so that she can find them quickly.

"The spots in the chocolate campaign have not been bought at the highest rate. However, we do book them into a designated break on the breakchart. If someone else pays more for that time, the spots can be moved or, in some cases, preempted. If the Commercial Booking people tell us about a likely spot preemption, we phone the agency and give them a preemption warning. We tell them the rate they would have to pay to upgrade the spot. Usually the agency responds by asking us to upgrade the rate for the spot. Then we have to confirm the upgrade and give the Commercial Booking people a copy.

"Sometimes, we hear that a spot has already been preempted. In this case, we find a suitable replacement break, usually at a higher rate, for the preempted spot and tell the agency of this choice. In the rare cases when the agency does not want to pay the extra for the new spot, they send us a spot cancellation. When a spot is canceled, we update our campaign file and advise the Commercial Booking people.

"Selling airtime is a challenging job. The sales executives enjoy it, and tend to stay here for a long time. Occasionally we have to hire someone new. When that happens, Personnel sends us the details of the new recruit and we record them in the Sales Executive Register. We also have to assign agencies to the new executive. This business relies on personal dealings, and we have to be sure that we match the right person to the account.

"When a new agency becomes accredited, the agency advises us. We have to assign the new agency to an executive and record the name, address, phone number, and person to contact in the Agency Register. We also have to fill out a new agency form and send it to the Computer Department. They need it for billing.

"Well, I think that's it, except for the files over there. That's where we keep the ratecards and the programme schedules file. This file holds the programme transmission schedule and predicted ratings.

"I hope this gives you what you need to get started. I'll see you tomorrow after you've had a chance to think about all this. You're sure to have some questions for me, and you may want to talk to some of the sales executives. Or maybe you'll want to watch how they do their work."

Hints on How to Work

Stamford Brook has described how Piccadilly sells its airtime. Now you have to make sure that what you heard (or read) is what Stamford meant to tell you. The best way to do this is to build a data flow model of the Sales Department and show it to Stamford.

The model that you produce here has to balance with its parent: bubble 3 from Diagram 0 in Chapter 3.3 (Figure 3.3.1). You can start by drawing all the flows that enter or leave the parent, and then connect them to the processes that Stamford described.

As you work through Stamford's statement, try to identify each data flow mentioned and each identifiable process. You will find that most of the paragraphs yield some data items and/or some processes. Draw each item as you come to it. Don't worry for the moment if you have many pieces that do not appear to be connected. By the time you finish the interview, there should be enough hanging data flows to connect everything. If your model turns out to have too many bubbles, group some of the processes and level upward to make a better partitioning.

Think of the data flow model as a note-taking device. As the users talk to you, draw what is being said. Remember that neatness beyond legibility is not important. You are trying to record what is being said. If your model doesn't make sense, then perhaps what you were told doesn't either. As soon as a user stops speaking, turn the model around and talk your way through it. This usually helps each user to clarify and re-explain the system.

Alternatively, you may prefer to start by making an assumption about a group of processes, and then leveling down to show the details. For example, Stamford gave you information about Dollis Hill and her advertising agency's campaign. You could add a process called PLAN A CAMPAIGN or something similar to your model, draw all its inputs and outputs, and return to it later using a lower-level diagram to show its details.

Sometimes, it's useful to annotate your data flow diagrams with comments enclosed in asterisks, like so: * Comment *. The typical sorts of comments concern

- the person who is carrying out a process
- the location where a process is being done
- the name of a form that is carrying information

43

- the color of a report (if it adds to the understanding of the model)
- the name of a file (if it is different from the name in your model)
- anything else that helps you to communicate with your users

It is unlikely that your users will ever present information in neatly partitioned paragraphs, as you have in this book, but you can overcome this handicap by asking them questions. What questions to ask? Follow the data. You have the data flows that enter or leave this area of the business. (Get them from the parent diagram.) Ask the users what happens to each of these flows: "How do you use them? What do you do to produce them?" As the users describe each process and its related data flows, ask enough questions so that you know what happens to every one of the flows. Use the model to identify the areas where you need to ask more questions.

What to Do

Re-read Stamford's statement. Build one or more data flow diagrams to model the Sales Department as he described it. Ensure that your diagram(s) balances with Diagram 0 in Chapter 3.3. Remember that any data flow in your diagram has to enter or leave its parent bubble.

When you find new data, define the data in your data dictionary. Compare your answer with the sample diagrams in Chapter 3.5.

Trail Guide

If you arrived here after Chapter 3.5, find your next destination:

● Easiest, ■ More Difficult, and ◆ Most Difficult: Go to Chapter 1.7 *Strategy: Focusing on the Essentials.* Now you'll leave the physical model to move on to a more logical view of the system.

✳ Promenade: You may rejoin your trail in Chapter 2.8 *Current Physical Viewpoint,* where you will learn more about the change in direction the Project is about to take.

STRATEGY: FOCUSING ON THE ESSENTIALS 1.7

Before You Reached Here...

Those of you on the ● ■ and ◆ Trails have built up your knowledge of Piccadilly by building a current physical model. Now you can take a more logical view of Piccadilly's requirements. This chapter reviews your progress and discusses the strategy of essential requirements modeling.

Before proceeding, though, make sure you've inspected the most up-to-date versions of the models that show how Piccadilly is currently implemented in Chapter 3.6.

● Easiest and ■ More Difficult: You'll need to have read Chapter 2.3 *A Variety of Viewpoints.*

◆ Most Difficult: This is a good place for a checkpoint. If you have any problem with this chapter, try taking a temporary detour to an easier trail.

❋ Promenade: This chapter is not on your trail, but you may care to read the first bit. However, if you haven't read Chapter 2.3 *A Variety of Viewpoints,* this chapter may not make much sense.

Your Strategy

As you know, building current physical models teaches you about the Project and about the system you are about to change. Your context diagram defines the boundaries of the Project; the boundary data flows tell you precisely where the domain of analysis begins and ends. They also show you all the data that the system needs to remember. Coming down one level, Diagram 0 partitions the context so that you have an overview of how the organization is currently constructed. This partitioning provides convenient slices of the system for you to study and model. For example, you should have built Diagram 3 Sales Department when you studied how Piccadilly sells its airtime. Details of the selling activity are now found in

Diagram 3.5 Make Changes to Spots and in Diagram 3.1 Plan Campaign. Along with the models that you built yourself, Chapter 3.6 contains a lower-level current physical model for each of the other Piccadilly departments.

Your data model shows a different viewpoint. This model sees the system from the point of view of what it has to remember. You build data models to help you understand the system (if you know what a system remembers, you know what it does). The data model is the first step toward the essential model, since it largely ignores how data are stored and instead concentrates on what has to be stored. The data dictionary gives discipline to the specification by defining all the data used by the data flow diagrams and the data model.

You could continue to analyze Piccadilly from the current physical viewpoint, but please don't. Modeling everything that happens in great detail would take far too long. Also, it's not necessary. Your basic understanding of Piccadilly's current business is enough to take a simpler, clearer, and less time-consuming view of the system.

Changing Your Viewpoint

The current physical system has served its purpose of helping you to get started. We suggest that you collect your models in a folder or clip them together and label the collection "Current Physical Viewpoint." Keep them close at hand; they contain useful information, and you will need to refer to them.

The current physical viewpoint does not necessarily represent the system you wish to implement in the future. After all, technology changes, and you don't want to be tied to obsolete equipment. To specify the future system, you will need to find its underlying policy—the logic behind the current activities. Why does the system do what it does? This is called the essential policy, part of the essential viewpoint.

The Essential Viewpoint

The essential viewpoint shows us the reason the system exists. To see the reason more clearly, ask, "What would this system look like if it operated independently of any implementation technology?" By stripping away all manifestations of the current implementation, you can see the underlying essential policy.

As you build the essential viewpoint, remember that it is an abstraction of the system. You have to disregard the implementation and extract only the essential policy. The essential viewpoint portrays a theoretical system; after all, it couldn't exist in the real world without some kind of technology.

To build the most appropriate future system, you must not inadvertently reimplement any obsolete or inappropriate technology that happens to be part of the current way of doing business. (We will discuss this concept in Chapter 2.10 *Essential Viewpoint.*)

Your data model has already revealed some of the system's essential policy when you looked at the essential data requirements. Instead of modeling how data are or will be stored, the data model simply defines logical groupings of business data required by the system. You will revisit the data model as you build models of the system's essential process requirements.

✚ Ski Patrol

You are about to go down some very steep slopes—the precipitous though exhilarating runs through essential modeling territory.

If you are not completely familiar with essential event-response modeling, we urge you to equip yourself for what lies ahead. That's why the ✚ Ski Patrol refers you to Chapter 2.10 *Essential Viewpoint,* which explains why you want to take the essential view, and, following that, Chapter 2.11 *Event-Response Models,* which describes how to build models of the essential requirements.

What to Do

We suggest that you start a new collection of models and call it the "Essential Requirements Viewpoint." Include in this collection a copy of the context diagram from the current physical model. Although you are changing the way you look at the system itself, you are not changing anything outside the system (the terminators), nor are you changing the content of the data that the system receives or provides to the outside world (the boundary data flows). So the context remains the same.

Also include in your collection a copy of the current data dictionary. While some of the data it describes are dependent on the current system's implementation, some other data are essential. We like to think of the data dictionary as an evolutionary document that you add to over the life of the analysis project as you model different viewpoints.

Naturally, the data model is included in the collection of models. The data model is the most implementation-free viewpoint you've had so far. It is the beginning of the system's essential requirements for stored data, and you will develop it further by the essential modeling effort you are about to do.

Trail Guide

● Easiest and ■ More Difficult: Go to Chapter 2.10 *Essential Viewpoint* for an explanation of the new viewpoint that you are about to use.

◆ Most Difficult: Tighten your ski boots one notch and launch yourself into Chapter 1.8 *Identifying Events*.

❋ Promenade: If you haven't already read Chapter 2.3 *A Variety of Viewpoints*, as well as Chapters 2.4 *Data Viewpoint* and 2.5 *Data Models*, now is the time. Otherwise, the most appropriate place to pick up your trail is at Chapter 2.10 *Essential Viewpoint*.

IDENTIFYING EVENTS 1.8

Before You Reached Here ...

You have built current physical models for Piccadilly. You will need to refer to some of them to complete this chapter.

● Easiest: You have already worked through Chapters 2.10 *Essential Viewpoint* and 2.11 *Event-Response Models*.

■ More Difficult: You have read Chapter 2.10 *Essential Viewpoint*. If the going gets rough, take a short detour through Chapter 2.11 *Event-Response Models*.

◆ Most Difficult: Keep on going, but be forewarned: Here we ask you to build the event list for Piccadilly Television. If you have problems, try detouring through the ● Easiest Trail.

✷ Promenade: This chapter is certainly not prescribed reading. However, if you are interested in this topic, may we suggest that you abandon your comfortable stroll and join one of the other trails? Picking up the ● Easiest Trail would be appropriate. Make sure that you have done the required reading before attempting this chapter. Otherwise, Chapter 2.10 *Essential Viewpoint* is the best place to rejoin your trail.

Your Strategy

Your current physical models and data model of Piccadilly Television provide you with a basic knowledge of the system. You are about to change your viewpoint and begin the search for the essential requirements by repartitioning the system into event responses.

To repartition the system, you first need to identify all events. An event list is an inventory of all the events affecting a system. You can make your event list more useful by paying particular attention to how you name the events. Event names should describe what happens in the terminator to generate the data flow to the system and why the system cares.

An analysis rule of thumb says that external event names should begin with the terminator's name. The second part of the name gives the reason that the terminator sends the data flow to the system (why the system is interested in the data flow). For example, *Customer decides to pay for services* tells you that the system is interested in the incoming customer payment. It is interested because receiving payments is a fundamental activity of the system. Moreover, this event name strongly indicates what the system's response is going to be. *Aircraft arrives in airspace* also indicates the reason for sending the data and the likely response from the system.

Don't name your events after the data flow. Such names as *Customer payment arrives* or even *Aircraft position notified* are not really descriptive. They give you no guidance as to where to start looking for the appropriate responses. *Aircraft position notified* almost suggests that the correct response is to simply record the position and wait for something else to happen, whereas *Aircraft arrives in airspace* suggests that the controller does whatever is necessary to route the aircraft safely through the airspace.

Temporal event names describe what it is time for the system to do. For example, *Time to analyze accounts* and *Aircraft is ten minutes from landing* tell you what processes form the response to the event.

The payoff for choosing good names comes when you build your event-response models. When a descriptive name is a clear guide to the system's response, to *why* the system is reacting, it is easier for you to find all the details, and to know whether all the details you find are really part of the response.

For each event on the list, append the incoming and outgoing boundary data flows for the event. One event can cause any combination of data flows into and out of the context—you must find enough events to account for all the flows on the context diagram.

What to Do

In this part of the Project, you will identify all of the events you can from the flows in the context diagram. To build the event list, you will need all the models that you've already built.

Use the complete current physical model in Chapter 3.6. You'll also need the interviews and business descriptions from Chapters 1.2 *Start with the Context* and 1.4 *The Piccadilly Organization* as reference material. Build an event list by writing down all the events you can identify from the context diagram. Choose names according to the rules of thumb described above.

In your event list, show the data flows associated with each event. Then compare your event list with the one in Chapter 3.7. Return to this point and read the following section before finding your next destination.

A Strategic Point

Recall, as we suggested earlier, one good reason to build a current physical model is that you don't have any previous experience with the business. (We chose the case study for this book for the reason that very few people could have known about the British television industry.) A second reason for building the model is that the background material about Piccadilly was both fragmented and complex.

Now that you have traveled along the safe and steady path through the current physical model, you are in a better position to answer this question: "Could you have developed an accurate event list without first building the current physical model?" Your answer depends on whether you have had experience with similar systems. Now that you know what is involved, could you have built the event list with less current physical modeling? Or would you feel happy starting the event list and retreating to build a physical model whenever you needed more information?

Whether you can complete the event list earlier in the project depends mainly upon the amount of user knowledge, your own understanding of the business being analyzed, and your access to expert users. If these factors all are favorable, then you could probably start the analysis by building a context diagram, a rough data model, and an event list with very little current physical modeling. If you know the right questions to ask, your expert users can help you build event-response models just as they can help you build a current physical model.

You'll naturally want to get the essential requirements for the system as quickly as possible. Do not spend any more time than absolutely necessary on the current physical model. Start your essential modeling as soon as you can and begin to produce an event list.

Trail Guide

After you have read the discussion and compared your event list to ours in Chapter 3.7, proceed onward:

● Easiest, ■ More Difficult, and ◆ Most Difficult: Go to Chapter 1.9 *Modeling an Event Response*. With your event list, it's time to start work on the individual event–response models.

✳ Promenade: If you did the exercise in this chapter, we suggest that you stop being a promenader and start being a worker. Following the other trails to Chapter 1.9 *Modeling an Event Response* is the most logical step. If you wish to resume promenading, Chapter 2.10 *Essential Viewpoint* is the appropriate place to rejoin your trail.

MODELING AN EVENT RESPONSE 1.9

Your Strategy

We want you to build an event-response process model for the current physical response to event 9 *New agency wants to do business.* Your model should include all of the processes that are part of the current system's response to the event. Make sure that each process has all the data that it needs to do its job. Use the necessary current physical data stores in this model.

Note that we are suggesting you build this first Piccadilly event-response model in two stages. In the first stage, concern yourself only with the current system's response to the event. Do not worry about refining it or turning it into an essential model yet. The second stage comes in Chapter 1.10 *Refining an Event Response,* when you will refine your event-response process model by eliminating the implementation-dependent processes and replacing the physical stores with their logical equivalents.

We will review the models after each stage. However, if you feel that essential event-response modeling holds no dangers for you, you may build the refined essential model and skip the first-stage review. If you plan to do it this way, read Chapter 1.10 *Refining an Event Response* before starting, then compare your answer with Chapter 3.9's. However, because this is the first event, we suggest that you proceed cautiously and use the following directions.

What to Do
Refer to the current physical data flow diagrams and the data dictionary in Chapter 3.6. Build a physical event-response process model for event 9 *New agency wants to do business.* Remember your model may contain implementation-dependent characteristics. Check your model with the sample in Chapter 3.8.

Trail Guide
● Easiest, ■ More Difficult, and ◆ Most Difficult: Go to Chapter 1.10 *Refining an Event Response,* where you will remove the physical characteristics of the model you have just built.

❋ Promenade: Either accompany the others to Chapter 1.10 *Refining an Event Response,* or resume your trail in Chapter 2.10 *Essential Viewpoint.*

REFINING AN EVENT RESPONSE 1.10

Before You Reached Here ...

The ● Easiest, ■ More Difficult, and ◆ Most Difficult Trails have brought you through Chapter 1.9 *Modeling an Event Response,* where you have built an event-response model. Early in the Project (in Chapter 1.3 *What About the Business Data?),* you have built a first-cut system data model. Now you are in a position to considerably simplify your event-response model by removing all of its nonessential requirements.

● Easiest: This chapter will call on your knowledge gained from Chapters 2.5 *Data Models* and 2.9 *Data Dictionary,* as well as Chapter 2.11 *Event-Response Models.*

■ More Difficult: Your required reading is Chapter 2.10 *Essential Viewpoint.*

◆ Most Difficult: You probably haven't bothered to read this paragraph and are already racing into "Your Strategy" below, so keep up the good work.

❋ Promenade: This event-response material seems to fascinate you even though it isn't on your trail. Skim "Your Strategy" below and follow the others to the sample model in Chapter 3.9. Alternatively, if you want to return to your own trail, Chapter 2.10 *Essential Viewpoint* is the place to do it.

Your Strategy

The event-response process model that you built in Chapter 1.9 *Modeling an Event Response* portrays the way business policy is currently implemented. Not surprisingly, the model contains some implementation-dependent processes and data. These are not part of the essential system and may obscure the business policy.

Examine your event–response model and remove any processes or data flows that are not part of the business policy and that exist purely because of the way Piccadilly currently does the job. While you are doing this, replace physical data stores with essential data entities and data-bearing relationships. The work you have already done with the system data model will help you to identify the essential data needed to support this event response.

Another aid for you to identify the essential data is to build an event-response data model. This model shows only the essential data stored or accessed by a particular event response, and its limited scope helps you to focus on the details for this event. Naturally, as before, the system data model that you already have will help you, but you should concentrate on establishing all the details of the data accessed or stored by this particular event response.

While you are building the event-response data model, consider how the essential processes use the data. Annotate each of the data entities and relationships with the appropriate CRUD (create, reference, update, delete) operators. When working at the event-response level, you can see exactly what kind of use is made of the data. After you have modeled all the event responses, you'll combine all the fragmented data models into the system data model. At that stage, you'll use the CRUD check to ensure that you have captured all the event responses. In other words, every entity and relationship has to have sufficient CRUD processes to make a working system. Missing CRUD processes mean missing events or redundant data. You'll learn about the CRUD check a little later in the Project in Chapter 1.15 *CRUD Check* and its associated review Chapter 3.14.

As you work with the data, remember that if you need to create any new data flows, define them in the data dictionary. Similarly, if you discover new attributes, add them to your definitions of the entities and relationships.

What to Do

You'll need three items: the event-response model for event 9 *New agency wants to do business* that you built in Chapter 1.9 *Modeling an Event Response;* the data dictionary from Chapter 3.6; and the data model in Figure 3.2.8.

Refine the event-response process model so that it shows only essential processes and data stores. This is easier to do if you know what data are essential for this event. Build an event-response data model of the essential stored data used by this event.

Update your data dictionary with any new data items you discover or need. Check your model against the sample answer in Chapter 3.9.

Trail Guide

In case you return here after reviewing Chapter 3.9, find your next trail.

● Easiest: Now that you have an essential event response, it's time to describe its processes. Go to Chapter 2.12 *Mini Specifications.*

■ More Difficult: Since you already know how to write mini specifications, proceed directly to Chapter 1.11 *Writing Mini Specifications.*

◆ Most Difficult: Proceed straight to Chapter 1.11 *Writing Mini Specifications.*

❊ Promenade: Writing mini specifications is probably a bit out of your line, but you should know at least what they are. Go to Chapter 2.12 *Mini Specifications* to read about them.

1.11 WRITING MINI SPECIFICATIONS

Before You Reached Here...
Following the ● ■ and ◆ Trails, you have built an event-response model. This chapter asks you to write a mini specification for that model.

● Easiest: You will need the skill acquired in Chapter 2.12 *Mini Specifications.*

■ More Difficult: Your knowledge of how to write mini specifications means that you can proceed.

◆ Most Difficult: You should already know about writing mini specifications, and you are in the right place to exercise that skill.

✳ Promenade: Writing mini specifications is something you probably don't want to do but you should at least know about them. We suggest you rejoin your trail in Chapter 2.12 *Mini Specifications.*

Your Strategy

In the Project Chapters 1.9 *Modeling an Event Response* and 1.10 *Refining an Event Response,* you have built the essential event-response process model and data model for event 9 *New agency wants to do business.* You've also updated your data dictionary to reflect the essential data. Now you tie all the pieces together by writing a mini specification, a one-page description of the essential requirements for a process.

Note the unique role of the mini specification: It specifies only what is not specified elsewhere. The essential activity model, the essential data model, and the data dictionary have specified different aspects of the system. The mini specification completes the picture by defining the processing within the bubble.

We asked Stamford Brook, manager of the Sales Department, to explain how he assigns a sales executive to every new agency: "We want to make sure that all of the executives have a reasonable and equal work load. The maximum number of advertising agencies that an executive can service is ten. When a new agency applies to do business with us, we assign it to the executive with the least number of agencies. If there is more than one qualifying executive, then we give it to the one who has not had a new agency for the longest time."

Your task is to write the mini specification for the essential activity ASSIGN AGENCY TO EXECUTIVE. Choose whichever mini specification method you think is most appropriate for this activity. As you write the mini specification, keep a list of any questions that occur to you.

What to Do

To do this exercise, you will need several pieces of the Project that you have already completed. They are the essential event-response process model for event 9 (Figure 3.9.2 in Chapter 3.9); the essential event-response data model, also in Chapter 3.9 (Figure 3.9.2); and the data dictionary from Chapter 3.6. You'll also need Stamford's statement above.

Write a mini specification for ASSIGN AGENCY TO EXECUTIVE. Compare your specification with the one in Chapter 3.10.

Trail Guide

● Easiest, ■ More Difficult, and ◆ Most Difficult: You all go straight ahead with the Project. Your next assignment is waiting in Chapter 1.12 *Another Event Response.*

✳ Promenade: If you want to see more event-response modeling, go with the others to Chapter 1.12 *Another Event Response,* otherwise rejoin this trail in Chapter 2.13 *Modeling New Requirements.*

1.12 ANOTHER EVENT RESPONSE

Before You Reached Here ...

The ● ■ and ◆ Trails all have taken you through Chapter 1.8 *Identifying Events*, where you have built the event list for Piccadilly. That list includes

1. Agency wants to run a campaign   CAMPAIGN REQUIREMENTS (IN)
                                              SUGGESTED CAMPAIGN (OUT)

In this chapter, you will build the complete event-response model for this event, including an updated data model and data dictionary and the appropriate mini specifications. The required reading for this event is the following:

● Easiest: Be sure you've read Chapters 2.5 *Data Models*, 2.9 *Data Dictionary*, 2.10 *Essential Viewpoint*, 2.11 *Event-Response Models*, and 2.12 *Mini Specifications*.

■ More Difficult: Your required reading to this point is Chapter 2.10 *Essential Viewpoint*. If you are finding this trail too tough, feel free to jump to the ● Easiest Trail for a while. However, if all is well, then keep on going.

◆ Most Difficult: If you have successfully come this far, we salute you. It probably gets a little easier from here on, but don't relax just yet.

❋ Promenade: This chapter asks you to model another event response. If you've been following some of the Project, we urge you to get the most out of this book by continuing what you are doing. Otherwise, rejoin the promenaders in Chapter 2.13 *Modeling New Requirements*.

Interview with Dollis Hill, Sales Executive

You are about to build the event-response process and data models for event 1:

1. Agency wants to run a campaign   CAMPAIGN REQUIREMENTS (IN)

     SUGGESTED CAMPAIGN (OUT)

Turn to Chapter 3.6 and look at the Piccadilly current physical model, specifically process 3.1 PLAN CAMPAIGN (Figure 3.6.5). It contains the data flows and stores that the system uses to respond to this event. Their definitions are in the data dictionary. The interview below with Dollis Hill describes the response to the event. Use the information from it and the current physical model to build your event–response models.

When you interviewed Stamford Brook, he mentioned that Dollis Hill is the sales executive assigned to the agency with a new campaign for liqueur chocolate bars. Let's talk to Dollis and find out more about how she planned the campaign.

"I had a phone call from my contact at Totteridge & Whetstone. They're the agency with the account for the liqueur chocolate bars. They want to advertise their product on television, and will spend £250,000 with Piccadilly. The remainder of the £1,000,000 budget is to be spent with some other television companies.

"During my phone conversation about the campaign, they said they want to advertise from the beginning of December to the end of February. This chocolate has an alcoholic filling, and so their target audience is adults. They also told me how many television rating points they want from the Piccadilly area. From the ratings, they will know how many adults have seen their commercials.

"The commercials they're going to use sound really interesting. Each one is thirty seconds long.

"To begin, I allocated the campaign a unique number (this is for identification purposes), and recorded their requirements in my campaign file. Next, I planned a campaign for them. I looked at the breakchart to analyze the available time. For each break, I found out how many seconds were available for sale and at what moveability rate. Then I identified which times would be suitable by looking at the predicted ratings on the programme transmission schedule. Remember, I want breaks close to the programmes that have the highest predicted ratings for adults, those most likely to buy the new chocolate bar.

"When I had identified the suitable amount of time, I used the ratecard to price each spot. The price for a spot depends on the moveability rate.

"Next, I went back to the campaign requirements the agency had given me, and compared the priced time against their requirements. I finally selected a mixture of spots that I thought would satisfy the media buyer at Totteridge & Whetstone, and I recorded my suggested campaign in my file. Then I phoned the agency and told my contact my suggestions for the campaign. She'll call back this afternoon with the decision."

If you need any more background to build the models, re-read Stamford Brook's statement in Chapter 1.6 *Selling the Airtime*.

Your Strategy

The system's response to event 1 is more complex than to event 9. To make your task easier, divide the event modeling activities into two steps, as we suggested in Chapter 1.9 *Modeling an Event Response.*

The first step is to model the current system's response to the event *Agency wants to run a campaign.* Your model should isolate all the processes that can be connected to the response. Make sure that each process has all the data that it needs to do its job. Show all the appropriate current physical data stores on this model.

Your first-step model is intended to reflect the way the business is currently being done. However, you know there are some features that are present in the model because of the existing implementation. If you can easily recognize them, you may choose to leave them out of your model. If you are undecided whether a process is essential or implementation dependent, then play it safe and include it in your model. The second step will take care of it.

When you are happy with your physical event-response model, compare it with the sample in Chapter 3.11 (Figure 3.11.1). The discussion there will give you some feedback on the first step before you plunge into the essential waters in the next step.

The second step in event modeling is to refine your model so that it reflects the essential view of the system. This view excludes anything to do with the implementation, as implementation-dependent features are not part of and often obscure the essential policy.

Finally, the model is refined by removing any processing or data that exist because of the implementation, and replacing the physical data stores with their essential equivalents. There are several ways to do this. Perhaps the easiest way to refine your model is to start by replacing physical data stores with the appropriate essential data entities and relationships. The first-cut data model that you have already built will help you identify them.

Look at the data dictionary definitions of the incoming and outgoing data flows for this event response. For each data element in a flow, ask, "Which entity is described by this element?" You will find the entities in the first-cut data model, or, in some cases, you may need to create a new entity if no suitable one exists. Make each entity a data store in your event-response model in place of the existing physical files.

Alternatively, you could look at the definitions of the physical files, and determine which entities can be combined to make the logical equivalent of each file. Draw those entities on your model. When you have replaced the physical files, relate the entities that you have used to each other to make the event-response data model.

Another alternative is to build an event-response data model first. Remember that the event-response data model is a collection of stored data that is private to the particular event response. The data for the data model come from the data flows that are input to and output from the event response. Use the entities and relationships to replace the physical files.

Another approach that we should mention is instead of building a detailed model of the current event response, you build a model showing only one bubble and all the data flows and physical data stores. Then, instead of refining the model by eliminating implementation-dependent processes, you refine it by replacing the physical data stores with the essential data stores that correspond to the appropriate entities. If the single bubble, known as an essential activity, is too complex for one mini specification, you level downward to decompose the single bubble into essential processes that are small enough to specify in detail. Once you have captured the event response's private data by building the event-response data model, then the essential processes are those that store or retrieve these data.

Whichever method you choose, annotate the event-response data model with the CRUD operators. In other words, show which entities and relationships are created, referenced, updated, or deleted by this event response. Doing so will help you later on when you are ready to correlate all of your event-response models.

If you want to break the task into smaller stages, check your progress frequently against the models in Chapter 3.11. Do this exercise in the way that is most comfortable to you; but whatever you do, don't wait until you have finished everything before checking. You must ensure you are developing the models along the right lines before you get too far.

Your final concerns at this stage include entering any new or altered data items in the data dictionary and writing mini specifications for each of the processes in the essential event-response process model. Remember, the work that you've done already on your event-response data model will simplify the task of writing your mini specifications.

What to Do

You will need the interview in this chapter, plus the current physical data flow diagrams, data dictionary, and data model, all from Chapter 3.6.

First, build the current physical event-response model for event 1 *Agency wants to run a campaign.* At this stage, the model will still contain all the physical processes and data stores. Compare your model with the physical sample in Chapter 3.11.

Next, transform your physical event-response process model into an essential event-response process model. You do this by building the event-response data model. As you proceed, update the data dictionary and write the appropriate mini specifications. Check the two models as you go with the samples in Chapter 3.11.

Trail Guide

If you return here from Chapter 3.11, choose your appropriate trail:

● Easiest, ■ More Difficult, and ◆ Most Difficult: Go to Chapter 1.13 *More Events,* where you will model the remaining events for the Piccadilly system.

❋ Promenade: If you are a genuine promenader, you have just come here for a stroll through the subject and are probably not interested in doing as much work as for the other trails. As you have already seen the essential models, Chapter 2.13 *Modeling New Requirements* is an appropriate destination.

MORE EVENTS **1.13**

Before You Reached Here ...

The ● ■ and ◆ Trails have brought you this far and so we needn't remind you of the required reading—we assume you have the necessary skills. In this chapter, we ask you to model the remaining events in the Piccadilly system. This is an important chapter, for it gives you necessary practice in essential modeling. If at any stage you want to refresh your knowledge, don't hesitate to visit the appropriate parts of the Textbook.

❋ Promenade: This chapter is not going to hold your interest. If you have looked at the essential modeling part of the Textbook, turn to Chapter 2.13 *Modeling New Requirements* as the next logical place to go.

Your Strategy

This chapter asks you to produce models for sixteen event-response models. We want to give you the best possible opportunity to practice and perfect your essential modeling skills. The strategy that you use depends on you.

What problems are you having with essential event-response models? If you are having trouble identifying all the processes that respond to an event, try thinking of yourself as a data tracker. For an external event, track the triggering data flow as it ripples into and out of processes that may be relevant. To decide whether a process is relevant, use the name of the event as a guide. Ask, "Does this process fragment respond to the event I am modeling? Or is it part of another event response?" The track of an external event response ends when all the resulting data flows have either been stored or been sent to the terminators. Tracking a temporal event response is done in the same way except that it starts at the data and ends outside the system at a terminator.

If you are having trouble coping with a particularly complex event, stand back and apply a top-down approach. Try starting with an essential activity model. Draw a bubble and identify all the boundary data flows and stores for the event.

This provides you with a context for the event response. Refer to the latest version of the data model to help you identify the data entities and relationships that are relevant to this event. Use these entities and relationships to build an event-response data model for each event before you attempt to partition the event response into smaller processes.

Sometimes, you will need to vary your strategy. If your knowledge of the event response is fragmented, for example, you might start by developing an event-response process model that contains many processes. Then refine the model to remove any implementation-dependent data or process, leaving only the essential requirements.

Building the event-response data model first often provides you with insights that will help you with the event-response process model. To assist you in building this data model, we suggest that you use the boundary data flows (shown in the event list) for each event. Every data element in the boundary data flows is either an attribute of an entity or a relationship, or it is calculated by your event response. This means that the data elements in the boundary flows help you to identify the entities and relationships that are relevant to the particular event response.

If you have trouble deciding what is essential and what is implementation dependent, try thinking in a different way. Our Atlantic Systems Guild partners Steve McMenamin and John Palmer propose the notion of "perfect technology." They explain this concept in their book, *Essential Systems Analysis,* as follows: Imagine, within your context of study, that you have access to devices that are infinitely fast, cost nothing, are totally reliable and infallible, and have infinite data storage. Now that you have this technology, for each process, flow, and store in the physical model, ask, "Do I need this process, flow, or store now that I have perfect technology? Does this do something essential, or does it support the old, less-than-perfect technology?"

As you build your models, update the data dictionary to include any new data flows, entities, and relationships. Finally, write a mini specification for each essential process.

✚ Ski Patrol

Attention, all readers: Study the following tips and guidelines before reading the "What to Do" section and before starting the exercise.

To begin, choose the easiest event response to model first. Which one is the easiest? Whichever one appears the simplest to you. In your work with the Piccadilly Project, you probably consider some aspects to be simpler than others; that's usually because you've had experience with similar subject matter in another project. If you still cannot decide how to start, use the complexity of the event's input and/or output data flows as an indication of the event's complexity.

If you are having trouble modeling one event response, leave it and try another. Each event will teach you something new about event-response modeling. Give yourself a chance to learn from your own success.

If your chosen modeling approach is not working, change it and select a strategy that you have not tried. Be prepared to vary your approach. The commentaries in Chapter 3.12 will provide you with illustrations of a variety of strategies. These new insights will help you to retry the events that you have trouble with.

For help on how to build event-response models, refer to Chapter 2.11 *Event-Response Models*. If you had any trouble with modeling the data, review Chapter 2.5 *Data Models*. These chapters both provide explanations and exercises to improve your modeling skills. The rationale for building these models is in Chapters 2.10 *Essential Viewpoint* and 2.4 *Data Viewpoint*. Refer to these chapters if you have difficulty understanding any of the event-response explanations in Chapter 3.12.

What to Do

Refer to the event list that you have produced (a sample list is in Chapter 3.7). There are sixteen events that you have not yet modeled. You may also need to refer to the descriptions of Piccadilly in Chapters 1.2 *Start with the Context*, 1.4 *The Piccadilly Organization,* and 1.6 *Selling the Airtime,* as well as to the current physical models in Chapter 3.6 *Complete Current Physical Model.* You will definitely need the data dictionary in Chapter 3.6. The event-response model in Chapter 1.12 *Another Event Response* and its commentary in Chapter 3.11 may also provide some useful material.

Your procedure is as follows: First, build an event-response process model and an event-response data model for each of the sixteen events.

Update your data model and data dictionary with any necessary changes, and write a mini specification for each of the essential processes. Most important, do not wait until you have attempted modeling all sixteen events before verifying that you are on the right track. As you finish each event, compare each model with the sample in Chapter 3.12. Solve any problems as you go. There is also a commentary for each event response in the chapter. Make sure that you can reconcile your answer with our reasons for our models before moving on.

Trail Guide

When you are satisfied with your models, you will have completed gathering all of the essential requirements from the current system. We suggest that you celebrate this milestone in an appropriate manner. Now you are going to look at a different type of requirement.

● Easiest and ■ More Difficult: Go to Chapter 2.13 *Modeling New Requirements* to see how the next stage of the Piccadilly Project unfolds.

◆ Most Difficult: Go to Chapter 1.14 *Some New Requirements.*

❋ Promenade: Chapter 2.13 *Modeling New Requirements* is a good destination for you.

SOME NEW REQUIREMENTS 1.14

Before You Reached Here ...

The ● ■ and ◆ Trails have led you to successfully build the essential requirements model for Piccadilly. Now there will be some changes. Piccadilly management has new requirements that were not part of the original context. To model these new requirements,

● Easiest: You need to have read Chapter 2.13 *Modeling New Requirements* before proceeding.

■ More Difficult: You also have read Chapter 2.13 *Modeling New Requirements.*

◆ Most Difficult: You are continuing with the Project without any preparation (macho, huh?). Good luck.

❈ Promenade: If you have already visited Chapter 2.13 *Modeling New Requirements,* stick around. This may be interesting for those of you managers who want to introduce new requirements into your own projects.

Piccadilly's New Ideas

One of the secondary goals of your analysis effort is to provide better information more efficiently to Piccadilly management. So now the managers are looking at ways to use their current business capabilities to increase their market share, and they've come up with some new requirements to be part of the system delivered by you. Read through this statement from Blake Hall, the Piccadilly sales director:

"We have to compete with all the other television companies for a share of the TV advertising revenue. We have managed to maintain a share of thirteen percent of the revenue, but our business aim is to push that up to fourteen percent. To do this, we need to add value to our advertising by providing a better service than the other companies. This morning in our management meeting, we came up with a great new idea. We're going to provide a new facility, one that isn't offered by any of our competitors.

"We'll offer the agencies a way to get immediate feedback on their advertising campaigns. We'll install a forty-line answering service so that viewers of a commercial can telephone immediately with orders or with requests for more information about the product. Our response-handling service will record the viewer responses and pass them on to the agency every morning. We can tell them which spots attracted the responses. This service won't be applicable to all campaigns, but the agency can tell us if they want it when they book the campaign.

"Sorry to make changes at this late stage of the Project. However, as you know, the television industry is always changing. Our system must be able to cope with that."

Your Strategy

You know a lot about Piccadilly, so you know that these new requirements will affect those you've already defined. However, give yourself a chance to understand the new requirements before attempting to integrate them. Start by treating the new requirements as a "mini project." Specify these requirements by building stand-alone models. This approach gives you the chance to raise questions and to assess the effort to implement the requirements before proceeding.

Modeling new requirements is like modeling any other requirement. You use event-response data and process models, and write data dictionary entries and mini specifications to support them. The main variation from the events you have modeled already is that this time there is no current system to observe. This time you will build only essential models.

Let's suppose you have modeled the requirements, the users have answered all your questions, and your project manager has given you the go-ahead to add the new requirements to the Project. Now you can integrate each part of the new requirements model with the other event responses within your context. Some of the new requirements will result in a change to existing event-response models, while others will become completely new event responses. You will also need to reflect the new requirements by making any necessary changes or additions to the context diagram, event list, data model, data dictionary, and mini specifications.

What to Do

You will need the statement of the new requirements from this chapter, as well as the data model (Chapter 3.12), event list (Chapter 3.7), event-response models (Chapters 3.11 and 3.12), and data dictionary (Chapter 3.6).

Build event-response models of all the new requirements you can identify from the above statement. Note any changes to entities, relationships, and attributes that are needed for the new requirements.

Identify all changes to the context. These affect the effort needed to complete the Project.

Update your data model, context diagram, event list, and data dictionary in order to take the new requirements into account. When you are happy with your models, compare your results with the sample answers in Chapter 3.13.

Trail Guide

● Easiest, ■ More Difficult, and ◆ Most Difficult: Proceed to Chapter 1.15 *CRUD Check* to verify the integrity of your essential models.

❊ Promenade: The others are going off to do some hard work. That's not why you're reading this book. You should go to Chapter 2.14 *New Physical Viewpoint*.

1.15 CRUD CHECK

Before You Reached Here...

The ● ■ and ◆ Trails have led you through the Project chapters that had you build the complete essential requirements model for Piccadilly. Or was it complete? The CRUD check tests the completeness and integrity of your model by ensuring that no events are missing from the Piccadilly event list, and that the stored data are both necessary and sufficient to support the event responses. If you have been through the Project chapters, read on.

❉ Promenade: This chapter really isn't for you. Pick up your trail in Chapter 2.14 *New Physical Viewpoint*.

Your Strategy

There are eighteen events in the event list. How can you be certain there aren't any more, and what are the implications of missing an event? Failure to find all the events during analysis means implementing a system that does not fulfill all its requirements. That's the reason for a CRUD check.

A CRUD check ensures that every attribute of stored data is created and referenced and, when appropriate, updated and deleted. If every attribute is not correctly CRUDed by the system, that means either that there are undiscovered event responses or that the stored data are redundant.

To perform the CRUD check, you build a table that cross–references the entities and relationships to the events:

Entity/Relationship	Create	Reference	Update	Delete
Advertising	1			
Advertising Agency	9			
Advertising Campaign	1	11		

The above sample is presented simply to show you what we mean by a CRUD table and to get you started. It is not yet correct or complete.

The first column of the table lists the Piccadilly entities and relationships. The next four columns show the numbers of the event responses that create, reference, update, or delete that entity or relationship. In this example, event 1 *Agency wants to run a campaign* creates two entities, one of which is referenced by event 11 *Agency wants to upgrade a spot*. Some of this information is already in the event-response data model that you have built for the event.

Your task is to work through all the events in the list, and to fill in the CRUD table. The input for this exercise is the event-response process models, event-response data models, data dictionary, and mini specifications. Check each of these models for access to the stored data, and fill in the event numbers against the accessed entities or relationships.

When you have exhausted the existing models, inspect the table for gaps. Is there anything that is referenced without ever being created? Is anything deleted without ever being created? Any such gap points to a missing event response that stores the data. All stored data must be created. If created, the data must be referenced. Every cell of the "Create" and "Reference" columns of the table should have at least one entry.

Are there any entities or relationships that are created but never referenced? This could mean you missed one or more events that reference the data. Perhaps the storing or referencing event response has been incorrectly modeled. Alternatively, there may be no reason within your context for referencing that data. In that case, you have data that are redundant from the point of view of your system, and you can remove those data from your data model.

Beware that some data are not meant to be deleted, and some may never be updated. Also consider the custodial processing, which maintains the stored data. A missing CRUD action often points to a missing custodial process. Once you identify the missing custodial process, then consider whether it should be part of an existing event response, or whether you need a new event response. Add any new events to the list, and update your context diagram and data dictionary with any new data flows. Starting at this higher level is the best and least time-consuming way of checking your context.

The final and most detailed check is to complete a CRUD table for the lowest-level data; that is, not just for the entities and relationships, but for each of the data elements attributed to each entity and relationship. This check ensures the complete integrity of all your system models at the lowest possible level. When an event-response process model is made up of more than one process, it is useful to put the process number, rather than the event number, in the CRUD table.

That's enough about building a CRUD table. Try building one yourself.

What to Do

You will need the data model, event list, and event-response models packaged in Chapter 3.12. You will also need the data dictionary from Chapter 3.6.

Draw a CRUD table listing all the entities and relationships in the data model. Use your event-response models to fill in the cells in the CRUD table. Identify any missing events and, if there are any, update the context diagram and data dictionary.

Compare your results with the sample CRUD table in Chapter 3.14.

Trail Guide

This Trail Guide repeats the one at the end of Chapter 3.14.

● Easiest and ■ More Difficult: Go to Chapter 2.14 *New Physical Viewpoint* to learn how to turn the essential requirements into a new implementation.

◆ Most Difficult: A word of warning: You are about to enter an area of systems analysis that has undergone many changes. If you feel confident about your analytical skills, plunge ahead to Chapter 1.16 *Strategy: Toward Implementation*. Otherwise, it's no loss of face to follow the others to Chapter 2.14 *New Physical Viewpoint*.

✱ Promenade: Proceed to Chapter 2.14 *New Physical Viewpoint* to see how all this becomes an operational system in the real, physical world.

STRATEGY: TOWARD IMPLEMENTATION 1.16

Before You Reached Here ...

The goal of the Piccadilly Project is to build a specification that contains all of the essential requirements. This chapter illustrates how to use the models you have built to design the new system.

● Easiest, ■ More Difficult, and ✱ Promenade: You all should have come here from Chapter 2.14 *New Physical Viewpoint*.

◆ Most Difficult: You have come here from Chapter 1.15 *CRUD Check*.

Your Strategy

So far, your work on the Piccadilly Project has resulted in a complete set of analysis models. The correctness criteria for these models ensure that you have completely specified the essential requirements by doing the following:

- You have defined the context of study.
- You have specified the events with event-response data and process models.
- You have added any new essential requirements to the models.
- You have run a CRUD check to verify the context and to ensure that there are no missing events.
- You have defined every component of every data flow and data store in the data dictionary.
- You have specified every functional primitive with a mini specification.

It's time to look at the system from a new point of view. Now you have to think as a systems designer. The objective is to make the most efficient and most creative use of the available technology.

The modeling effort has looked at the system from the essential point of view. The reason for using this viewpoint is to understand and define the essential purpose of the system without being influenced by the implementation. You have a clean sheet—

75

there is no leftover, obsolete technology, and you have the perfect start for selecting the new implementation.

When you start considering the implementation details, you must think about how to map the essential requirements into a program hierarchy, what programming language to use, how the screen handlers will work for this system, what effect the database management system will have, what the network protocols will be, what transaction volumes will the system have to cope with, what security will be needed, what are the end users' job descriptions and manual procedures—pretty soon you are drowning in the details.

Instead of drowning in details, let's use the tried-and-true analysis technique of decomposing things into manageable pieces. Let's break up the design task so that it restricts the users' and developers' involvement to the parts that they need to know about. We can break the problem into external and internal design.

External versus Internal Views

Of the two main groups of people involved in the design, the users are primarily concerned that you have specified all the essential business policy. A secondary concern is how the technology affects their jobs. Users are interested in the appearance and behavior of the screens, the reports they get, and their interaction with other devices in the implementation. The users take an *external view* of the system: how it appears to someone on the outside of the hardware. They are not interested in how the database management system stores their data, nor do they care what happens on the network to provide a screen of information. Such matters are the concern of the software developers.

The software developers take an *internal view* of the system. They need to see how the operating system works, how to structure their programs, how to use the programming languages to best advantage, and how all the other factors affect the software implementation.

While the internal and external are different views of the implementation, remember that the two views are tightly linked. Decisions made regarding the external behavior of the system influence the internal design. Similarly, the internal design of the system affects the external appearance. Your design strategy needs to separate the internal and external views, but it must provide a connection between them.

Design Strategy

Figure 1.16.1 illustrates the tasks involved in doing the preliminary design for your system. Let's look at each of them:

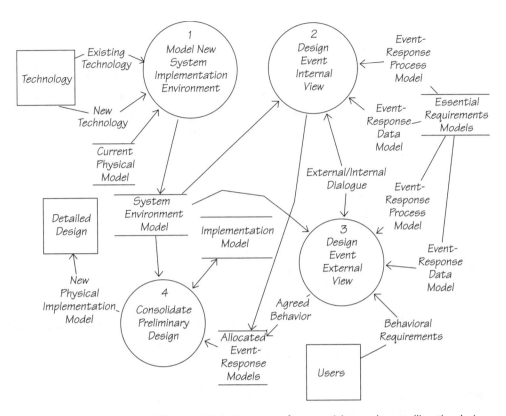

Figure 1.16.1: A strategy for organizing and controlling the design activities. Notice that the essential requirements models are the input to systems design. "Design" is finding the best way to implement each event response with the available technology.

√   *Model New System Implementation Environment.* Consider the existing technology within the organization, and the new technology that can be purchased. The existing technology includes the people working for the organization. Also refer to the current physical model to see how the existing technology is currently used. The output from this process, the system environment model, is a physical model of the processors that you can use to implement the system.

For each of the events,

√   *Design Event External View.* The design of the event's external view specifies how the users want the event to behave and how they want the data to appear. Include any expectations concerning the physical implemen-

77

tation; for example, users may want a screen, an ATM, a fax modem, or any other such technology.

√ *Design Event Internal View.* In this task, you assign each of the essential processes to a processor. Assign all of the data associated with the event response to data carriers and data containers. Annotate the event-response models with the design decisions.

√ *Consolidate Preliminary Design.* This task brings together all of the allocated event-response models, and summarizes the design decisions onto an implementation model. This model becomes part of the final new physical model.

Note that the design constraints discovered during external design will affect the internal design and vice versa. This dependency, identified by the data flow EXTERNAL/INTERNAL DIALOGUE, means that the external and internal design for each event will be iterative.

After completing the preliminary design, you proceed next to detailed design, when you design the internal structure of the software and the databases, and specify the manual procedures. During this step, you impose more and more implementation constraints and their consequent complexity on the essential requirements. To help control this complexity, take advantage of event partitioning by designing the software event by event.

If you work event by event, you are freed from the tyranny of the waterfall model of systems development.* Instead, you can choose to do the detailed work on one part of the system and then implement that part of the system independently of other parts. Suppose you are particularly concerned with a key event. You can do external design, build a prototype, complete detailed design, and do implementation for that event response. Since your event partitioning defines the interfaces with other parts of the system, you have the security of knowing which other parts of your design are affected. This phased approach puts you in control and, as a result, you can gradually implement and test your system piece by piece, rather than in one daunting uncontrollable chunk.

Trail Guide

The next two chapters will provide examples of how to build the new physical model for Piccadilly Television.

● ■ ◆ ❋ All trails: Go to Chapter 1.17 *Piccadilly's New Environment.*

*This waterfall model is a systems development methodology that dictates the completion and freezing of one activity before beginning the next. All analysis and design must be done in detail before any part of the system can be implemented. The method uses the "big bang" approach to implementation, in which every part of the system is implemented all at one time.

PICCADILLY'S NEW ENVIRONMENT 1.17

Before You Reached Here ...

● ■ ◆ and ❋ All trails have brought you here from Chapter 1.16 *Strategy: Toward Implementation*. In this chapter, you will see how the strategy outlined there is used to link your essential requirements specification to the tasks of preliminary and detailed design.

Interview with Stamford Brook:
Implementing the New System

The pure analysis is coming to an end. The people at Piccadilly are satisfied that you have captured all the essential requirements, and now they are getting excited about the new implementation for their commercial airtime sales system. The sales manager, Stamford Brook, wants to give you his thoughts on the new implementation.

"As you know," Stamford explains, "our main reason for having a new computer system is to maximize our revenue from selling airtime. We want to make it easier for our sales executives to develop campaigns that suit the agencies' requirements. The current system uses manual modeling, which makes it very difficult for the executives to consider all the implications when they select spots for a campaign. We want to replace the old breakchart with a central computer system so that the executives can have on-line links. This way, they can look at all the available time and model their advertising campaigns before making suggestions to the agency.

"The Commercial Booking Department will also need to be on-line so that the people can add new breaks and programmes to the computerized breakchart. We hope you can also provide the Research Department with some facilities for modeling their predicted ratings. It's a very time-intensive task at the moment, and the predictions are not as accurate as they could be.

"We have decided that the Programme Transmission Department will not be linked to the central computer. Instead, the Commercial Booking and Research departments will continue to communicate with the Programme Transmission

Department by phone, at meetings, and with printed reports. We'll be installing fax machines, so that every evening, Commercial Booking will fax the next day's break transmission schedule to Programme Transmission. This is an improvement on the method that we currently use.

"Every day, the Programme Transmission Department will fax the transmission log to the new Accounting Department. Accounting will be responsible for entering the actual transmission times into the computer system. Also, all the invoices will be printed in the Accounting Department, and the people there can take care of sending them to the agencies every day.

"Piccadilly management would like the new system to provide an up-to-the-minute analysis of our airtime sales. This will replace the weekly revenue report that is produced by the current computer system. However, the managers still want to retain some of their weekly summary reports. Management will also continue to have daily sales target meetings with the sales executives.

"Here's a guide to the number of workstations we think is appropriate for each department or area:

Sales	20
Commercial Booking	5
Research	5
Accounting	2
Management	2

"Most important of all, we must be sure that all of the essential requirements are implemented exactly as you have specified them. We're looking forward to hearing your ideas for the implementation."

Applying the Design Strategy

In Chapter 1.16 *Strategy: Toward Implementation,* we discussed this phased design strategy:

- Build a system environment model of the part of the organization affected by the new system.
- For each event,

    √ Design the external and internal views.

    √ Annotate the event-response models with the design decisions about how the essential processes and data are allocated to processors and data carriers.

    √ Summarize the design decisions on the implementation model.

The first step in applying the design strategy to Piccadilly is building a system environment model to define the technology available for the new system.

Piccadilly's System Environment Model

Using the system environment model as a way to partition the system allows you to focus on the technology. This is a different view of the system from what you've been working with. The first step to build this model is to copy all the terminators from the context diagram to the system environment model.

The next step is to identify all the processors that will be available to implement the requirements. The interview with Stamford Brook mentions a central computer and a total of 34 workstations at various locations. This is a fairly high-level model of the system, so you can treat each of the locations as a processor. Figure 1.17.1 shows the known terminators and processors.

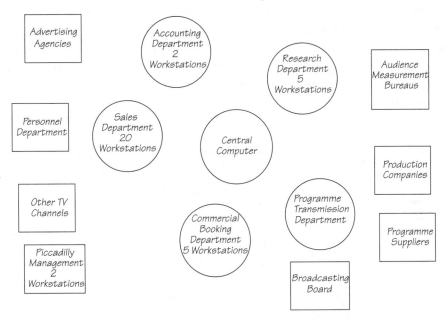

Figure 1.17.1: The first step in building the system environment model is to identify the processors that will carry out the essential requirements. The terminators are other systems that will communicate with the system.

Having identified the processors, you need to consider how they will communicate with each other and with the other systems outside the context of your study (the terminators). For each of the terminators in the system environment model, consider what technology is both available and appropriate to carry data between the terminator and the processors inside your system. For instance, what technology is available to carry data between Piccadilly's Programme Transmission Department and the programme suppliers? If the answer is phone and mail, then add these two data carriers to your system environment model.

You also must consider whether the system has an existing arrangement about technology to communicate data to and from a terminator. (The current physical model will be very helpful here.) For example, Piccadilly expects to receive commercials from the production companies on automatic cassette recordings (ACRs). While there are other technologies available for commercial recordings, this is the one used by the network of independent television stations. There does not appear to be a case for changing it, so ACRs and an accompanying letter can be added to the system environment model.

You also know that the Piccadilly management terminator will have two workstations installed. So there will be an interactive dialogue flow between management and the central computer. Also, since management wants weekly summary reports, there will be a one-way reports flow to the management terminator. Another less obvious technological flow is the daily meeting between management and the Sales Department. If you view a meeting as a way of transferring data, it is just as much a part of the system's technology as an interactive dialogue that is implemented on a computer.

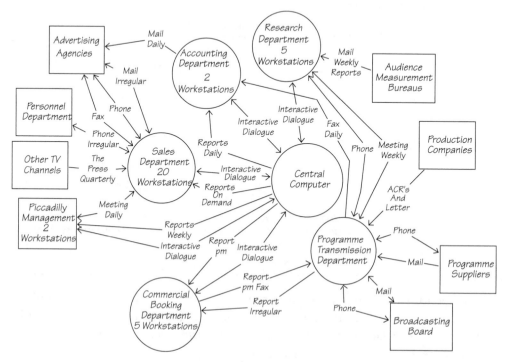

Figure 1.17.2: The system environment model augmented by the data carriers. It summarizes the proposed technology that is available for the implementation of the Piccadilly system.

A guide for naming the data carrier flows on the system environment model is to use the method of carrying the data. For example, PHONE, MAIL, and INTERACTIVE DIALOGUE are all reasonable names. Sometimes, adding the expected frequency makes the system environment model more specific to a particular environment. A REPORT PM FAX flow between the Commercial Booking Department and the Programme Transmission Department emphasizes that there will be a fax transmission each afternoon.

Allocating Event Responses

The next step is to specify how the technology will be used to implement the essential requirements. It would be extremely difficult, if not impossible, to think about the details of every requirement simultaneously. So we will continue to use event partitioning, and to organize the design activities one event at a time.

The objective is to *allocate* each of the fragments of processing and data to the processors and data carriers on the system environment model. You allocate by first looking at the event response from an external, or user-oriented, view. However, the users' vision of how the event response should behave might not be possible with the allocated computer technology. Therefore, before the allocation is final, you must take an internal, or developer-oriented, view of the event response.

Let's see how this works using event 1 *Agency wants to run a campaign* as an example. Figure 1.17.3 shows the response to this event.

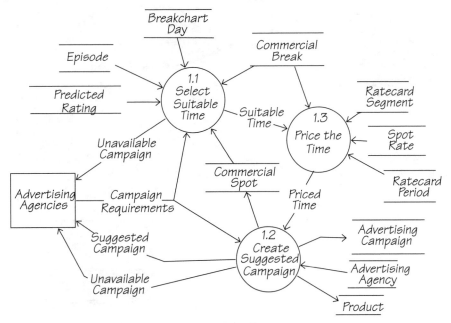

Figure 1.17.3: The essential event-response process model for event 1 Agency wants to run a campaign. *The essential requirements must be fitted into the environment as specified by the system environment model.*

There are three components of the event-response model that must be allocated to the processors and data carriers in the model:

- the essential data stores
- the essential data flows
- the essential processing policy

We will organize the allocation around these components.

Allocating Essential Data Stores

Consider each of the essential data stores in your event-response model and decide which processor will be responsible for storing that data. In the interview, Stamford Brook told us that Piccadilly wants to automate campaign modeling and breakchart maintenance. Looking at our system environment model, we can see that all the data stores in the event-response process model in Figure 1.17.3 will be allocated to a storage device inside the processor called CENTRAL COMPUTER.

Before finalizing our choice for the essential data stores, we need to consider both the external and internal views. The solution shown in Figure 1.17.4 satisfies

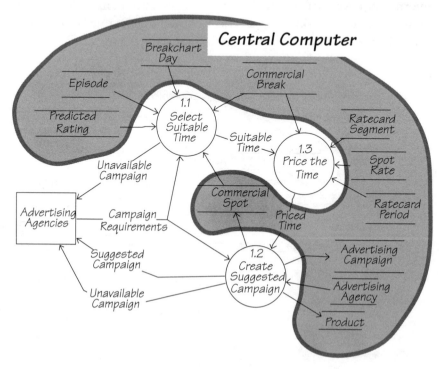

Figure 1.17.4: A boundary line drawn to indicate that all of the essential data stores for this event have been allocated to the central computer.

Stamford Brook's external view of the data storage. He wants the executives to use centralized breakchart information. However, consider the internal view of this allocation. Does the data storage facility have the capacity to store the expected volume of data? Can the database management system provide a suitable response time for Stamford and the executives? Is it feasible to provide the requested kind of modeling facilities on a centralized basis?

Naturally, at any stage of the allocation, if you find for some reason that the internal view of the system will not support the external view, you will need to renegotiate the allocation.

Allocating Essential Data Flows

Next, consider each data flow in the event-response model and ask,

- Which processor or terminator produces the data?
- Which processor or terminator receives the data?
- How are the data carried?

Let's say that the data flow SUGGESTED CAMPAIGN is produced by the Sales Department, and is received by the advertising agencies. The system environment model (Figure 1.17.2) shows that a mail, phone, or fax data carrier is available. So we can have a SUGGESTED CAMPAIGN PHONE/MAIL data carrier. It isn't necessary to spend any time worrying about the internal and external views for this flow. The data content of SUGGESTED CAMPAIGN is defined in the data dictionary, and the telephone and mail services are well established and well understood by the people involved. This would not be the case if we had chosen, say, a microwave link via satellite.

CAMPAIGN REQUIREMENTS is another data flow in our event-response model. Let's say that the advertising agencies send this flow to the Sales Department by mail or telephone. However, the flow does not stop in the Sales Department. Knowing the sales executive wants to model the campaign using the central computer, you must provide a data carrier to carry CAMPAIGN REQUIREMENTS from Sales into the computer system. Because the sales executives want on-line workstations, you need a CAMPAIGN REQUIREMENTS DIALOGUE to record the campaign requirements in the central computer. This data carrier is capable of fragmenting data and hence supporting more varied behavior patterns than phones and the mail. Now you will have to spend some time working on the internal and external behavior of this data flow.

The users (that is, the sales executives) probably have expectations and questions about how the dialogue should work. For example, "Will there be a menu for choosing the campaign requirements screen? Will it matter if we enter the product name or agency name first? Will there be a screen that links to other campaigns being run by this agency? What other historical information is useful to the executives entering the campaign requirements?"

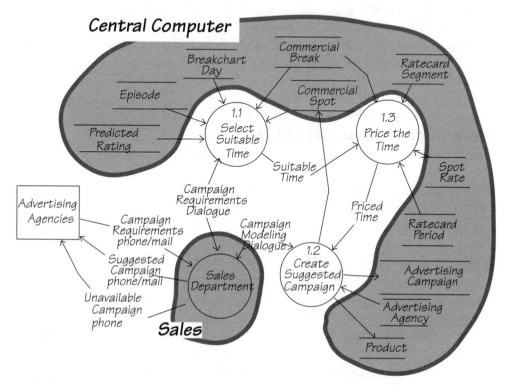

Figure 1.17.5: *The event-response model after allocating the essential data stores to the central computer and the essential data flows to data carriers. Some interfaces between the Sales Department and the central computer are also identified.*

You can model the external behavior, including the displayed data, using either transaction synchronization models developed on paper, or prototypes built with an automated prototyping tool, or both. As a starting point you can build a transaction synchronization model, to plan the communication between the sales executives and the system while the campaign requirements are being entered. Campaign requirements is one of the most vital interfaces in the system so you can bring it to life even more effectively by using your transaction synchronization model as the input to building a working model or prototype. Your aim, whatever approach you use, is to simulate the interface with the users so that they can agree on the interface behavior before the system is implemented.

Once the external behavior of the data flow is established, you should examine the internal technology of the central computer and the workstation to verify the users' expectations can be met.

At the center of Figure 1.17.5, you can see that the essential processes and the data flows between them are not yet allocated. That is our next task.

Allocating Essential Processing Policy

Your final task is to verify that every fragment of the essential policy has been allocated to a processor. For each process, ask, "Is all of this process allocated to one processor?" If the answer is yes, then the boundary of the designated processor will be redrawn to include the process. If the answer is no, then the boundary will dissect the process. Figure 1.17.6 shows the system after the essential processes have been allocated.

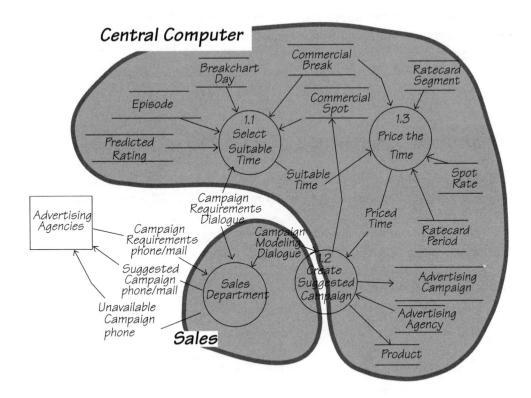

Figure 1.17.6: Every component of the essential event-response model has been allocated to the appropriate processor, and each of the interprocessor data carriers has been defined.

All of the processes SELECT SUITABLE TIME and PRICE THE TIME have been allocated to the central computer. However, the boundary through CREATE SUGGESTED CAMPAIGN indicates that some of the policy is allocated to the Sales Department and some to the central computer. The mini specification for this process says that the selection of suitable time depends on the experience of the sales executive. In other words, some of the policy is subjective and is not feasible to be automated.

The CAMPAIGN MODELING DIALOGUE between the Sales Department and the central computer indicates that the sales executive will be able to interactively model a campaign before suggesting it to the agency. In this case, you build a model that synchronizes the system's pattern of behavior during the time that the CREATE SUGGESTED CAMPAIGN processing is active. Call this behavioral model CAMPAIGN REQUIREMENT AND CAMPAIGN MODELING DIALOGUE. Such a model provides a cross-reference between the allocated event-response model and the behavioral model. Chapter 2.14 *New Physical Viewpoint* discusses a variety of behavioral modeling techniques.

So far, we've talked about just one event response. Naturally, you will follow the same procedure for each of the events in the system, though it gets easier as you go on. Some events will be so alike that one allocation can serve as a pattern for others. Similarly, once the internal view of some interfaces is designed, you can use the same design for other events.

Summarizing the New Physical Model

As you allocate each event, you can summarize the new implementation by adding your design decisions to an implementation model. This model illustrates how the technology defined by the system environment model will be used to implement the particular system. Figure 1.17.7 shows the allocation for the first event.

The implementation model is a project management tool that summarizes your design decisions. (The design details are contained in the collection of allocated event-response models.) The model can be used to coordinate the work of people concerned with the internal design.

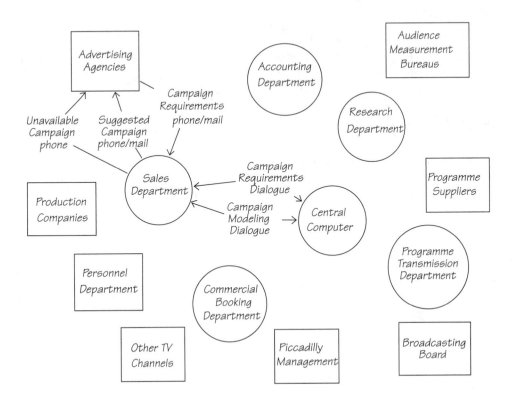

Figure 1.17.7: The beginning of the implementation model for the Piccadilly system. The design decisions for event 1 Agency wants to run a campaign *have been added to this model.*

What to Do

Review the models in this chapter. Consider whether the system environment model (Figure 1.17.1) represents the most appropriate technology. Given your knowledge of technology, are there any changes you would like to suggest to Blake Hall or Stamford Brook? Do you like the allocation of the event? Can you think of any improvements?

Trail Guide

There is a fork in the trail here, as you pause to read Chapter 2.15 *Object-Oriented Viewpoint*. However, if your installation does not use object-oriented languages, you might leave it for future reference and continue on with the Piccadilly Project in Chapter 1.18 *Analysis Strategy*.

● ■ ◆ ✳ All trails: Go to Chapter 1.18 *Analysis Strategy*.

or

● ■ ◆ ✳ All trails: Go to Chapter 2.15 *Object-Oriented Viewpoint*.

ANALYSIS STRATEGY 1.18

Now That You Have Reached Here ...

You have reached the end of the analysis. Well done! On the way, you have built the requirements model for a substantial project. Now, while the model is still fresh in your mind, we want to take a moment to review what happened, to discuss the strategy that you used during the Piccadilly Project, and to see how you might use this strategy as a template for your future projects.

Your Analysis Strategy for Piccadilly

Your first analytical task for the Piccadilly Project was to identify the context and to make the first-cut data model (process 1 in Figure 1.18.1). The purpose of the *context diagram* is to define precisely the scope of the project. Remember that the boundary data flows define the limits of your analysis and determine the system's data. The *first-cut data model* confirms the context, and begins to reveal the business policy by uncovering the reasons for storing data.

The role of the users at this stage is to provide you with a high-level description of the business and to confirm as far as possible that your context diagram and first-cut data model are an accurate statement of the domain of the system to be studied. We feel this review is so important that we consistently refuse to proceed with our own projects until the management and the users have all agreed that the products of this first step are correct and that the areas of uncertainty have been identified.

In the Piccadilly Project, the users' knowledge came to you in the form of recorded interviews. In the real world, it is not this easy. Users are not always as lucid as you'd like them to be, and they aren't always available to spend time with you. It's also difficult to know whether you are asking all the right questions. The *current physical model* (output from process 2 in Figure 1.18.1) helps you to understand the system, identify problems, and assess the impact of proposed changes. The purpose of current physical modeling is to communicate with your users.

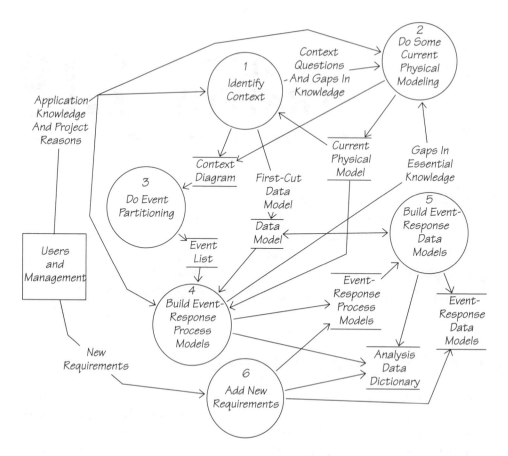

Figure 1.18.1: The analysis strategy that you used for the Piccadilly Project can act as a template for all of your future projects.

When you started the Piccadilly Project, you didn't have detailed knowledge of the British television industry (we deliberately picked something obscure to give you practice with this part of the strategy). When you built the context diagram, it must have raised many questions in your mind. Some of them were answered by building the current physical model. Often, the understanding you get from the current physical modeling process leads you in turn to make changes to the context diagram.

Please be clear about the purpose of the current physical model. It is not built for its own sake, nor is it a required phase of analysis of every project. If you already possess application knowledge or have access to users who thoroughly understand the system, you can minimize the amount of time spent on current physical modeling and go straight to event partitioning. This is not an irreversible decision, for you can always do some physical modeling if you find you don't understand parts of the system well enough to build the essential models.

You have experienced how event partitioning (process 3) logically divides the context into manageable, functionally cohesive pieces. The *event list* is a practical tool for keeping track of the system's events. Note that each event represents a discrete amount of processing. It is attached to the remainder of the system only by stored data. This functional detachment means the event-response partitioning continues to provide manageable units for working on the design and implementation of the system.

In the Piccadilly Project, you analyzed each event in detail by building *event-response process models* and *event-response data models* (processes 4 and 5) to specify the essential processing and the data required for the system's response to the event. This is the heart of analysis and the heart of the project. Although at times the analysis may seem endless, keep in mind that every requirement missed, and every requirement misstated, can mean months of extra work when the installed system must be changed to accommodate a requirement that existed all the while. The impact of your system on the business depends on the accuracy of your work in this part of the project.

Mini specifications, one for each of the essential processes in your event-response process models, define the essential policy carried out by the process. The *data dictionary,* which you added to throughout the project, defines all the data flows, data stores, entities, relationships, and data elements/attributes used by the system.

As the Piccadilly event-response modeling progressed, you amalgamated the various event-response data models into a *system data model.* The contribution of this model, apart from specifying the system's stored data, is that it enabled you to run a *CRUD check.* This check helped you to discover any missing events or redundant data. These inconsistencies raised questions about the context of your project and thus enabled you to correct the gaps by either additional event modeling or changes to your context of study.

As you saw with Piccadilly, your strategy is to model new business requirements when they occur, and then add them to your existing models (process 6). The important point is to control how you add new requirements to your context of study. You must adjust the project plans and estimates to accommodate each new requirement. Otherwise, your project plan will be based on a context of study that differs from reality, and your project will go out of control.

Your analysis has resulted in a number of deliverables: *context diagram, first-cut data model, current physical model, event list, event-response process models, event-response data models,* and *system data model,* all supported by *mini specifications* and a *data dictio-*

93

nary. These deliverables contain a precise collection of requirements for the Piccadilly system. Your analysis models have ensured that everyone understands the requirements in exactly the same way. In the Piccadilly Project, the participation of the users was simulated by interviews with the users and the reviews of your models that we have supplied in Section 3 of this book. In your own projects, you will use the same modeling approach to help you capture the users' requirements and to make it possible for the users to participate in the analysis and to take responsibility for decisions about business policy. Once you have captured their essential requirements, the users' involvement is limited to external design feedback. They can stand back while you and the other designers concentrate on the best implementation for the requirements.

Preliminary Design

You have already learned about some aspects of the preliminary design. Chapter 1.16 *Strategy: Toward Implementation* provided a strategy for the preliminary design, and Chapter 1.17 *Piccadilly's New Environment* illustrated the strategy using one of the Piccadilly event-response models.

The aim of preliminary design is to do enough of the external and internal design to reveal the design tasks and the interfaces between them. To illustrate, turn to the Piccadilly example in Chapter 1.17. The preliminary design for event 1 *Agency wants to run a campaign* resulted in a definition of the interfaces between the tasks carried out by the central computer and the tasks carried out by the Sales Department. To make the design decisions about the interfaces, you did some external design by looking at the system's behavior from the users' point of view. You also did some internal design by determining whether the required external behavior is possible given the internal technology of the central computer.

Your preliminary design work adds some components to your specification. Notice we said "adds"—you do not redo anything that has already been specified as part of the essential requirements analysis.

In this design task, you first model the technology (process 1 in Figure 1.18.2). The output from this process, the *system environment model,* defines the technology that may be used to implement the essential requirements.

The external view of the system (process 3) defines the appearance and behavior of each event. The internal view (process 2) describes how the technology will support that behavior. The output from these processes, the *allocated event-response models,* are annotated copies of the event-response process models. Each model identifies design tasks by defining the interprocessor interfaces.

In process 4, the allocated models are consolidated to form the *implementation model.* This summarizes the connections between processors, referred to as interprocessor interfaces, for all the events in your system. It is a project management guide and a tool for familiarizing outsiders or new developers with the project.

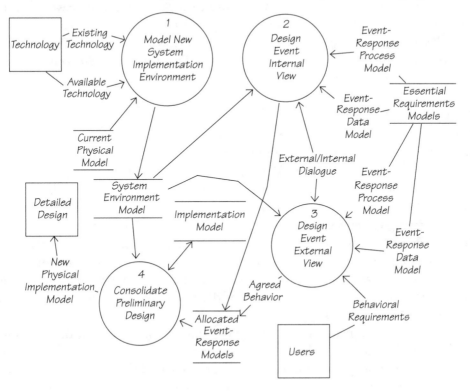

Figure 1.18.2: You can use this preliminary design strategy as a template for planning the preliminary design activities in all your projects.

As a result of the design activities, the physical characteristics of each interprocessor interface are added to the data dictionary. Design decisions are recorded in the data dictionary by recording the physical characteristics of each flow. These physical characteristics are

- planned frequency (for example, daily, continuous)
- estimated volume (for example, 200 per day, 2 per month)
- medium for carrying the data (for example, interactive transaction, paper report)
- mechanism for triggering the data (for example, user request, end of month)
- physical identifier (for example, batch number, user identifier)

95

Note that some interprocessor interfaces will be the existing essential data flows with the addition of physical characteristics. However, in the cases where one essential process has been allocated to more than one physical processor, the fragmentation makes it necessary to add a new physical interface to the data dictionary.

Behavioral models specify the external behavior of interfaces. These can take the form of transaction synchronization models or automated prototypes.

Detailed Design

The objective of design is to use the available technology to implement the essential requirements so that the partitioning of the implemented system is as much like the essential system as possible. When you achieve this objective, your system looks like the problem it is solving. Hence, it is more understandable to the users and easier for the technicians to maintain.

During preliminary design, you allocated the essentials to processors, data carriers, and data containers. Detailed design makes the essentials work within the constraints imposed by the characteristics of the system environment. As different processors and data carriers have different constraints, the detailed design path varies depending on the characteristics of the processor. A detailed design strategy is shown in Figure 1.18.3.

The system is made up of many parts; and, as you've seen in the preliminary design, not all of them are allocated to the same processor. Different processors may require different design techniques. Sometimes, much of the system is allocated to a single processor (a computer), but not all of the system's tasks are implemented using the same programming language and design technique. The purpose of distributing the design tasks is to ensure that each task is designed using the appropriate technique.

Suppose your preliminary design determines that part of your system is to be implemented on a computer, and the preferred language for that machine is one of the third-generation languages such as COBOL, PL/I, Pascal, or Ada. The design path for that part of the system indicates that you design the implementation using hierarchical structure charts. (We refer you to Myers, Page-Jones, or Stevens in the Bibliography for more information on designing for third-generation systems.)

If an object-oriented environment has been chosen for some of the requirements, your design tasks are quite different. Object-oriented design involves selecting suitable classes from the system environment's class hierarchy, designing any additional classes, and deciding the message flow within the system. There is a brief introduction to object-oriented technology in Chapter 2.15 *Object-Oriented Viewpoint*. (We also refer you to Booch, Meyer, Rumbaugh et al., and Shlaer and Mellor for a more detailed treatment of this subject.)

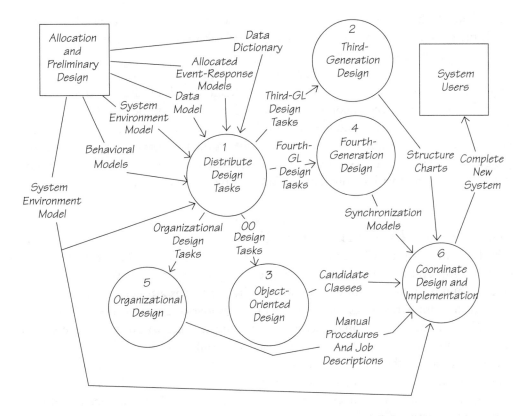

Figure 1.18.3: The detailed design strategy is determined by the system environment. If parts of a project follow different design paths, the results must be coordinated to produce the complete new system.

Fourth-generation languages (4GL) provide powerful high-level macros. The advantage of these macros is that you can considerably accelerate the process of implementation because you do not have to spend time writing any low-level code. The power of 4GLs may tempt you to sidestep design, but beware of this temptation. Failure to design a 4GL system, as in the case of any other system environment, means that you can end up with long-term maintenance problems—except with the power of a 4GL, you can build a bigger mess more quickly. Businesses that use 4GLs successfully go through the same analysis and preliminary design activities that we have outlined, and then design for their 4GL using a combination of allocated event-response models, hierarchical structure charts, and transaction synchronization models.

The system fragments that are allocated to the organization (processes performed by people in departments) differ since those tasks must be designed to

97

accommodate human skills and capacities. Concepts and tools borrowed from the fields of organizational psychology, management strategy, and human communication are used to define jobs and procedures.

In our experience, the coordination of the design paths is a danger zone for most projects, mainly because this activity is not treated as a real job, leaving the results of the different design paths unconnected. This failure to control the details of interfaces always means a fragmented implementation of system pieces that never quite work together. Even the simplest system that we have worked on has involved at least two design paths. There is always a necessity to coordinate the various design paths to produce a new system that has coherent working connections between the pieces.

Using Your Models As Estimating Tools

Way back in Chapter 1.2 *Start with the Context,* when you defined the Piccadilly context, you could estimate how long your analysis would take. We gave you the rule of thumb that every input data flow in the context is a potential event. So the context diagram determines (more or less) the number of events to be analyzed.

Now suppose you could measure how long it takes to analyze the essential requirements for one event. Knowing that, you could multiply by the number of events to produce the total amount of time needed for analysis. If we knew how long it would take your organization, we'd certainly tell you. Unfortunately, the time spent on analyzing an event response varies widely from project to project. Some of the factors causing these variations are

- the analyst's skill
- the users' skill
- the availability of the users
- the newness of the system
- the amount of organizational red tape
- the degree of management commitment
- the level of project management's skill
- the degree of team cohesion
- the geographical distribution of the project team
- the physical support for getting work done (offices and technology)

There are some things that you can do to control the effect of these various factors. The first is to analyze a few event responses, and to use the average analysis time as a benchmark for the remainder of the events. However, you must choose events that are complex enough to be representative of the rest of the system. Use the above list to check for variations in your working environment, and reflect these in the benchmarking guideline. A rough guide to the time needed for analysis is the benchmark time multiplied by the number of events:

Estimate of Total Essential Requirements Hours = Estimated Number of Events
x Estimated Hours Per Event

When you do the detailed requirements analysis for some of your event responses, keep track of how long it takes you to finish the analysis of a functional primitive. Remember that the analysis of a primitive is finished once you have written a mini specification of the process. Use the resulting metrics to provide a more detailed estimate:

Refined Estimate of Essential Requirements Hours = Estimated Number of Events
x Average Number of Primitives Per Event
x Estimated Number of Hours Per Primitive

Models As Management Tools

Whatever variation on our strategies that you choose to use, every strategy produces defined deliverables, all of which can be used to measure progress. It's useless to say that the analysis is 75 percent complete without some metric as proof. However, it is very useful to say that using a measured number of person-hours, 75 event-response models are completed, 200 mini specifications are written, and the data dictionary contains 4,000 entries. These numbers represent reality, a volume of work that has been completed and are therefore a real measure of progress.

The partitioning of your system into controllable pieces means that you can employ a *spiral development* strategy. Instead of doing all the analysis, then all the design, and finally all the implementation, you can spiral through the development activities piece by piece. Part of the project team may decide to work on one or a group of data-related events all the way through to implementation. Meanwhile another group can be working on the analysis and design of other parts of the system. Spiral development, because it fragments the project and the development activities for each part of the project, requires a high degree of management and coordination. The models built by the project team members can provide the basis for management measurement and control.

The best thing about having measurability is that managers can monitor the project by comparing the project's deliverables with the specification. This means that time normally reserved for endless progress meetings can be devoted instead to doing real work. All this presupposes that the managers understand the specification and are prepared to find their way around it. (How can you manage something unless you can understand it?)

If you are using a CASE tool to document your model, progress is easier to measure. The specification is measurable, we have certainly demonstrated that. If the model is automated, it is a relatively simple matter to automate the measurement and provide a variety of detailed statistics to assist project management. (We refer you to DeMarco's *Controlling Software Projects,* listed in the Bibliography, for more information on this subject.)

Models As Presentation Tools

Your analysis models can serve as the basis for presentations about your project, but which models you use depend on the audience with whom you are trying to communicate.

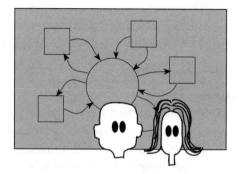

Figure 1.18.4: *The choice of models to use in presentations depends on your audience.*

We once had thirty minutes to explain to a board of directors, first, the scope of a system and, second, the justification to continue the project. To achieve these goals, we chose two models: the context diagram to illustrate the boundaries of the project and the complexity of the interfaces with other organizations and other systems; and the first–cut data model to illustrate the project's subject matter. Soon, we could see that the directors understood from the questions they asked. Indeed, they understood so well that they saw the benefits of the project and decided on the spot that it should continue.

When making a less formal and more detailed presentation to your fellow developers, you can use an event-response model as the focus of your discussion. As you proceed with the project, the data dictionary and the mini specifications help reviewers to do more detailed reviews.

In most analysis projects, team members must periodically report to their management on the state of the project. In such a report, always include the measurable components of your analysis. For example, one of the project groups we work with produced a management report to summarize the project plans. The team wrote an introduction to the report and included the context diagram. Then the team used the event list to show how long they estimated the analysis would take and how many events they had already analyzed. Once they had demonstrated their measures of success, with actual numbers of completed events and mini specifications, management became far more supportive of the project.

The moral of these real-life stories is that models are pretty pictures and are worth at least a thousand words. Use them well to convey your own ideas and aspirations.

Your Next Project: Reusing Analysis

We have been analyzing systems for the past twenty or so years. Most of our projects began with the users saying, more or less, "Our system is different." What they meant was, "We do things differently here." Essentially, there is very little difference between systems that deal with the same subject matter but you must think abstractly to see these similarities. As you develop your essential modeling skills, you will see that an event you analyze today is similar to the one you analyzed yesterday, so similar that you can simply modify the first model to produce the second one. The more systems you analyze, the more similar patterns you will discover, especially if you are consciously looking for reusable components.

Now that you are using a common analysis modeling language, you can take advantage of others' models. You'll find that work done by other analysts on your project provides a starting point for some of your analysis.

As the practice of requirements analysis matures, the reuse of analysis will not be restricted to models produced within one project, or even within one organization. If the work of other projects is accessible and if you can identify system patterns that are applicable to your project, then you can avoid work that has already been done, perhaps countless times before.

An industry-wide reusable approach to analysis means that analysis models could be shared between organizations. You could use pattern books in the same way that dressmakers use pattern books. You could start a project by looking at available patterns and selecting those most applicable to a particular project. (See Arango in the Bibliography for the story of how one organization built its own reuse pattern books.)

Before reuse can happen on a wide scale, however, the software industry needs a project development culture that values reuse above reinvention. The benefits will be twofold: reduced development times, and quicker reaction to new user requirements. So when you finish your next project, keep copies of your models. They will give you a head start for the next project.

✚ Ski Patrol

The ✚ Ski Patrol is not here to dig you out of the snow. If you got this far, you must be skiing safely and with style. We are only here for a few last words.

Analysis demands an enormous variety of skills. Your human communication skills must be good enough to deal with people and politics. Your technical skills must be precise enough to systematically record the results of your work. You must be able to communicate with a wide variety of people and to look at systems from many different viewpoints, particularly the other person's. Most important of all, you must be able to use your models to ask the right questions and to record the answers.

The Piccadilly Project has taught you about viewpoints, models, and strategies. We are confident the exposure to the analysis technique in this book will equip you for your own world of projects.

You will meet ski patrollers there. You will meet many people who know these techniques. Listen to them carefully. When you have a few projects under your belt, you will find yourself picking fallen skiers out of the snow and steering them to the correct trail.

What to Do

You will need all your analysis models. Review the Piccadilly Project and the models you've built. Note which models have given you the best insight in the shortest time. Think about the order in which you built the models. Will that order work on your next project? Consider how to improve any weaknesses in your analysis skills.

Resolve to work with at least one model every day. Use the models to jot down notes of phone conversations or to explain fragments of systems to your colleagues. Practice using them as shorthand to communicate ideas and to verify understanding.

Use one or more of the analysis models to describe a system. It can be any system, from a real-world computerized system to the system for arranging your next birthday party. Only by working with these models can you learn more about them.

Trail Guide

● ■ ◆ ✳ All trails: This final Trail Guide points you into the world. Take along your tools, viewpoints, and strategies to use on your own projects. Good luck! May your users be as knowledgeable, tolerant, and friendly as those we've met over the years.

SECTION 2

The Textbook

ANALYSIS MODELS 2.1

Before You Reached Here ...

You have done the required reading: Chapter 1.1 *Your Project Starts Here* explains the Trail Guides you are about to see. If you don't know what they mean, then return to Chapter 1.1.

● Easiest: If you've already read Chapter 1.1 *Your Project Starts Here,* you're now in the right place.

■ More Difficult: Your trail assumes you already know about analysis models, and you may skip this chapter. If you have already been to Chapter 1.2 *Start with the Context,* then we suggest you pick up your trail in Chapter 2.3 *A Variety of Viewpoints.*

◆ Most Difficult: Your trail doesn't bring you into the Textbook, as your experience indicates you don't need it. However, as this Textbook updates the subject, you may want to stick around. If not, pick up the ◆ Most Difficult Trail in Chapter 1.2 *Start with the Context.* On second thought, if you are following this trail and are reading this chapter, perhaps you'd better go to Chapter 1.1 *Your Project Starts Here* for a refresher on the Trail Guide system.

❋ Promenade: You've read Chapter 1.1 *Your Project Starts Here.* After reading this chapter, you may skip the exercises and go directly to the Trail Guide at the end of the chapter.

Analysis Models: A History of Sorts

At the dawn of time, men and women lived in caves. During the day, they emerged to hunt for food and to club the other tribes senseless. At night, in their caves, they wrote programs for their computers. Unfortunately, their programs were not the well-organized set of instructions befitting the technological masterpieces of their new computing machinery. Instead, their programs were tangled and confused. Some of the programs were so bad that no one in the tribe ever understood them. Those of the tribe called maintenance programmers suffered many troubled days and sleepless nights trying to make these programs work correctly.

The elders of the tribe knew something had to be done. They reasoned that if everyone wrote programs in the same clear way, then everyone in the tribe would be able to understand, and thus modify, all the programs. With this in mind, the elders set out on a voyage of discovery. For many years, they traveled through lands of confusion until lo and behold! In a far-off land, they found understandable programs. The people of this land wrote programs using only two constructs—selection and repetition—to join their statements.*

The elders introduced the method to the tribe. They called it *structured programming,* and it was good. It improved the code of the tribe immeasurably, but it also led to a new problem. The tribe, now able to write understandable code, fearlessly tackled ever larger and more complex programs, but they couldn't organize the larger programs they could now write. So the tribe asked the elders to undertake another voyage of discovery. This time, they brought back a revolutionary idea: functional hierarchies, made up of small modules with known interfaces. They called this *structured design,* and it was good.

The tribe produced elegant, well-designed, and maintainable systems. But the elders noticed another problem. Often, the elegant systems solved the wrong problem! The elders realized that what the tribe needed was a way to make sure they knew the exact requirements before designing a solution.

This time, the elders stayed home and studied the tribe's behavior. They found that those of the tribe called users spoke a different dialect from those of the tribe called analysts. The only way, the elders reasoned, that the two could understand each other is if they had a common language. One more voyage of discovery and the elders returned with a set of graphic models belonging to neither the analysts nor the users. Now everyone could communicate effectively by building graphic models of their systems.

Those called analysts and those called users sat on the floor of the cave and together they drew their models in the sand. Soon, they made a major discovery: To build a model of the system, they had to understand the system. At the same time, they found that building the models helped them to understand the system. The models were successful—they made it much easier to gather the correct requirements. Now the tribe could always build the correct system.

Once they had solved the urgent problem of inventing a suitable systems analysis technique, the tribe moved forward. Next, they invented fire.

*Many years later, the Italian explorers Böhm and Jacopini came across remnants of stone tablets inscribed with the two constructs. Their paper, published in 1966, was an immediate hit. See the Bibliography for a reference, and the Glossary (under "structured programming") for a serious explanation.

The Cavemen Build a Model

The elders wanted an explanation of the tribe's hunting system. To show them, the users and analysts built a model, which is presented in Figure 2.1.1.

The Working Model

Like the cavemen who modeled their hunting system, an architect draws up final building plans, which are referred to as "working drawings." Using these drawings and a written specification, the builder constructs the building. Why are the plans called working drawings? Because they represent what will be a working system.

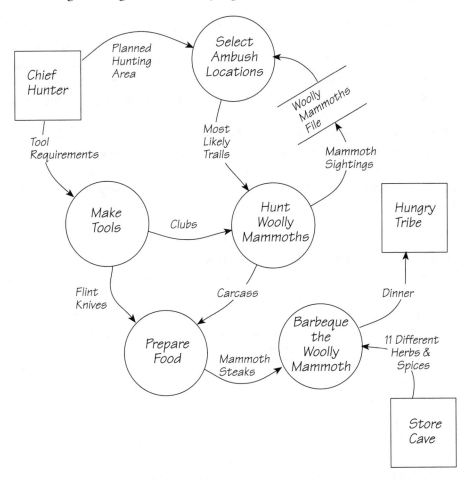

Figure 2.1.1: The cavemen's hunting system. Each bubble represents an active process of the system. Arrows between the processes indicate the flow of information or material produced by each process.

If the construction engineer erects the building exactly as the working drawings show, the building will function correctly. Indeed, before construction can begin, the architect uses the working drawings to obtain approval and to demonstrate to the proper authorities that the building will be safe and meets code. The architect also uses the working drawings to demonstrate to the owner of the building that it will function as desired.

Because it represents a working system, the analysis model is also called a *working model* of the system. How do you know from studying the model that the system will work? You can tell from looking at the data that enter and leave each of the processes. Ask, "Is there enough data entering a process for it to produce its output data?"

For example, in Figure 2.1.1, the bubble HUNT WOOLLY MAMMOTHS receives the incoming data CLUBS and MOST LIKELY TRAILS. These data are sufficient for the cavemen to hunt for food. When the process of hunting is finished, the cavemen return with MAMMOTH SIGHTINGS and, if they were sufficiently agile, a CARCASS or two. In this case, we can say that the process has enough input to produce its output. In other words, it works.

When you can say the same thing for the other processes in the system, then you know you have a working model of a correctly functioning system and therefore the correct specification of the system. How else can this kind of model help in the task of systems development? To answer, let's first look at how you understand systems.

How Do You Understand Systems?

Understanding a large and complex system is difficult. You can only understand it by controlling the amount of information you take in at a time.

Let's leave the cavemen's system for a moment and find an example in the twentieth century to illustrate. Suppose that you, a systems analyst, need to understand an automobile. Assume that you have only the most rudimentary knowledge of how a car works and that an expert is available to explain the car. Although cars are made of hundreds of individual components, initially you do not have to know anything about any of them.

In the beginning, the most useful information your expert can give you is an explanation of how the major components of the car interact with the others to produce the desired result. This means, of course, that the major components are fairly large things such as the engine, transmission, suspension, and body. For the moment, you must accept that each component works as the expert says it works, and not be concerned with how it accomplishes its task.

Once you understand the complete system in terms of the interactions between the major components—the engine powers the wheels, the body rides on the suspension, and so on—then you can begin to investigate how each of the components at this level works. Again, you must accept an explanation of the interaction between subcomponents, and believe for the moment that the subcomponents all work. Ensure that you understand this level of detail before investigating the subcomponents.

Each of the subcomponents is in turn broken into its subcomponents, and this partitioning continues until you arrive at a level of detail such that any component is so understandable you don't have any more questions. Its function is obvious, or it can be described without benefit of any further partitioning.

How far do you go with the partitioning of a system? With the automobile example, you can continue dividing and subdividing until the components are single pieces of metal, or until you get to the molecules of the alloy, or perhaps until you reach the fundamental particles that make up the atoms. The answer is that it depends on your purpose when you study a system.

> The system is divided into components, and the components are divided until they are readily understandable, head-sized pieces.

In systems analysis, we use the convention of *head-sized pieces:* pieces of the system that comfortably fit inside an analyst's head and are readily understood. In the automobile example, the engine of the car is larger than head-sized, whereas the carburetor is just about right. For information systems, head-sized pieces are those that can be satisfactorily specified in a page or less of text description. These descriptions are called *mini specifications,* and they are the topic of Chapter 2.12 *Mini Specifications.*

The alternative to a leveled, successively detailed approach to understanding systems is too awful to contemplate. Suppose that in trying to understand the car, you started at the nuts-and-bolts level or the sub-sub-subcomponent level. How many sub-sub-subcomponents are there? Too many to let you easily understand how a car works. By starting at this level, you'd be swamped with details to the point that you'd probably never understand the system. The advantage of a leveled approach is that you move into the details as you choose. When you are controlling the amount of detail, you are far more likely to succeed with your goal of understanding the entire system.

Now look back at the cavemen's hunting system. Hunting was just part of their lives. If you were studying anthropology and wanted to understand the ways of the cave people, you'd not likely first study their recipe for the barbeque sauce they used on the woolly mammoth. Instead, you'd first look at a higher level to study all of their activities: hunting, gathering, painting the cave walls, clubbing other tribes, and so on. Then you would study each activity that you identified.

You would produce the hunting system diagram for study at this level, and then you might go on to look at even lower-level details, such as how they made their tools, how they mixed the paint for the cave walls, and so on. Eventually, by progressively descending into more and more details, you would capture all the activities as well as the links between them. This method is called a *top-down approach*.

Making Functional Pieces

The tactic of breaking large systems into progressively smaller components seems wise. However, you must ensure that you do produce useful components. There is little point in randomly chopping up systems in the way that Lizzie Borden partitioned her parents. When you divide a system, the resulting components must have some rational relationship to how the system works. In other words, each component must be a *functional piece* of the system.

How do you tell if you have functional components? First, a component is functional if it can be easily and informatively named, and if the name makes sense in the context of the system. Second, if you can honestly name a component using a verb and an object, it is a function. For instance, a name like SELECT AMBUSH LOCATIONS indicates that the component has a single function. It can be recognized by the user, and described by the analyst. A name such as PROCESSES BEGINNING WITH "M" does not pass the test and indicates that the partitioning didn't produce functional processes.

> The best partitioning is the one that makes a system's interfaces as narrow as possible.

We can also tell a lot about the functionality of a process by inspecting its *interfaces*—the data flows that enter and leave it. These flows should carry as little data as possible, and thereby make narrow interfaces.

Functional components need less data than do nonfunctional ones. Imagine the data flows for the two processes mentioned above. The data for SELECT AMBUSH LOCATIONS are reasonable (Figure 2.1.2). However, in any modern system, a process called PROCESSES BEGINNING WITH "M" would have an absurdly large number of data flows and, as a result, be meaningless to users. While functional components make

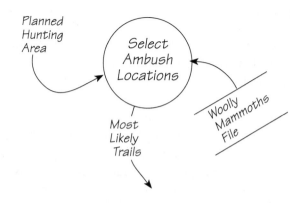

Figure 2.1.2: *Part of a model showing the data for SELECT AMBUSH LOCATIONS.*

narrow interfaces (alternatively, narrow interfaces make functional components), both are easier to get if you find the right places to divide the system.

Cutting at the Natural Joints

Clearly, a woolly mammoth was too large to be cooked all at once, so the cave dwellers prepared it for cooking by cutting it into pieces. When they butchered the animal with their primitive tools, they cut along the soft natural joints, where it was easier to cut through than the hard bones. Similarly, many fruits, like oranges, provide natural seams for dissection.

The same is true in systems analysis. The natural joints in the system are along the *narrowest interfaces,* or where you find a narrow flow of data with two processes. If you try to divide the system by pulling apart functions, you'll get messy interfaces. By partitioning where you find the data interface at its narrowest, you'll deliver a model with functional components that are easy for the users to recognize.

Summary

The cavemen's model is one example of the many structured analysis models you will meet as you progress through the Piccadilly Project. For now, the important thing to remember is that modeling tools are complementary. Each shows one aspect of the system. Together, they make a complete working model of the system.

What to Do

The chapters in this section describe in detail each of the structured analysis models, and tell why and when to use each model. Examples and exercises in the chapters will give you good practice in using each model. The exercises are rehearsals in effect, and we urge you to do them so that you become proficient before using the model in the Piccadilly Project. First, try doing the three exercises below, and compare your answers with those in Chapter 4.1. When you feel comfortable about the results, you are ready for the next stage.

Exercise 1: Woolly Mammoths

The cavemen's hunting system model (Figure 2.1.1) shows five processes, two of which are MAKE TOOLS and HUNT WOOLLY MAMMOTHS. Can you think of a reason the analyst drew two bubbles and didn't combine them into one?

Exercise 2: Other Uses for the Model

The cavemen built their model to explain the hunting system to their elders. Can you think of other uses for this model?

Exercise 3: The System Remembers

The model shows a symbol called WOOLLY MAMMOTHS FILE. This is a file of locations where mammoths were seen during previous hunts. In other words, the system is remembering these data. What items of data do you think the system keeps here for later analysis?

Trail Guide

As you read in Chapter 1.1 *Your Project Starts Here,* the Trail Guide symbols represent degrees of difficulty in working through this book. Select the degree of difficulty that you feel most comfortable with and follow its trail. Good luck with your analysis modeling.

● Easiest: Go to Chapter 2.2 *Data Flow Diagrams* to learn how to use this type of model.

■ More Difficult: You didn't have to come here. However, if you want to know more about the data flow model as we use it, jump to the ● Easiest Trail and turn to Chapter 2.2 *Data Flow Diagrams.* If you are already familiar with data flow diagrams, proceed to Chapter 1.2 *Start with the Context.* If you have already been there, your next Textbook chapter is 2.3 *A Variety of Viewpoints.*

◆ Most Difficult: This is not on your selected trail and we cannot be sure why you are here. We suggest that you return to Chapter 1.1 *Your Project Starts Here* for information about using the Trail Guides, and select a destination from there.

❋ Promenade: Your destination is Chapter 2.2 *Data Flow Diagrams.* You already saw this model in the cavemen's hunting system.

DATA FLOW DIAGRAMS 2.2

Before You Reached Here ...

● Easiest: You have passed through Chapters 1.1 *Your Project Starts Here* and 2.1 *Analysis Models* in preparation for this chapter. The data flow diagram is a relatively simple model, and this chapter gives you enough information to use it in the Piccadilly Project. We have simplified some of the explanations in this chapter so that you're not kept away from the Project any longer than necessary. However, we will return to data flow diagrams in Chapter 2.6 *More on Data Flow Diagrams* to complete your knowledge of this model.

■ More Difficult: This chapter is not required reading for you, as we assume you already know about data flow diagrams. However, you may wish to test your knowledge by doing the exercises at the end of this chapter. Otherwise, if you haven't already been to Chapter 1.2 *Start with the Context,* that is your destination. Otherwise, Chapter 2.3 *A Variety of Viewpoints* is the appropriate place to be.

◆ Most Difficult: You have elected to work through the Project without coaching from the Textbook. Therefore, you should turn to either Chapter 1.1 *Your Project Starts Here* or Chapter 1.2 *Start with the Context.* As you are here, you may care to do this chapter's exercises before rejoining your trail.

❋ Promenade: This chapter describes how data flow diagrams are used to model systems. Read it without getting immersed in the details. After all, you don't intend to build your own analysis models, just understand how they are built.

A Graphic Model

Quite some time ago, travelers discovered they couldn't adequately describe the earth's surface with words alone. Text descriptions of travelers' routes were usually confusing, ambiguous, or incomplete. This led very quickly to the invention of maps. Maps are something we take for granted today, and it is very difficult to think of a better or more convenient way of describing the earth's topography.

A map shows a huge amount of information. To transmit the same amount of information verbally would take several hours of fast talking, and still not be precise or useful. It is the amount of information contained in a map that makes it so useful. A map shows the roads, bodies of water, campsites, public footpaths, park areas, and borders between countries. A map also shows the topography of an area, with contour lines revealing the steepness of the hills and the depth of the valleys. From that, you can tell how the land sheds its water. (See Figure 2.2.1.)

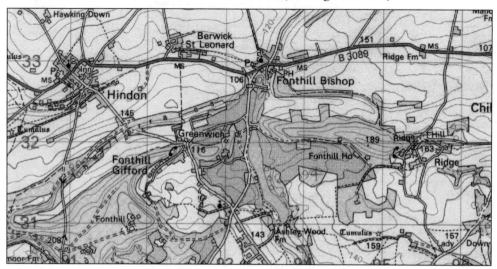

Figure 2.2.1: A map provides much information far more quickly and conveniently than a text description of the area. Not only does the map give you more information, but it also allows you to bypass information you don't need at the moment.

The systems that you will analyze are multidimensional. For example, look around any business office. What you see is a number of people, each performing a task that is a part of the office system's work. Note that they are all working at the same time. The people and machines don't stand and wait their turn to be active. The flows of information, the documents passed around the office, are simultaneously moving in many different directions. If you tried to describe this system with a one-dimensional medium, such as words, the nature of your tool would impose a serial order on the processes within the system. A multidimensional or graphic tool, such as a picture, provides a way of representing the parts of the system without imposing an artificial order. Hence, you have a more understandable representation of the system. Now let's look at a graphic model that can be used to describe any information system.

A Working Model: The Data Flow Diagram

The data flow diagram is more than just a passive map of the system. It is also a working model because all of its components work in the same manner as the real system. In other words, the data flows in the model imitate the movement of data in the real system, and the model's processes use the same input data as the real-life system to produce the same output data.

The pay-back for constructing a working model comes when you deliver the specification to the system builders. Instead of a description of the system, you are handing over a functionally correct scale model. The builder's task is to make a full-size, real-world version of your working model.

Let's examine this idea of the working model by seeing how a data flow diagram processes data. Consider the model shown in Figure 2.2.2. This is a working

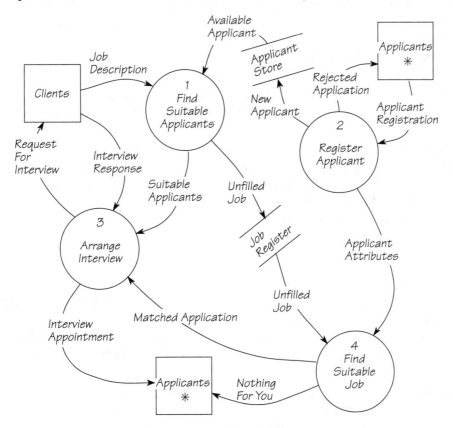

Figure 2.2.2: A data flow diagram is a working model of a system. In this system of the Nelson Buzzcott Employment Agency, the arrows represent the flow of data into and out of the processes (shown as bubbles). The data stores (indicated by parallel lines) hold data for later reference.

model of an employment agency system. For it to work correctly, each processing component in the system must be able to produce its outputs from its inputs. So each of the processes must receive enough data by way of the incoming data flows and data stores to construct the outgoing data flows. This convention is known as the *Rule of Data Conservation*. This is how it works.

The process REGISTER APPLICANT in Figure 2.2.2 receives the incoming data flow APPLICANT REGISTRATION, and uses the data from that flow to assess the candidate, reject unsuitable ones (REJECTED APPLICATION), add a successful applicant's details to the APPLICANT STORE, and produce the outgoing flow APPLICANT ATTRIBUTES. For this process to obey the Rule of Data Conservation, the incoming data flow APPLICANT REGISTRATION must contain all the data necessary for the process to do its job. To ensure that the content of the incoming flows is sufficient, you need to look up the definitions of all the appropriate data flows and the data store in the *data dictionary*. This is a dictionary like any other, except that it restricts itself to definitions of the data in your model. (The data dictionary will be discussed in detail in Chapter 2.9 *Data Dictionary*.)

You also need to see the *mini specification* for the process to ensure that the transformation policy explains the rules carried out by the process. (A mini specification is a one-page description of the processing in a bubble and is the subject of Chapter 2.12 *Mini Specifications*.)

One bubble does not a system make, so the Rule of Data Conservation must be applied to all the processes in the system. Once you have proved that every bubble can produce its outputs from its inputs, then you can be sure that the data flow diagram is a working model of the system being studied. Now let's see how each of the components contributes to the working model.

Terminator

Terminators provide the data for the system to process. They also receive the data when the system has finished with the data. A *terminator* is a person, department, company, system, machine, and so on that exists outside the system being studied.

> A terminator is outside the system. It interacts with the system by giving data to, or receiving data from, the system.

For example, APPLICANTS is a terminator for the employment agency system in Figure 2.2.2. It is shown as the square in the top right corner, as well as at the bottom, of the diagram. (Disregard the asterisk, as it is only a notation to say this icon appears more than once in the diagram.) An applicant is anyone who applies for a job through the agency. Applicants are interesting to you because they provide data for your system, and receive data from it. However, you are not interested in finding out why they provide data, or what they do when they get something from the system. Nor are

Figure 2.2.3: A terminator. Think of it as another system to which your system has a data connection.

you interested in modifying the applicants' actions, so there is no need to study them. Provided you can be reasonably sure that your analysis study and the implemented system will have no effect on the behavior of the applicants, then you can declare APPLICANTS a terminator, and therefore outside the scope of your study. From the point of view of your system, your only interest is the data connection with the terminator.

Consider the fragment of the data flow diagram shown in Figure 2.2.4. The *boundary data flow* APPLICANT REGISTRATION enters the system from the outside world. It is sent by the terminator APPLICANTS. In this piece of the diagram, the analyst is declaring, "I know all about the data flow APPLICANT REGISTRATION because I have declared its definition in the data dictionary. I don't know what is happening inside APPLICANTS because by showing it as a terminator, I am saying that it is outside my system. So my system's processing, and therefore my interest, begins when the data flow arrives from the terminator."

Figure 2.2.4: The data flow APPLICANT REGISTRATION from the terminator determines the limit of the system being studied. We call it a boundary data flow.

Let's think about the role of a terminator. Imagine a typical applicant. Call her Sarah Typical. Sarah wants a job. Before the agency can do anything for her, it needs to have some information about Sarah: her name, address, date of birth, employment history, relevant skills, and so forth. Naturally, this collection of data must flow into the system before it can be processed. We show this data flow in the model by the arrow labeled APPLICANT REGISTRATION.

Data Flow

A *data flow* is a roadway for data, traveled by "trucks" carrying loads of data. Of course, they travel in the direction of the arrow. Each truckload contains the same collection of data elements. Naturally, the values of the data elements will change, but the elements themselves are constant.

Figure 2.2.5: A data flow is a roadway for moving a collection of data of known composition.

The APPLICANT REGISTRATION data flow carries information to the process REGISTER APPLICANT. Like Sarah Typical, each applicant must supply the required information (her name, address, date of birth, skills, employment history, and so on). For the system to work, the data flow must carry the same collection of data elements each time it appears in the system. For you to make sense of this part of the system, you must know exactly what that collection contains. This is how the data dictionary fits into the model. The data dictionary is where you define the data items that make up the data flow.

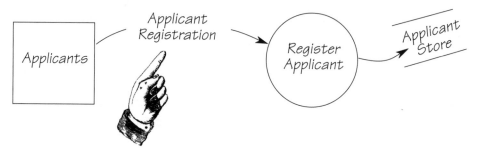

Figure 2.2.6: The data flow contains the information that the process needs to register an applicant.

Applicant Registration = Applicant Name + Applicant Address + Date Of Birth + Skills + Employment History

Figure 2.2.7: Sample data dictionary definition. Every data flow in the model is of known composition and is recorded in the dictionary.

The name you give to the data flow has a lot to do with the readability of the diagram. By way of an example, go back to Figure 2.2.2 and replace all the data flow names with, say, the names of your friends. Your friends may have nice names, but they make the model unreadable. So it pays to give careful thought to the names you assign to data flows. The more precise the names, the easier it is for your readers (and you) to recognize the system.

The medium that carries the data is not important; you are interested only in the data content of the flow. The data in the flow APPLICANT REGISTRATION may be telephoned or written. Sarah Typical may have come into the agency in person and told the agency personnel what they need to know. Any of the above will make the system work. The data dictionary entry for APPLICANT REGISTRATION shows only the data Sarah delivers. It makes no mention of Sarah's pleasing personality.

However, simply moving data around the system is not enough. The system has to process the data. (After all, it is a data processing system.) So, how does it do this?

Process

A *process* transforms the incoming data flows into the outgoing flows. Consider what happens to Sarah's application when it enters the process shown in Figure 2.2.8.

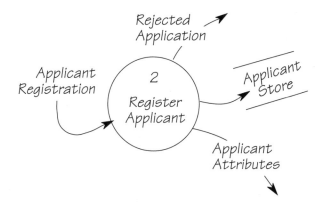

Figure 2.2.8: *The* REGISTER APPLICANT *process changes the* APPLICANT REGISTRATION *into a data flow called* APPLICANT ATTRIBUTES, *and an entry labeled* NEW APPLICANT *in the* APPLICANT STORE.

The APPLICANT REGISTRATION is transformed by the process REGISTER APPLICANT. Part of the process checks the application. You know this because there is a flow called REJECTED APPLICATION. If the application is approved, the process uses some of the data from APPLICANT REGISTRATION to record the applicant in the data store

APPLICANT STORE. This store remembers the data for later use. Some of the same data make up the flow called APPLICANT ATTRIBUTES that is then used by another process. You describe the process in detail when you write its mini specification.

Data Store

A *data store* is a repository of information until the process needs it. The process REGISTER APPLICANT adds information about the applicant to a data store called APPLICANT STORE. The system has to remember every

> The data stores are the system's memory.

applicant, so there is a requirement to store data. If the applicant is not remembered, the process FIND SUITABLE APPLICANTS could not function correctly.

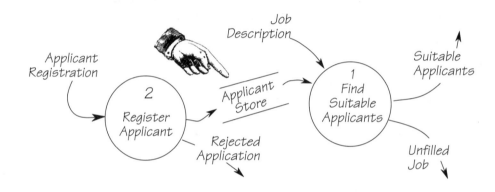

Figure 2.2.9: *Every new applicant is stored in the APPLICANT STORE. When new jobs arrive, the process FIND SUITABLE APPLICANTS searches the store of applicants to find people whose qualifications match the job requirements.*

Data stores are shown in the data flow diagram by a pair of parallel lines with the store's name between them. Some people call data stores "files," but please don't be one of them. You may show a data store in one of your models called, say, aircraft file. However, from a modeling point of view, this is a data store, not a file. The term "file" implies a computer file, and that excludes the other devices used to store data. Books qualify as data stores, as do filing cabinets, index cards, and sometimes people's memories. Again, the data being remembered, not the media, are what are important.

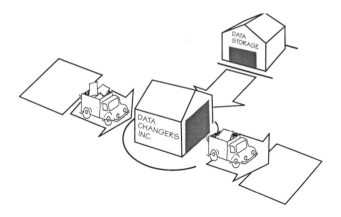

Figure 2.2.10: Putting it all together. Something happens in the terminator to generate a load of data. The data are driven along the data flow until the data reach the process. The process acts like a factory, converting the raw material (the incoming data flow) into the finished product (the outgoing data flow). During this process, the factory uses data from the store.

A Tour of the Model

Now let's look at the rest of the system, picking up with the data flow APPLICANT ATTRIBUTES as it carries data to the process FIND SUITABLE JOB.

FIND SUITABLE JOB searches the JOB REGISTER, looking for an unfilled job whose requirements match the attributes of the applicant. The company has established rules for matching applicants to jobs. In fact, all the processes on the data flow diagram have their own rules. Part of your analysis task is to discover these rules and record them in the mini specifications.

> Each process is described by a lower-level data flow diagram, except when it is small enough to be called head-sized, in which case it is described by a mini specification.

Sometimes, the process is too large and complicated to write a mini specification. In that case, think of the process as a lower-level mini system. Instead of writing a specification, you draw another data flow diagram showing the details of the process. If the processes of the lower-level diagram are still too complex, each of them in turn becomes the subject of yet a lower-level diagram. This decomposition process continues until each piece is small enough to be specified in a one-page mini specification. This technique is known as *leveling,* and Chapter 2.7 *Leveled Data Flow Diagrams* describes it in more detail.

In the example, the process FIND SUITABLE JOB is simple and concise enough to be described by a mini specification. In fact, all of the processes in this diagram are small enough to need no further breakdown. You can tell this by inspecting the data flows that enter and leave the bubbles, and realizing that you know or can imagine the processing that goes on inside them.

Once the process FIND SUITABLE JOB has carried out its task, either MATCHED APPLICATION or NOTHING FOR YOU is issued. NOTHING FOR YOU tells the applicant that there are currently no jobs requiring the skills contained in the APPLICANT

ATTRIBUTES. But let us suppose that Sarah Typical has the necessary attributes to be matched to a job. The diagram tells you that the data flow MATCHED APPLICATION activates the process ARRANGE INTERVIEW.

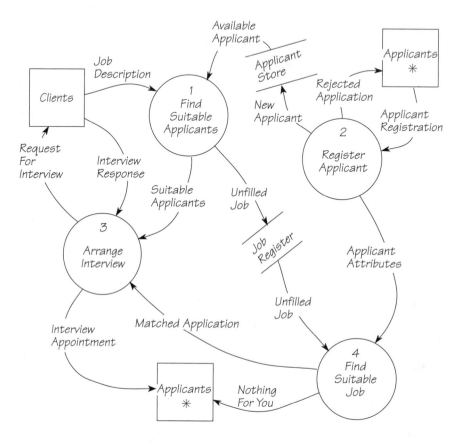

Figure 2.2.11: The Nelson Buzzcott Employment Agency system.

When either MATCHED APPLICATION or SUITABLE APPLICANTS arrives, ARRANGE INTERVIEW sends out a REQUEST FOR INTERVIEW. This contains some of the information about the candidate, and some suggested times for interviews. The CLIENTS review the information to decide if they wish to interview the applicant, and their intentions are communicated via INTERVIEW RESPONSE.

You do not know, nor do you care, how the clients decide. The clients are shown as a terminator, thus indicating they are outside the domain of your analysis. However, you are concerned with the data the clients need in order to make their decision. The data content of REQUEST FOR INTERVIEW must be acceptable to the terminator. When a client responds favorably, an INTERVIEW APPOINTMENT is sent to the applicant.

> Each process is activated by the incoming data, and has no knowledge of the other processes in the system.

We've not described all of the system shown in Figure 2.2.11, and you should look through it before reading further. As you do so, note that each process is activated by the arrival of an incoming data flow. When the data flow arrives, it provides the material to be processed. A process has no notion of where the data flow has come from, nor where the outgoing data flow goes after it leaves. When a process has finished, it is dormant until the next flow arrives.

A data flow diagram is an *asynchronous model* in that it represents the system as a network of independent processes. Because it is a network, there is no synchronization of the system as a whole. There is no centralized controller of the processes, telling them when it is their turn to be active. Why is this so? The reason is you are using the model to describe the *requirements* for the system. The way in which the current system is controlled is not neces-

> The data flow diagram models the system purely from the point of view of the data, and ignores any control or management structures the system may currently happen to have.

sarily a requirement for any future system. You do not yet know what the implementation of the future system will be, so you certainly can't be concerned with its control structure.

Context Diagram

We have talked about modeling a system by breaking it into its functional pieces. How do you get started? And how do you know what the system is in the first place? How big is it? Where does it start and end? Before beginning to analyze, you must know the scope of the system. The best tool for this purpose is the context diagram.

The context diagram is like the markings on a tennis court. Once the boundaries of the playing area have been established, the game can commence. The *context diagram* defines the boundaries of the system being studied by showing how it connects to the outside world. It is largely built before you begin to break the system into its functional pieces.

The context diagram shows how the system under study connects to the systems surrounding it. It is these connections, or boundary data flows, that tell you the precise scope of your study. You use the context diagram to demonstrate the scope of the systems analysis to your users. Without an agreed context, there is no way of telling if the correct system is being analyzed. There is no real place to start the analysis, and no real place to stop.

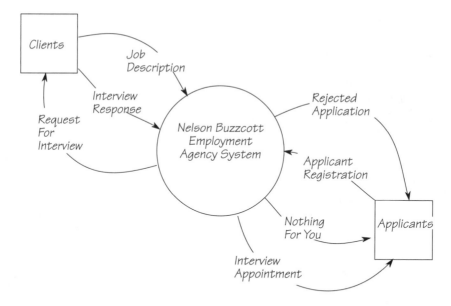

Figure 2.2.12: *The context diagram for the Nelson Buzzcott Employment Agency system. Here you declare that you do not intend to study how* CLIENTS *react to a* REQUEST FOR INTERVIEW. *Similarly, the processes of the* APPLICANTS *are not the concern of the analysis study. The context diagram states that the study will begin and end with the flows to and from the terminators.*

The context diagram is a data flow diagram just like any other, but it differs in that all the system's processes are collected into one bubble. This bubble is like all of the others that you have seen. It transforms the incoming data flows into the outgoing ones. The Rule of Data Conservation applies to the context bubble as well: It must receive all the data flows that are necessary to produce the system's output.

A common problem with analysis projects is that the context is not large enough. How big should it be? Most systems analysis projects are undertaken because a computer system will eventually be built. Should the context include only the computer system? Generally, the system's scope defined by the context diagram must be larger than the anticipated computer system. In fact, it must be large enough to include the manual and mechanical tasks that surround the computer system. The rule of thumb to use here says that the terminators should be unaware that your new computer system has begun operating.

> The context diagram must include the anticipated automated system and the surrounding manual tasks.

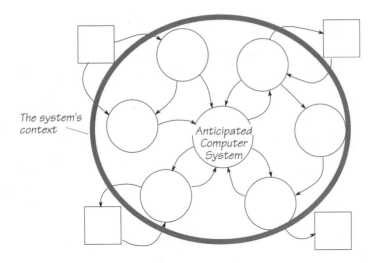

The system's context

Anticipated Computer System

Figure 2.2.13: The context of study includes the anticipated computer system, as well as the surrounding manual processes.

When the ultimate objective is to build a computer system, isn't it a waste of effort to study the activities that are outside the computer? It's not wasted effort because these activities may affect the precise content of the data entering and leaving the computer system. You must understand the manual processes that produce the input and receive the computer output so that you can ensure the exact content of both the input and output. Additionally, the manual processes may change because of your new computer system. If they do, you need to know how they will be affected, and who has to be trained for their new tasks.

The exception to this rule is when the terminator is another computer system, which supplies data directly to the new one. In this case, there is no manual process between the two computer systems.

Summary

A data flow diagram is used to build a model of a system. You do this by partitioning the system into the flows of data (hence the name data flow diagram) and the processes and stores that use those flows.

The data flow diagram is a working model of the system being studied. It works because you can prove that each process in the model can produce its output from the input data flows. The value of building these models is that they are readily understandable by the users of the system. Moreover, when you have finished studying the system, the data flow models become part of the specification of that system.

Enough reading! Let's do some work.

What to Do

The three exercises that follow will give you some practice using data flow diagrams before you apply them in the Piccadilly Project.

Do each exercise and look at the sample answer in Chapter 4.2 before going on with the next one. Also, read the commentary provided there and make sure you understand it before proceeding. If necessary, re-read the appropriate part of this chapter.

Keep your answers. It will be illuminating to look back on them as your modeling skills improve over the course of this book.

Exercise 1: Nelson Buzzcott's Employment Agency

You have seen the data flow model for Nelson Buzzcott's system already. Have another look at it (Figure 2.2.14), then answer the questions.

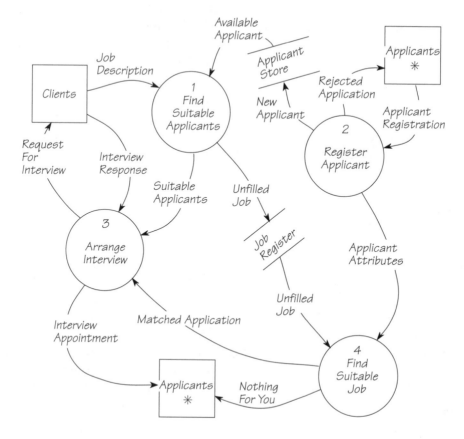

Figure 2.2.14: The employment agency data flow diagram discussed earlier.

a)   What is the most probable data content of MATCHED APPLICATION? Write down the data items that you think are necessary for this flow.

b)   What kind of information would you find in JOB REGISTER? Consider the data needed to support the process that uses this data store.

c)   Which is processed first: process 1 FIND SUITABLE APPLICANTS, or process 2 REGISTER APPLICANT?

d)   Why isn't CLIENTS shown as a bubble?

e)   What happens to the data flow UNFILLED JOB? Why?

f)   How do we know that a job has been filled?

Exercise 2: The FastBuck Book Company

The FastBuck Book Company was formed a few years ago as a partnership between a failed bookmaker, a snake oil seller, and the nephew of a mob chief. Despite police attention, the company has managed to remain in business for a number of years.

Here's how the company does business: Orders are sent in on coupons that customers clip from magazines such as *Slug Lovers' Monthly* and *The Gravel Journal*. (The company tends to advertise in magazines with a readership profile showing high gullibility.) Naturally, this is a cash-only business. Any orders received without payment are discarded.

The company doesn't actually publish books. It simply has a supply of different covers. The appropriate cover for each order is glued to the generic book, which is sent to the hapless customer. Long-standing customers may receive several copies of the generic book with different covers. This is regarded by FastBuck as a hazard of being a customer.

The generic book was written in 1908 by a prominent bore. It rambles over such a wide range of topics that it is almost impossible for customers to complain of irrelevance. The book is also so stunningly soporific that it is doubtful anyone has ever read it. Complaints are few at FastBuck, and those that do arrive are ignored.

Repeat customers are in for a special treat at FastBuck. Those who attempt to deal with the company on more than one occasion are thought to be naive enough to fall for anything. An invoice for the previously ordered book is generated and sent in the hope that the customer will pay the invoice, even though cash arrived with the original order.

Benedict Shady is an analyst who had an unfortunate relationship with the daughter of one of the partners—unfortunate because the daughter is now in a delicate condition, and doubly unfortunate because her father is the one with mob connections. In return for not being thrown off the Brooklyn Bridge, Benedict has agreed to do an analysis of the business.

Benedict is presumably a better lover than analyst. His model is shown in Figure 2.2.15. Write down all the errors that you can find in Benedict's model, then compare your list with the one in Chapter 4.2.

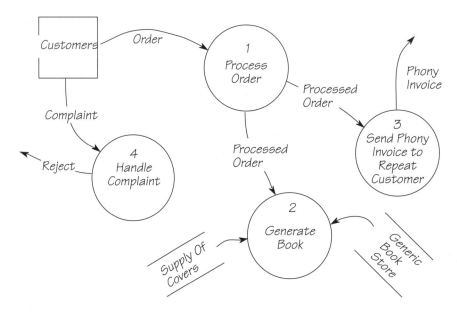

Figure 2.2.15: The data flow model, which Benedict built, appears to have some problems.

Exercise 3: The Government Research Paper Clearing House

The federal government maintains the Government Research Paper Clearing House for unclassified research and development papers. The Clearing House sends scientific research papers to any citizen who asks for them. So that citizens know what is available, indexes of recently published papers are sent out each month.

Citizens may request to be added to the subscribers list for no charge and, upon being added, will be sent all the indexes currently on file.

For a fee, a citizen can request to have the full paper sent to him. This request must first be checked to see that the citizen is a subscriber and the index is correct, then the paper is sent. The invoicing for this service is handled by the Accounting Office.

The procedure for mailing has been in place for some years. When a paper is requested, the clearing house generates an address label with a tear-off strip identifying the paper to be sent. Each weekday evening, an order for the required papers is sent to the Papers Store. The store sends the papers the following morning, and the Clearing House workers match the papers to the labels and ship them. The tear-off portion of the label is discarded before shipping.

The indexes are maintained and shipped by the Index Department. This department is run by two women who use personal computers to maintain their list of subscribers. They are also responsible for shipping the indexes to subscribers. The women do their jobs very well. In fact, they have won awards several times for their efficiency.

Draw a context diagram for the activities of the Government Research Paper Clearing House. Remember that a context diagram has only one bubble to represent all of the system you intend to study. Remember also that the data flows around the boundary of the system are the most important element of your diagram. Check your diagram against the one in Chapter 4.2.

Trail Guide

You may finish here or in Chapter 4.2 after completing the exercises. For your convenience, we duplicate this Trail Guide in Chapter 4.2.

● Easiest: Now that you know about data flow diagrams, go to Chapter 1.2 *Start with the Context,* where you will build a context diagram for the Piccadilly Project.

■ More Difficult: If you made it here, perhaps you should be following the ● Easiest Trail. If you have not already built the Piccadilly context diagram in Chapter 1.2 *Start with the Context*, that is your destination. Otherwise, go to Chapter 2.3 *A Variety of Viewpoints.*

◆ Most Difficult: Any chapter number beginning with a "2" is not part of your trail. Try picking up your trail in Chapter 1.2 *Start with the Context.*

❋ Promenade: The purpose of data flow diagrams is to model the processes and the flows of data between the processes. The data in a system are just as important as the processes, so the next logical step for you is to learn about the data. Jump to Chapter 2.9 *Data Dictionary.* Don't worry about the chapters you pass over to get there. Eventually, you will return to learn about the models in those chapters.

2.3 A VARIETY OF VIEWPOINTS

Before You Reached Here ...

● Easiest and ■ More Difficult: Your last official chapter was 1.2 *Start with the Context,* where you built Piccadilly's context diagram. You could also have come here from Chapter 3.1, where we gave our answer.

◆ Most Difficult: This chapter is part of the Textbook, which is not part of your trail. Your trail runs through the Project chapters. Of course, you can always jump to an easier trail; they all pass through this chapter. To rejoin your own trail, go to Chapter 1.3 *What About the Business Data?*

❄ Promenade: You have come here from Chapter 2.9 *Data Dictionary.*

Models and the Need for Different Viewpoints

Systems analysis is the craft of understanding systems by building models of them. However, today's systems are large and complex, and while analytical models need to be accurate, they need to avoid becoming as large and complex as the systems they seek to represent. Viewpoints are the way that systems analysts can conquer the complexity and size of today's systems.

Viewpoints are a justified distortion of the reality being represented. Using viewpoints, analysts can build models that include only as much information as they need to see. This doesn't mean that models are a false representation of the system. It does mean the models are more usable because they show what the analyst needs at the time, and such models don't burden analysts with details that can be delayed or shown in another model.

A well-known example of justified distortion is the map of the London Underground shown in Figure 2.3.1. Anyone who has visited London has probably seen, if not used, the map of its Underground. Each of the "tube" (subway or train) lines is in a different color running horizontally, vertically, or at a 45-degree diagonal. This map uses a viewpoint. It shows the way the stations are connected, and discounts their actual geographical location. Why? Consider the people using the map. They want to travel from one tube station to another. What matters to them is the way their two stations are connected. By straightening out the lines, the map designer makes it easy for travelers to see which lines and which interchange stations must be navigated before they arrive at the intended destination.

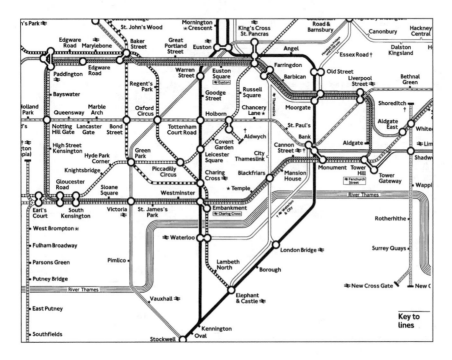

Figure 2.3.1: *This representation of the London Underground* is displayed throughout the trains and stations, and frequently appears on postcards and T-shirts. The map distinguishes the lines of the network with different colors.*

In reality, the Underground is not at all like the colored map. A topographical map showing the true layout of the lines would be almost unrecognizable to Londoners. The topographical map represents reality, but it has not been used since the other map was produced in 1933. Why not? The reason is simply that it shows a viewpoint that is of little use to Underground passengers. When you are a hundred feet below the streets, you have little interest in knowing whether you are passing under the Old Bailey or Nelson's Column.

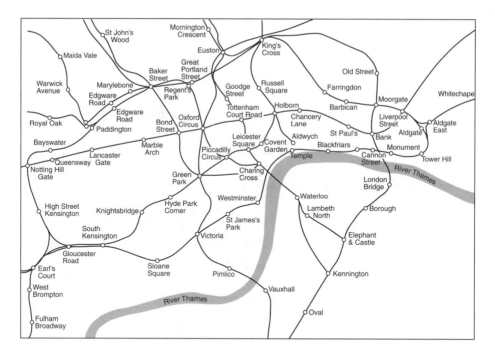

Figure 2.3.2: Topographical map of the London Underground. This map shows the correct geographical location of the lines and stations.

The stylized map in Figure 2.3.1 takes great liberties with the geography of London, but the distortion is entirely justified. By filtering out unnecessary information, it presents a viewpoint that is far more useful for its intended audience.

Today's information systems are big, and more and more systems serve large areas of the users' business. In the past, it was sufficient to build a computer system

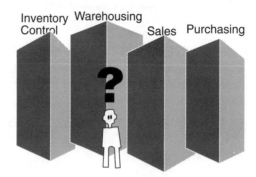

Figure 2.3.3: Yesterday's inventory report is no longer enough. Today, the computer system must cover inventory control, warehousing, sales, purchasing, and more.

that produced a single report. Those days are gone. Today's computers are more powerful and capable of accommodating elaborate systems. Increasingly sophisticated users are expecting integrated systems that bring together the different activities of their businesses, as well as higher levels of specialization and functionality within each application.

Filtering Information

The challenge for today's systems analysts is to deal with these complex systems and users' expectations. The solution may be to erect a filtering mechanism as a defense against the barrage of information.

For example, if you live in a large city, chances are that each day, you are exposed to several thousand pieces of advertising material. The average American sees more than eighty television commercials each day. Sunday editions of the *New York Times* have carried up to 350 pages of advertising. Italian and French drivers negotiate a forest of eye-catching billboards erected beside their roads and on conspicuous buildings. Add to this the magazines, newspapers, subway posters, shop displays, handbills, and packaging that everybody sees. Each day, we are exposed to an amazing barrage of advertising material urging us to buy something or other.

Figure 2.3.4: Did you take careful notice of all the advertising you saw or heard yesterday? Were you impressed by all the claims of bigger, better, newer, or brighter? You probably ignored most of them or gave them at best a cursory glance before continuing on your way.

What is it that keeps you from absorbing this oversupply of advertising material? It is your *viewpoint*. You filter out the advertisements you don't need or don't want to see at the moment, and look at those you do. You ignore the advertisements for coffee if you are a tea drinker. You skip the ads for bottled water if you like tap water. Most men disregard the pitches for women's cosmetics and most women ignore the electric shaver commercials, except at Christmas and birthdays. Your viewpoint is saying, "I don't need to see that now." You keep your sanity by filtering out unneeded information.

Similarly, systems analysts must filter information to be effective. As systems get larger and extend to more areas of the users' business, the number of people who have knowledge of part of the system increases. More people means a wider variety of attitudes, skills, and vested interests confronting analysts. To deal with the increasing number of user groups, analysts can respond by building system models that show the viewpoints of greatest interest, or most relevance, to each audience.

Analysts also need to filter out what is not needed at the moment. When analyzing a large system, they need to distort the reality by modeling the viewpoint of the system that contributes most to their understanding. The viewpoints we've found most successful for our own projects, and the ones we'll introduce in this book, are the *current physical, essential, data,* and *new physical.* Let's meet them.

Current Physical Viewpoint

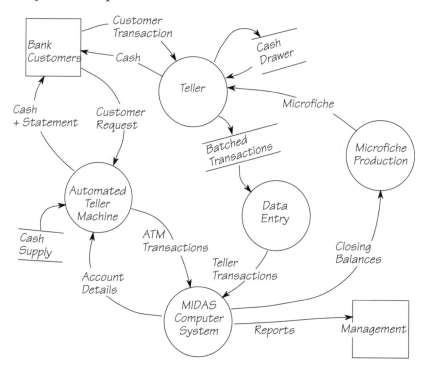

Figure 2.3.5: *The current physical model focuses on a system's current implementation. While it shows the system requirements, the model also includes the processors that do the work. This view usually shows departments, people's jobs, existing computer systems, machines, and so on.*

The *current physical viewpoint* is used in a limited way at the beginning of a project to establish the system's context of study, and to provide users with recognizable models. Users are initially more comfortable if they see models that show the people and machines currently performing the tasks. It also helps the analyst to identify problem areas, to get to know the sources of the system's information, and to assess the impact of the future system. We discuss this topic fully in Chapter 2.8 *Current Physical Viewpoint*.

Essential Viewpoint

The *essential viewpoint* is considered the "perfect" view of the system; it shows only the requirements and intentionally excludes anything that exists because of the way the system is designed and/or implemented. Filtering out the system's current technology is desirable if you are to select the best possible future implementation. The essential viewpoint is necessary for any project. We've found it to be the most useful, and therefore the most used. You will meet this viewpoint in Chapter 2.10 *Essential Viewpoint*.

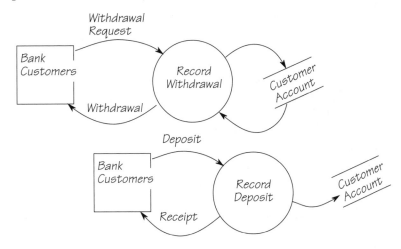

Figure 2.3.6: The essential viewpoint, which shows only the business policy of the system and ignores the machinery that does the work.

Data Viewpoint

The third viewpoint we find useful focuses on the system's information. This *data viewpoint,* as represented by the *data model,* ignores the processing part of the system, and instead concentrates on modeling the information that is essential to the system. This viewpoint and notation will be introduced in Chapter 2.4 *Data Viewpoint*.

135

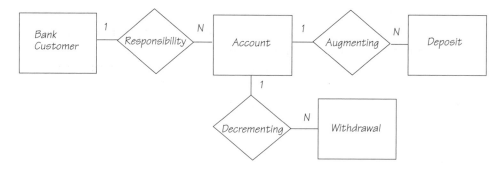

Figure 2.3.7: A data model of the system. This abstract viewpoint looks solely at what the system has to remember.

New Physical Viewpoint

The *new physical viewpoint* and its associated model is used to illustrate, negotiate, and define the implementation of the new system by computers, humans, and machines. This viewpoint is also known as the *new implementation environment* or the *preliminary design model.* The new physical model uses the same notation as the current physical. You will meet this viewpoint in Chapter 2.14 *New Physical Viewpoint.*

Using Viewpoints

The usefulness of each of these four viewpoints depends on the particular audience and the situation. All of the viewpoints are described in the Textbook, and individual chapters discuss when each viewpoint is most effective and what type of audience and situation calls for a particular viewpoint.

Unfortunately, life is not so simple as to allow analysts to build a complete model of one viewpoint, verify it, and then move on to the next one. The information required for each viewpoint, as well as the users who supply the information for the viewpoint, will appear at random times.

As the project analyst, you cannot expect to get all the information you need when you need it. You have to be prepared to capture it whenever it becomes available. As a result, you will typically have several models, each showing a different viewpoint, in various stages of completion. This does not hinder the analysis effort. Your models, though incomplete, enable you to continue gathering the requirements.

Summary

Although not every project will use all of the four viewpoints introduced here, we suggest that you use them all in the Piccadilly Project. The context diagram, which you have already built, is a starting point for the current physical, essential, and data models. The context diagram is used to set the scope of the project and to give you some understanding of an unfamiliar business. As you will see, the current Piccadilly system is largely manual, so you'll need to filter out the manual features to find the underlying essential policy. Finding the policy is made easier by first studying the system's stored data, so very early in the project you will construct a data model of Piccadilly's information. Once that is underway, the complete essential requirements are modeled.

For most of the time, you'll model the Piccadilly system's essence. Toward the end, when the essential viewpoint is largely complete, you will select the new operating environment and construct a new physical viewpoint of your implementation recommendations.

The Trail Guide will lead you to the viewpoint chapters when you need them and to the chapters that discuss the various models that represent the viewpoints.

Trail Guide

● Easiest and ■ More Difficult: Our first viewpoint in the Piccadilly Project is that of the data remembered by the system. We discuss this topic in Chapter 2.4 *Data Viewpoint,* your next destination.

◆ Most Difficult: Unless you want to follow the easier trails and read about stored data, you should already be in Chapter 1.3 *What About the Business Data?* If you find yourself having difficulty with modeling Piccadilly's data, you can always backtrack to Chapter 2.4 *Data Viewpoint.*

❇ Promenade: We are taking you in a different direction. While it may be a large jump in chapters, it is conceptually a small leap for you to go on with process modeling. Continue your promenade in Chapter 2.8 *Current Physical Viewpoint.*

2.4 DATA VIEWPOINT

Before You Reached Here ...

● Easiest and ■ More Difficult: You've arrived here from Chapter 2.3 *A Variety of Viewpoints,* where you have read about several different ways to see a system. In this chapter, you'll look at the system from the viewpoint of its stored data.

◆ Most Difficult: This chapter is not on your trail. However, understanding data is an important part of systems analysis, so consider staying here now that you have arrived. If you intend doing the case study without any coaching from the Textbook, Chapter 1.3 *What About the Business Data?* is the place to be.

❄ Promenade: You've come here from Chapter 2.8 *Current Physical Viewpoint.* Now you will shift perspective and look at the system from the point of view of what it remembers.

Adventures in Data Modeling

Mysterious things are happening at the NanoSoft Corporation. Its people are completing their software projects on time and within budget.

Unfortunately, you work for PicoSoft. Naturally, your boss wants to know how NanoSoft does it, and has selected you to go on a spying mission. You must penetrate their office sometime after the NanoSoft systems developers go home (they are always able to leave on time) and discover their secret systems methods. If you are caught or captured, the Secretary will deny any knowledge of you and your mission.

Sometime after dark, you enter the office through an air shaft. It's very dark inside, and you have to feel your way. Blind groping gives way to limited sight as your eyes become accustomed to the dark. You discover you're in the part of NanoSoft where the analysts work. (You spot a sign saying "Analysts' Section," and you have terrific powers of deduction.) There is a book on each of the desks. You grab one and stuff it into your trench coat pocket. (Of course you're wearing a trench coat. What else could you wear on a spying mission?) More wandering in the dark and you find a whiteboard attached to the wall. The board bears the heading "New Projects." You presume this has something to do with their new projects (powers of deduction again), and you unscrew it from the wall. As you grope your way toward the exit, you pass a notice board. The printing on one notice is barely discernible in the dark, but the heading "Analyst" catches your interest. You remove the paper from the notice board, then make your escape through the unlocked basement.

NEW PROJECTS   REUSE

		REUSE
MIDAS Currency		97,104,22,76
Foreign		24, 77, 112
DEALING Bidding		80, 56
Stockbroking		85, 14, 109

Figure 2.4.1: What's a nice analyst like you doing in a place like this?

Back in the office the next morning, your workmates and the boss gather 'round as you show off your ill-gotten plunder. The first two items are revealed (in shameful daylight) in Figures 2.4.2 and 2.4.3.

"Can you make sense of all this?" your boss asks.

ANALYST NAME: C. HAWKINS

Model Number:	Business Area:
42	Sales
142	Sales
63	Accounting
79	Stockbroking
85	Stockbroking
97	Currency Trading

Figure 2.4.2: A page from one of the analysts' books.

Figure 2.4.3: The analysts' whiteboard.

	NEW PROJECTS	REUSE
MIDAS	Currency Trading	97, 104, 22, 76
	Foreign Loans	24, 77, 112
DEALING	Bidding	80, 56
	Stockbroking	85, 14, 109

"No problem," you say. "While nobody's here from NanoSoft to tell us what they do, I know what they remember. After all, they're writing this stuff in books and on notice boards because they have to remember it. *Once we can understand what they remember, we'll understand what they do."*

And then you tell them how to do it. "The book belongs to an analyst. All the analysts have one. Using the facts they're recording, I can build a data model for part of what they are doing. It looks like this," you say as you draw Figure 2.4.4.

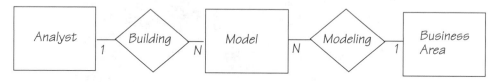

Figure 2.4.4: *The diagram you drew to explain NanoSoft's working policy.*

"Each of the boxes represents whatever information they need to remember about a subject. For example, I know they have a unique number for the model, so models must be important to them. I've drawn a box for MODEL, and one of the data items inside the box is a model number. I may find more information about the model later, and I will store it in that box.

"The analyst's name is on the book, which indicates that the analyst is important, so I have another box to store any data about ANALYST. The book tells me that analysts are interested in models and business areas, so I add another box for BUSINESS AREA.

"The diamonds give the reasons that they keep the information. You can see here where it says that an analyst has a BUILDING relationship with a model. The analysts write down all the models that they build, and this indicates there is some reason to remember which models are built by which analysts. Of course, an analyst can build more than one model, so I show the relationship as one analyst is building N, or many, models. While I know there is a reason to remember which analysts build which models, I don't yet know what it is.

BONUS LIST for this year	
C. Hawkins	$40
J. Coltrane	$30
L. Hampton	$50

Figure 2.4.5: *The notice removed from the board.*

"Now look at the whiteboard. Each new project covers a number of business areas. Under the heading 'Reuse' and beside each business area are certain numbers. Look, some of the numbers are the same as the model numbers in the analyst's book. This has to mean there's a connection between the new project and the models. Or, to be more precise, I can add a diamond to link BUSINESS AREA to MODEL.

"Now for the paper from the notice board." (This is shown in Figure 2.4.5.)

"This name at the top of the list is the same as that on the book, so that establishes a connection. If I add the connection to the model, I get Figure 2.4.6."

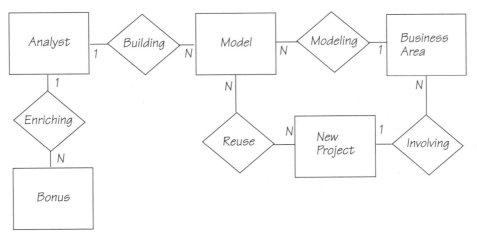

Figure 2.4.6: The finished model.

"What does it mean?" asks your admiring boss.

"It means," you begin, "that we now know how NanoSoft is completing software projects on time and within budget—the analysts are reusing their models. Look at this model. It tells you they keep track of which models have been built by which analysts, and which business areas the models cover. When they start new projects, they look at the business area of that project, and decide which models from previous projects can be reused. That must save them a lot of time. No wonder they always finish their projects within budget.

"Now here is the best part. Whenever they reuse a model," and here you pause for dramatic effect, "the originating analyst receives a ten-dollar bonus."

Your boss and workmates are now looking at you with admiration. "Could this be the time," you wonder to yourself, "to ask for a raise?"

Learning What a System Does By What It Remembers

In your secret mission to NanoSoft, you learned about their policy of reusing models. (Reusing models is a good idea, and we'll discuss this topic later in this book. For the moment, however, forget their policy and concentrate on how you learned it.) You analyzed NanoSoft's memory and found out that it was in the form of the data they stored on notice boards and in books. When you modeled these data, you revealed the policy, or the reason, for storing the data.

The task of analysis is to understand a system. The data flow model helps you to understand the system by breaking up its functionality into small, digestible pieces and by revealing the data flows and stores used by those pieces. The data model, by contrast, reveals the system by connecting the pieces of stored data used to support the system's processes.

> The facts remembered by a system reveal what that system does. Your task as an analyst is to understand the system. By investigating why the system's facts are remembered, you will uncover the system's reason for existence.

The memory of any system reveals almost everything about it. For example, what is stored in your own memory? You have a unique set of information collected from your experiences, work, family, education, environment, and so on. There is no other human with a memory identical to yours because there is no other human who has done exactly the same things you have. People of different cultures and in different occupations have different facts to remember. The memory of an Alaskan oil worker is substantially different from that of a South Sea island fisherman. Similarly, you would know if someone is a brain surgeon or a boat builder by knowing the contents of his or her memory.

Why Analyze the Stored Data?

There are certain advantages to analyzing a system by inspecting its memory and building a data model of it:

1. The data model gives a global view of the system. The system's memory is shared by most of the processes in the system and is not owned by any single process. By building a data model, you understand the policy in a broad, system-wide context, not just in isolated fragments.

2. The data model is an abstraction. It looks at the information without regard to how the data are kept and without regard to whether or not the data are intended to be part of any future computer system. This means that your view of the business is in no way limited by technological constraints.

3. The data model specifies the requirements for stored data, but they are independent from the technology that uses the data. This device-independent specification means that you can subsequently design files that insulate the

users' applications from the storage medium. The result is a system that will be more easily adapted to future technological changes.

4. The data model represents the real purpose, or business policy, of the system. Thus, by building this model, you ensure that you ask all the right questions about the policy. The model then gives you a vehicle for ensuring that all the answers you get provide a consistent view of that policy.

A commonly (but falsely) held belief is that you can define the stored data requirements by simply providing data that are sufficient to produce the output of the current computer system. This working-backward-from-the-reports approach makes the erroneous assumption that today's computer output meets the essential requirements for the business system and will continue to meet them for any future incarnation of the system. Such an approach also means that the analysis is being unnecessarily restricted to the current computer system. The computer output is only part of the story, limited by the technology available at the time the output was first generated.

To discover the rest of the story, you must understand the underlying reason for keeping the data in the system. The reason for storing or retrieving the data is as important as the data itself.

Summary

The data model is an abstraction. It shows the system's stored information only; and because of its viewpoint, it disregards the processes that use the information. It also shows the information without any regard to the way that the system happens to implement the information.

For example, the model doesn't consider whether the data will be stored in a computer database, in a book, remembered by a person, or whatever. When you look at the data separately from the storage technology, you ensure that you see the real data, not data organized for the convenience of some database or filing arrangement.

The data model is another analysis tool that you'll use when you are studying systems. In this case, you are studying what the system remembers and, by knowing what it remembers, you are able to determine what it does.

Let's go on to build some data models.

Trail Guide

● Easiest: You will find out how to build an abstract data model in Chapter 2.5 *Data Models*.

■ More Difficult: This trail is about to become more difficult, as it goes straight to the Piccadilly Project and to Chapter 1.3 *What About the Business Data?* Be forewarned: The Project yields a substantial model, and if you are not well-versed in data modeling, you may need to follow the ● Easiest Trail for a few chapters as a refresher course.

◆ Most Difficult: This is not an obligatory part of your trail, and you should turn to Chapter 1.3 *What About the Business Data?*

❅ Promenade: Now that you know the reasoning behind the data model, you can read about how they are built. Your next chapter is 2.5 *Data Models*.

DATA MODELS 2.5

Before You Reached Here ...
● Easiest and ❈ Promenade: Chapter 2.4 *Data Viewpoint* has discussed how to understand a system by looking at its memory or its stored data. This chapter discusses the technicalities of building such a model.

■ More Difficult: This chapter deals with how to build a model, and your trail bypasses it. As you are here, take the time now to look through this chapter. You may find some new variations on the data modeling technique. If you want to rejoin your own trail, turn to Chapter 1.3 *What About the Business Data?*

◆ Most Difficult: You should be at Chapter 1.3 *What About the Business Data?* but feel free to stay here to review this material. A guide at the end of this chapter will steer you back to your chosen trail.

Role of Data Models

A fundamental task of information systems is to store and retrieve data. Regardless of whether the system uses computerized files or manual ledgers to keep its data, the system must remember facts about subjects that are important to the system.

So far in the Piccadilly Project, you have modeled only the processes that transform and/or store data. Apart from adding data stores to the data flow diagrams, you have not modeled any of the data. This is like forgetting to breathe because you happen to be eating. It is necessary in systems analysis to model both the data and the processes, and to model them at the same time. If the stored data are not understood, there is little chance of understanding the processes that store the data.

> Without knowing what the system remembers, you cannot know what the system does.

To build the data model, you use the idea of partitioning as you did for the data flow diagram. However, instead of using flows and processes, you use *entities* and *relationships* to partition the system. Data models are also known as entity-relationship diagrams, Chen diagrams, and information models, among other aliases.

In this chapter, we'll demonstrate how to build a data model to reflect the system's business policy. We'll also discuss how definitions of the existing stored data may help you to build a data model. Neither of these methods is perfect. After we have taken you through event partitioning, we'll show you another strategy. However, you first need to understand the role of the data model and how it represents the policy of the system.

Entities

An *entity* is a rational collection of data elements. It describes something from the real world that is important to the business.

For an example of a business, let's look at an annual event held at the Richmond Arms. The Richmond Arms is a public house built beside the millstream at West Ashling in the county of Sussex. Each year, the pub's landlord raises money for charity by holding a duck race.

"Foul!" cry the readers (pun intended). "Ducks are not orderly enough for organized racing!" This fact of nature was indeed apparent to the landlord at the time of the inaugural race. So he hit upon the idea of using wooden ducks.

Figure 2.5.1: A Richmond Arms duck. Made of wood, the ducks are available in the public bar of the Richmond Arms in return for pledging money to charity.

The ducks are placed in the millstream, and they float on the current from the starting gate to the far side of the Southbrook Bridge. Customers of the pub who wish to enter a duck in the race (the landlord calls them "sponsors") must pledge a sum of money before the duck is handed over. The duck must float in a proper upright position, and it has to be modified below the waterline for this to happen. Modifications to improve floating and use of the current are encouraged, with the exception of any form of motorization or other power.

The sponsor may not handle the duck during the race. The landlord very sensibly introduced this rule after several over-enthusiastic, and possibly over-inebriated, sponsors entered the water to assist their ducks without realizing they were unable to swim (the sponsors, not the ducks). They (the sponsors) were rescued, safe but not exactly sound, on the downstream side of Oakwood Weir. The ducks are still missing.

The sponsorship money, the total of all the pledges, is presented to the charity of the winner's choice during post-race ceremonies in the public bar.*

What entities can you find in the above description of the business? First, recall that an entity is a rational collection of data elements that describes something important to the business. The landlord has to remember who the sponsors are, as they play an important role. Sponsors pledge money to enter the race. The system has to remember such facts as the sponsor's name and telephone number. (The term "remember" here means that the system stores that data. It doesn't matter if the landlord writes the information in a book, keeps it in his head, or uses a computer. The point is that the system retains the information.) Because the system has to remember each sponsor, you can say that SPONSOR is an entity.

The landlord also keeps information about each duck and charity. The sponsors may independently raise money to increase their pledges. This means that a duck, charity, and pledge are all real-world things, and the system needs to remember them. DUCK, CHARITY, and PLEDGE, thus, become entities.

SPONSOR   DUCK   PLEDGE   CHARITY

Figure 2.5.2: Entities from the West Ashling Duck Race.

An entity has a unique purpose and a definable role in the enterprise. For example, a SPONSOR promises a PLEDGE and nominates the CHARITY that he wants a duck to represent. The DUCK is an entrant in the race. The PLEDGE is an amount of money that may be increased from time to time. A CHARITY is the beneficiary of all the money raised by the event. Every entity in a data model must have a clearly stated purpose. If you cannot state the purpose, the proposed entity is not an entity at all; it is probably data that describe some other entity.

> Every entity must have a unique and definable role in the business.

Each of the entities—SPONSOR, DUCK, PLEDGE, and CHARITY—represents a

---

*This is not fiction. There actually is an annual West Ashling Duck Race. Skeptical readers are invited to attend the event at the Richmond Arms, West Ashling, Chichester, West Sussex, England. One or two details of the race rules have been altered for technical reasons, and the rule about the ducks entering the oil drum was omitted.

collection of data elements. For example, the name of the sponsor is recorded along with the address and telephone number. The DUCK entity contains its identifying number and the starting position. (Ducks are assigned their starting position from a draw. One side of the gate is released at the start. Ducks drawing low-numbered positions are on the advantageous side.)

Sponsor

- Name
- Address
- Telephone

Duck

- Number
- Starting Position

Figure 2.5.3: An entity is a data container, and is represented by the rectangle symbol.

An entity must have at least one attribute to describe it.

Each entity stores data. It is like a container in which you put data elements that are all about the same subject. Each container must have at least one data element. These elements, or *attributes* as they are known, describe one entity and no other.

An entity is the subject of some data. *It is not the data itself.* For instance, SPONSOR is an entity. You cannot give a value to SPONSOR, but you can have attributes to describe it. NAME is an attribute of SPONSOR, for example. On the other hand, NAME has a value but no attributes.

Figure 2.5.4: Each entity type represents all occurrences of the entity. For example, the entity SPONSOR represents all the sponsors taking part in the event.

148

There can be many occurrences of each type of entity. For example, there are many ducks. The race rules allow up to fifty of them. Similarly, there are many sponsors, many charities, and so on. Each entity type in the data model is not a single instance, but rather it represents every occurrence of the entity.

You may prefer to think of this situation by using a table in which each column represents one of the entity's attributes, and each row an occurrence of the entity. Figure 2.5.5 is an example of such a table.

Name	Address	Telephone
H. Purcell	Westminster Lodge	925 1658
E. Elgar	Broadheath House	262 1857
F. Delius	Bradford Cottage	768 1862
R. Vaughan Williams	Down Ampney	855 1872

Figure 2.5.5: This table represents all occurrences of the entity SPONSOR. The column headings are the attributes, and each row contains the unique data values for each sponsor. Extra rows are added as each new sponsor enters the race.

Each of the entities DUCK and PLEDGE also has such a table. If there are many occurrences of the entity, the table becomes quite long and unwieldy. So for the purposes of this book, we'll represent the system's entities using a rectangle on the data model, with its attributes listed in a data dictionary. If you prefer to use the table notation, that is your choice.

Because there are many occurrences of an entity, each one must be uniquely identified. For example, the entity SPONSOR has an attribute NAME that distinguishes a specific instance of sponsor from all the others. Each DUCK has a unique number, and so on.

While entities have numbers, names, and so on to identify them, they do not have codes that alter the business role of the entity. For example, at the moment, sponsors are people who frequent the Richmond Arms pub. Suppose that we change the system a little, and we now wish to include information about the course marshals who organize the race and oversee its correct running. Both the marshals and sponsors are customers, but it wouldn't be correct to model an entity CUSTOMER

Entities are uniquely identifiable.

> **Each attribute applies to every occurrence of its entity.**

and give it a code to indicate which role it is playing. Although they both happen to be customers, marshals and sponsors play different parts in the system and have different attributes. If you only had a CUSTOMER entity, some of the marshal's attributes would not be relevant to a sponsor, and vice versa. So you must ensure that each entity is a unique collection of attributes.

The entities you have seen so far are not connected to each other. Each one contains data that are relevant only to that entity. For example, the SPONSOR entity holds only data that directly pertain to someone willing to pledge money and float a duck. It does not contain anything else. The DUCK entity contains only attributes that describe a wooden duck. How do you know which DUCK belongs to which SPONSOR? You know because of the relationship.

Relationships

A *relationship* connects entities and, through the connection, expresses the business policy of the data model. In the duck race, the landlord needs to know which ducks are being sponsored by which sponsor. He also needs to know which sponsors have promised a pledge. By using a relationship to connect ducks to the relevant sponsor, and another to connect a sponsor to his pledges, the landlord can keep track of all the entities.

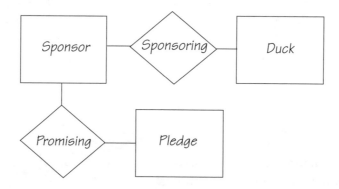

Figure 2.5.6: The relationship describes a business association between the entities. The relationship's name reflects the reason for the association.

Figure 2.5.6 shows the relationships that connect the duck race entities. The relationship is not just the simple association of entities; there must be some business policy that describes the circumstances for making the connection. You'll find the policy for these relationships in the rules for the race:

Only standard ducks issued by the Richmond Arms on a pledge of a minimum of £100 sponsorship may enter the race.

Ducks may be decorated in any way above the waterline, as long as they carry their official number in the standard position. Ducks may not, however, be modified structurally above the waterline.

All ducks must float in a proper upright position and modification below the waterline is required for this. Any other modification to aid floatation and efficiency may be used below water EXCEPT ANY FORM OF MOTORIZATION OR OTHER POWER. The sole means of propulsion of any duck shall be the natural water current of the stream.

> The relationship is a business connection between entities. There must be some business policy to establish the connection.

Looking through the rules, you can see that a pub customer may sponsor a duck if he promises to hand over at least £100 (which goes to charity), and agrees to modify and number the duck in the approved manner. Of course, a duck must be available in order to be sponsored. So for each willing sponsor, the landlord writes down the sponsor's details and the number of the duck being sponsored. By associating two entities according to the business rules, the landlord establishes the relationship.

You must be able to write the rules for associating the entities. If you can't, there may be no reason for the association. Normally, the processes that store or retrieve data make use of the relationships. Mini specifications describe processes (the topic of Chapter 2.12 *Mini Specifications*), and it is here that you describe each specific circumstance under which the relationships and entities are created, referenced, updated, and deleted.

> The rules for relating entities must be definable business rules. They are part of the mini specifications for the processes handling the entities.

Some relationships have attributes. For example, the landlord at the Richmond Arms records the date, time, and a witness' name when one of his customers asks to sponsor a duck. (He does this so that he can remind sponsors of their obligations if they forget to pick up their ducks on the day of the race.) However, the attributes of the relationship must be data that exist *because the relationship exists*. For example, the date, time, and witness' name are only important to the business because a sponsor has a SPONSORING relationship with a duck.

> The relationship may contain attributes provided they describe the relationship itself, and not the participating entities.

151

Note also that you cannot attribute these pieces of data to either the duck or the sponsor, but only to the SPONSORING relationship.

To make the model more informative, you must give each relationship a unique name to describe the reason for the relationship's existence. While you are thinking of a suitable name for a relationship (it's not always easy), try to remember that a relationship is not directional. It simply indicates which entities participate in the relationship and the reason for their participation. For example, Figure 2.5.6 shows a SPONSORING relationship between a duck and a sponsor.

While verbs in business descriptions tend to become relationships in data models, do not read the relationship as a verb. Do not say, "A sponsor is sponsoring a duck" because it doesn't work the other way. Wooden ducks don't have the money to sponsor one of the pub's customers. Instead, read "sponsoring" as the noun form of the verb (or gerund, as is its proper grammatical term). In other words, "The sponsor and the duck have a sponsoring relationship." There is a business link that the system needs to remember because of an act of sponsorship.

Similarly, the sponsor and the pledge have a PROMISING relationship: The pledge is promised, and the sponsor promises. Pledges are donated to a charity, the charity receives a donation, so a DONATING relationship is appropriate. The money raised by the duck race goes to the charity that is represented by the winning duck. Alternatively, the winning duck represents a charity. To call the relationship REPRE-SENTING describes the reason that the system needs to remember the relationship between a duck and a charity.

We suggested that when you name relationships you use the noun form of the verb. Typically, relationship names end with "ing," "ment," "tion," or "ance." Remember that you are describing why the relationship exists, not the action of one of the entities.

We built the model in Figure 2.5.7 by putting together everything we know about the West Ashling Duck Race. Since the data model is a reflection of the system's business policy, before continuing, be sure that it is as accurate as possible at this stage of the analysis. The model in Figure 2.5.7 seems to have all the necessary entities, and there is nothing else in the race rules to suggest other data that need to be stored. The system doesn't have to remember its owner (the landlord), its place of business (the Richmond Arms), or other information regarding the internal workings of the system.

Do you understand the necessary relationships? There is always a tendency when building data models to relate every entity to every other. However, show only the relationships that are necessary for the business in the model. For example, should the SPONSOR relate to the CHARITY? While a sponsor may be an avid sup-porter of one or more charities, it is of little interest to the system because it is the DUCK that determines whether the charity benefits (all the money goes to the char-

ity represented by the winning duck). The model shows only the direct and important relationships.

Similarly, should there be a connection between the DUCK and the PLEDGE? All the money raised by the pledges is given to a charity after the race. This means that the system has no interest in how much money has been pledged for any given duck, so there is no need to show a relationship between those entities.

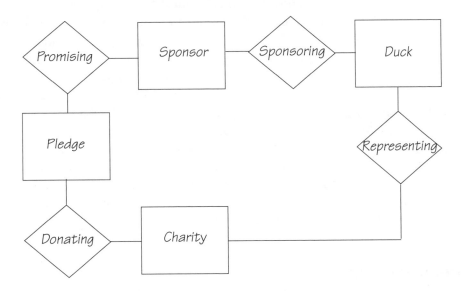

Figure 2.5.7: Partial data model for the West Ashling Duck Race.

No Foreign Keys

Before going on, we wish to exorcise a common misconception about relationships. The misconception concerns foreign keys.

A *foreign key* is one or more attributes that are included in one entity for the purpose of identifying another. For example, a SPONSOR has an attribute NAME to identify it. If the DUCK also held the sponsor's NAME and, on the other side of the relationship, the SPONSOR contained the duck's NUMBER, the two entities would be linked using foreign keys. The term "foreign" signifies that the attribute really belongs to another entity besides the one holding it.

> Foreign keys are a way of implement-ing a relationship. They are not the relationship.

So consider the purpose of the data model. It is a model of the system's busi-
ness policy, not its implementation. A relationship between entities is a business connection and deliberately avoids representing any possible implementation of this connection. In other words, this is not a model of a database, nor of any other data

storage system. Foreign keys break the rule about each entity containing only data that describe that entity, and no other.

If you don't have foreign keys, how do you know which DUCK belongs to which SPONSOR or, given a DUCK, who is the SPONSOR? The relationship shows that there is a *requirement* to connect the entities. In other words, it is the existence of a relationship in the data model that says, "I have a need to be able to remember that this entity has a connection to that one." Remember that you are dealing with an abstract model of the data used by the system and you are not designing a database. If you include foreign keys in the entities at either end of the relationship, you are saying that there is a business requirement to include those data elements in the database. That's not true: Foreign keys are one technological solution for the implementation of a relationship. They may or may not be applicable to the particular database management system (DBMS) or other storage technology chosen for your system. Right now, you need to focus on analyzing the business requirements for a system to be able to determine who has sponsored a particular duck. Later, after you've modeled the associated processes and you know the navigational requirements, you can design the appropriate implementation for the connection.

> The data model does not include foreign keys.

Cardinality

Cardinality tells you how many entities of each type participate in a relationship. For example, a sponsor may choose to sponsor several ducks. This happens when the sponsor is unsure whether a deep or shallow draught will be most effective at catching the stream's current. To increase his chances of being the sponsor of the winning duck, he opts to enter several ducks, each with a different modification below the waterline. To show this aspect of the policy, you add cardinal operators to the model.

Figure 2.5.8: Cardinal operators show that any one sponsor may have more than one duck in the race.

The operators in the model in Figure 2.5.8 show that for one instance of SPONSOR, there may be many instances of DUCK. The operator N (sometimes M) is used to symbolize one or more than one. This relationship is known as a one-to-many relationship. Note the way to read the cardinality: "For any one sponsor, there may be one or many ducks. For one instance of duck, there may be only one sponsor." To determine the cardinality, you have to ask this question from the entities at both ends of the model: "If I have a single instance of this entity, how many are there at the other end of the relationship?"

One-to-many is by far the most common type of cardinality, though you will come across others. For example, if the rules of the race are changed to limit a sponsor to a single duck, and a duck may have only one sponsor (as at present), the cardinality in Figure 2.5.8 would be changed to show a one-to-one relationship.

Sometimes, many-to-many relationships arise. For instance, suppose the rules were again changed to allow several sponsors to be involved with one duck, and each of those sponsors could sponsor several ducks. In this case, the sponsoring relationship would have a many-to-many cardinality. However, this rule change is unlikely to occur, as the landlord of the Richmond Arms does not want to chase more people than necessary to collect their pledges and certainly does not want to have to deal with multiple sponsors for each duck.

A final word on cardinality: The cardinal operators should be added to the data model as soon as you come across the cardinal policy. Note that in your models, you needn't be concerned about the exact location of the operators—above, below, left, right, or in the line—just make sure their placement is clear and unambiguous. Use cardinality as a way of raising constructive questions about your model. Eventually, to understand all your data, you will need to define all the cardinal operators.

As we've said, the data model represents the policy of the business being studied by partitioning whatever the system remembers into entities and relationships. Is the model in Figure 2.5.9 an accurate statement of the business policy? We have already examined the relationships to see if the business needed to remember any other links between the data. So now let us turn our attention to examining the meaning of the cardinality.

The relationship REPRESENTING says that several pub customers favor a CHARITY, and so it is represented by more than one DUCK. Note that each DUCK races on behalf of a single CHARITY. The cardinality for the relationship PROMISING says that any SPONSOR may make more than one PLEDGE. This is probably because he is sponsoring several ducks. We have already discussed and dismissed a relationship between DUCK and PLEDGE, and the model is saying that there is no way to tell which pledge is for which duck. So how does the system know which charities get which pledges?

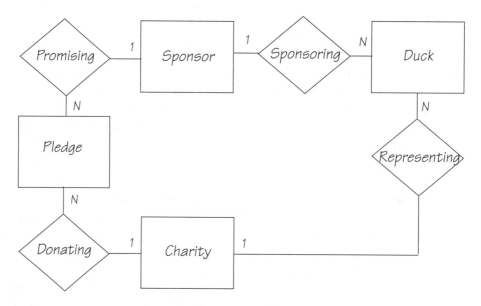

Figure 2.5.9: *This shows the Duck Race model with the cardinality added. It now states the complete policy of the race.*

It doesn't have to. All the money generated by the race is donated to the charity represented by the winning duck. The winning sponsor (the system can identify him from the SPONSORING relationship) officially gets nothing, but the landlord always stands him a few pints.

Building a Data Model

Let's look at how you go about building a model of the system's data. First, keep in mind that data modeling and process modeling are not exclusive, nor are they serial, activities. The book you are reading is a serial device; you read one chapter and presumably finish it before starting the next. In this book, we present the models in the order that we think is appropriate for most analysis projects. But the order of the book's chapters does not mean that one activity is completed before the other is started. That is not how it works. Data modeling and process modeling are done in parallel. Each model helps you to build the other by giving you different views of the same system, and thereby to build a complete and accurate model.

Finding Entities

You can build the first cut of a data model from the information that is usually available at the beginning of an analysis project. Remember the definition of an entity: a rational collection of data elements that describes something from the real world that is important to the business.

The definition helps you get started in your search for the entities. One or more of these rules of thumb will help you to recognize candidate entities:

- An entity has a defined business purpose. If you can't say what it means, you don't need to remember it.
- An entity holds at least one, and preferably more than one, attribute.
- An entity has more than one instance. If there is only one instance, it is probably a piece of constant information that is part of the policy of a process.
- Each entity must be uniquely identifiable. The identifier may be made up of one or a number of attributes, or it may be a specially generated code. For example, a person's name can be a satisfactory identifier, or in the case of the ducks, the landlord generated a number to tell them apart.
- An entity doesn't have a value. For example, a telephone number is not an entity, but is an attribute of an entity.
- A terminator from the context diagram may be an entity. If the system sends data to a terminator, it must at least remember the terminator's address.
- A report is rarely an entity.
- Column headings and names in reports often are entities. So are rows and columns in spreadsheets. But remember that entities are not calculated or derived.
- Nouns in business descriptions are often entities.
- Products (toasters, tomatoes, services, and so on) often are entities, providing there is more than one such product.
- Roles (such as the sponsor in the duck race) are usually entities.
- A repeating group in a data flow or data store is usually an entity or a collection of entities.

Where do you start looking for entities that are important to the business? The context diagram is a good place to start. All the data that enter and leave the system appear as data flows in this diagram. By analyzing these boundary data flows, you discover all the attributes that the system must store. By determining which entity each one rightfully belongs to, you identify all the entities. This process is called *attribution,* and is discussed later in this chapter.

Where else can you discover things that are important to the business? The documents used by the business are often good sources of entities and attributes. In

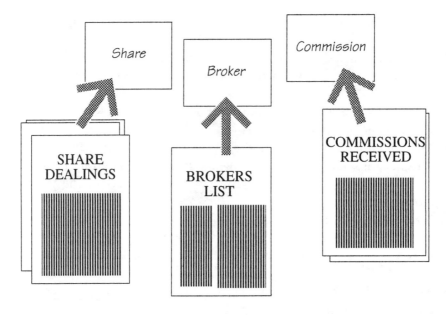

Figure 2.5.10: Most nouns become entities in the data model.

most reports, the entities usually appear as names or headings, but remember the rule of thumb that says entities are not usually calculated or derived.

Users' descriptions of the business reveal entities. As the users discuss their business, note the things they use in their business. The rule of thumb says that nouns are potential entities. Documents such as procedural manuals, job descriptions, and existing business forms all usually contain nouns that are potential entities.

Finding Relationships

The rule of thumb for finding the relationship between entities says that a relationship exists when a verb is used in a description of a business. Consider the following statement from the landlord of the Richmond Arms:

"The sponsors put a lot of effort into decorating their ducks, so we have a number of decoration categories, such as historical, topical, local humor, and so on. Local businesses have donated prizes for the different categories, and we award the prizes to worthy sponsors during the post-race ceremonies."

Not every verb you hear is a suitable candidate to become a relationship. A verb must describe some action that needs to be remembered, and it must describe some activity that includes two entities that form the subject and the object of the verb.

Because both the subject and the object are involved in the action ("The prizes are awarded to the sponsors"), there must be a business reason to remember the link between them. The name of the relationship should reflect *why* the connection exists.

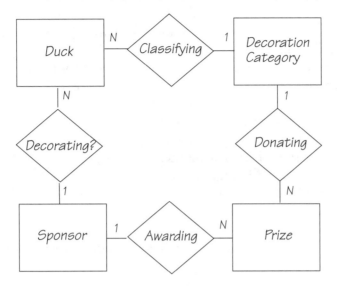

Figure 2.5.11: Verbs reveal relationships. The relationships in this model can be traced to the landlord's policy statement.

Verbs are not always reliable indicators of relationships. For example, the relationship DECORATING? linking the duck and the sponsor in the model shown in Figure 2.5.11 is probably not interesting to the business. It is unlikely that the landlord cares who decorated the duck, and if the duck has not been decorated, it will not win a prize in any of the decoration categories. You could omit this relationship from your model. Besides, you already know that there is a SPONSORING relationship between these two entities. However, if the DECORATING relationship is important to the business, and if it differs in meaning from SPONSORING, you could have both relationships between SPONSOR and DUCK.

The following rules of thumb can be useful when considering relationships:

- A relationship is an interaction between two (or more) entities that the system needs to remember.
- A verb often indicates a relationship.
- For a relationship to exist, there must be some business policy stating the circumstances under which it is established and used.
- A relationship's name reflects the reason for the association.
- A relationship may contain attributes, but it doesn't have to.

159

- If a data element describes an action, it may be an attribute of a relationship. For example, SPONSORING DATE cannot be attributed to either the DUCK or the SPONSOR. Therefore, it is an attribute of the relationship between them.
- A relationship may be needed if there are two adjacent attributes in a form or in a file, and if the attributes belong to different entities.
- Consider only those relationships that are necessary for the business within your context. Ignore those that are outside the scope of your system.
- A relationship is the requirement for a connection between entities; it is not the implementation of the connection.
- Relationships are usually of the one-to-many type. If you have a one-to-one relationship, there may be no need for the second entity. If you have a many-to-many relationship, you may have missed an entity.
- Relationships should not be directional. The noun form of the verb (or gerund) often works.

Keep in mind that these rules of thumb will only guide, but won't lead, you.

Subtypes and Supertypes

Most systems contain entities that are very similar to one another. In some cases, the characteristics of the entities suggest that perhaps there ought to be one entity instead of several. For example, earlier in this chapter, we talked about how customers could be either sponsors or marshals. Our rule said that their respective roles and attributes made them separate entities.

However, suppose you come across entities that have some identical and some different attributes. To illustrate, let's expand the West Ashling Duck Race to allow corporate sponsors. While both corporate and individual sponsors have the same relationship to a duck, the information the system needs to keep about each of them may be somewhat different. The corporate sponsor has a name and address that corresponds to the individual's name and address, and, in addition, has a contact name and an authorizing officer. Alternatively, let's now say that the system has to keep a record of the individual's favorite pint. This attribute cannot possibly apply to a corporation.

An entity must be a unique collection of attributes, and must not have a code to say, for any given occurrence, whether it is a corporate or an individual sponsor. As a data modeler, you want to have it both ways. You want to show the similarities between these entities, but you also want to preserve the differences. You can do this by using *supertypes* and *subtypes*. A *supertype* is a generalized entity. Its business role and its attributes apply to all its subtypes. A *subtype* is a specialized case in which the entity has its own unique characteristic as well as the characteristics of its supertype entity. An example of a supertype-subtype relationship is shown in Figure 2.5.12.

The arrangement in Figure 2.5.12 shows SPONSOR as the supertype. This entity has a relationship with DUCK, and the relationship exists for any kind of sponsor. The attributes of SPONSOR are needed for any kind of sponsor. The subtypes CORPORATE SPONSOR and INDIVIDUAL SPONSOR are specialized kinds of sponsors and have their individual data requirements. There is no theoretical limit to the number of subtypes, and you could add FOREIGN SPONSOR as another subtype to the model.

Sometimes, supertype-subtype relationships are confusing. Keep in mind that you can describe the true sub-supertype relationship as "is a." For example, the individual sponsor "is a" sponsor. The corporate sponsor "is a" sponsor. You must distinguish between this and a relationship that is an accumulation or composition. For example, TRANSMISSION, ENGINE, and BODY are not subtypes of an entity CAR. A transmission is not a car, nor is an engine. However, it may be useful to model the transmission, the engine, and the body as subtypes of CAR COMPONENT.

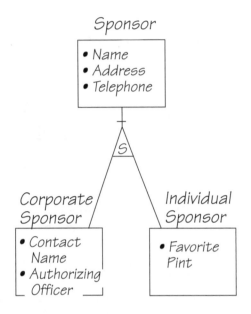

Figure 2.5.12: The supertype SPONSOR relates to DUCK as before. The two subtypes, CORPORATE SPONSOR and INDIVIDUAL SPONSOR, carry attributes that are unique to their specialized roles. The "S" in the triangle indicates the supertype-subtype relationship in the model.

Existing Files and Stores

The data model specifies the requirement for the system's stored data. This suggests that the files and data stores belonging to any current implementation may be a useful source of information. After all, if the data are currently being stored, there may be a reason for keeping the data as part of your new system.

Let's have a look at another example. Say there is an existing computer system that processes a file about animals in a zoo. If we write a definition of the file, we might learn about the system's need for stored data. Look at our definition in Figure 2.5.13. (The caption explains the notation. If this brief explanation is too brief for you, read through the relevant parts of Chapter 2.9 *Data Dictionary;* then return here.)

Animal File = {Species + Gender + Animal Weight + Date Of Birth
            + {Pen Number + Pen Location
            + {Keeper Identity + Pen Key Number}}
            + {Food Type + Food Schedule}}

Figure 2.5.13: The definition of an existing file. The notation, in brief, is as follows: The file is made up of everything following the =. The + means "together with," and anything enclosed in { } is repeating. Thus, the file is made up of repetitions of everything, while for one animal there is a number of pens, each having a number of keepers. The information about food applies to an animal.

In this example, braces { } enclose a *repeating group,* a collection of data items that occur more than once. Typically, the group repeats because its data items have a different subject than whatever preceded it, and the group therefore is a potential entity. The data elements within the repeating group are potentially the attributes of that entity.

In the definition in Figure 2.5.13, you can break out a new entity for each repeating group. The first lot of data elements applies to an animal, and there are several pens for each animal. This reveals two entities, with a relationship between them. Similarly, there will be a number of keeper entities that relate to the pen. The double right braces }} after the keeper information indicate that the repeating has ended, and that the next lot of data refers to the animal.

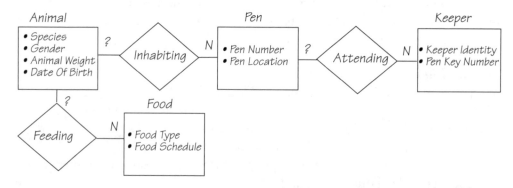

Figure 2.5.14: This model is derived from the data dictionary entry for ANIMAL FILE. Note that some of the attributes do not appear to be in the correct entity. The question marks are used to signify the unknown cardinality.

The data model in Figure 2.5.14 was translated directly from an existing file. You have no way of knowing whether the design of the file was based on logical groupings of data or not, and so you must regard it with suspicion. The cardinality resulting from this translation looks very suspicious. For example, an animal is relating to many pens, but an animal can't possibly be in more than one place at any one time. This may be historical information, so that the system keeps track of where the animal has lived. If so, the file doesn't appear to keep any information about when an animal was moved into or out of a pen. This may be deliberate, or it may be poor file design. A pen may be attended by several keepers, but do you need to know which keepers look after which pens?

The problem with using existing files is that if they are poorly designed, they may not always reveal the precise nature of the stored data, and you will need a lot of information from your users. However, such files do provide you with an alternative strategy for starting your data model.

Put your questions aside for the moment, and let's continue with the animal file. You now need to examine each of the attributes to determine if it is in the proper place.

Attribution

Attribution is best described as assigning data elements to the appropriate entity. An *attribute* is a data element that describes an entity or a relationship. For each data element, ask, "What entity or relationship does this element describe?" If it is describing the entity to which it is attached, then well and good, leave it alone. Otherwise, you need to find the entity that this element describes, or, to give the process its correct name, you attribute the element to another entity or relationship.

> Every data element stored by the system must be attributed to one, and only one, entity or relationship.

Sometimes, there is no suitable entity or relationship in the data model for an attribute. In that case, you must create a new one. Let's revisit the animal data model for an example of attribution.

Note the attribute PEN KEY NUMBER. The existing computer file stores this next to the identity of the keeper. The file designer did this to keep track of which keepers held keys to which pens. In other words, the physical placement of PEN KEY NUMBER is a way of implementing a relationship. However, an element called PEN KEY NUMBER cannot be an attribute of KEEPER. An attribute called PEN KEY NUMBER is describing something other than a keeper. What to do? Simply attribute it to the appropriate entity, in this case PEN. You also create a relationship that reflects the reason why PEN

Keeper

- *Keeper Identity*
- *Pen Key Number*

163

KEY NUMBER and KEEPER IDENTITY are adjacent in the physical file. In this case, there is already an ATTENDING relationship between KEEPER and PEN. If this relationship captures the required meaning, it will suffice (let's say it does in this case). If it did not serve the purpose, you would then add a relationship between the two types of entities. The rearranged fragment of the model is shown in Figure 2.5.15.

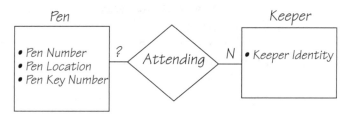

Figure 2.5.15: The process of attribution allocates each data element to the correct entity.

Sometimes, after attributing all the elements, you are left with an entity that has no attributes. When this happens, remove the entity from your data model. As our rule of thumb suggests, if there is no data to describe the entity, it is not an entity that your system needs.

A Sample Data Model

We will now show you a data model under construction with the aid of a travel agency example. Mallard Travel is a London-based company that provides low-cost flights to holiday and vacation destinations. Part of the business involves selling flights to the more obscure summer destinations in Europe. Our data model will be for this part of Mallard's business.

A fragment of the context diagram for this booking, or reservation, system is

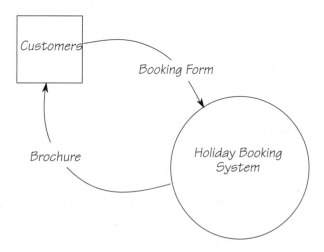

Figure 2.5.16: Part of the context diagram for the Mallard Travel system. The flow BROCHURE is constructed from stored data held within the system. The data from BOOKING FORM are remembered by the system.

shown in Figure 2.5.16. There are other boundary data flows, but the two shown will give us enough information about the system's stored data for our purposes. Most of the incoming data from BOOKING FORM is remembered by the system. The outgoing flow BROCHURE is made up of data from the system's memory. By analyzing these flows, and understanding the business policy they represent, you can build the data model. Instead of giving you a definition of the flows, we've taken some pages from the Mallard Travel brochure and partially reproduced them in Figures 2.5.17 and 2.5.18. Assume that all the data making up the two data flows are in these figures. The goal is to define the business policy in the form of a data model.

A data model is a statement of business policy. Building a data model from information about the business is a translation process. You first understand the business policy, then translate it into data model terminology. However, before you start interpreting this policy, you must disregard any knowledge of travel booking

---

Mallard Travel   Summer Flights

Holiday Code	Destination	Duration	Season	Price
ANM7	Andorra	7 days	May–Jun	£ 99
ANM14	Andorra	14 days	May–Jun	£119
ANJ7	Andorra	7 days	Jul–Aug	£109
ANJ14	Andorra	14 days	Jul–Aug	£139
LIM7	Lichtenstein	7 days	May–Jun	£ 89
LIM14	Lichtenstein	14 days	May–Jun	£109
LIJ7	Lichtenstein	7 days	Jul–Aug	£109

All flights depart and return on Saturdays. The applicable season is determined by the date of departure. You will be told your take-off times close to departure.

Car rental per day. (Prices include unlimited mileage, but exclude insurance and petrol, or gasoline.)

A Suzuki Alto	£16.50
B Opel Corsa	£18.50
C Opel Kadett/VW Golf	£20.50
E Opel Ascona/Ford Sierra	£24.50

Figure 2.5.17: Some of the Mallard flights and car rental options. (Not all flights are shown.)

Mallard Travel   Booking Form

How to book

Once you have selected your destination and dates of travel, contact your travel agent or Mallard Travel. When your dates have been confirmed, you will be given a reference that identifies your booking. Please use this reference for all future correspondence.

Holiday Code	Destination	Duration	Departure Date

Booking reference given on telephone: _____

Title	Initials	Family Name	Age (if under 12)

Address for correspondence: _____

_____

_____

_____

Day telephone: _____ Night telephone: _____

Car rental section

Car group: _____ Car type: _____

Pick-up date: _____ Drop-off date: _____

❑ **I wish to pay by credit card**

Card no.: _____ Expiration date: _____

Signature: _____

❑ **I enclose a check for:** _____

Payment number: _____ *(For internal use)*

Figure 2.5.18: Part of the Mallard booking form.

policy you've acquired from any other system. It is irrelevant because you are studying a different system. In this exercise, you'll have to interpret what you read about Mallard's business. Since there will be no users available to answer questions, this exercise will not produce a conclusive statement of business policy, but it will adequately demonstrate the process.

Begin with the booking form shown in Figure 2.5.18. Let's start with the obvious. This system is all about booking holidays, so there is a potential entity called BOOKING and one called HOLIDAY. HOLIDAY CODE and BOOKING REFERENCE GIVEN ON TELEPHONE are unique identifiers. One of the rules for entities says that they must be uniquely identified. Therefore, a unique identifier indicates the existence of an entity. Mallard Travel wants to know which bookings are reserving which holidays, so it is reasonable that they have a RESERVING relationship.

Figure 2.5.17 tells you that the company is offering flights to various destinations; so, you guessed it, FLIGHT is the next potential entity. When we add it to the model, the result is shown in Figure 2.5.19.

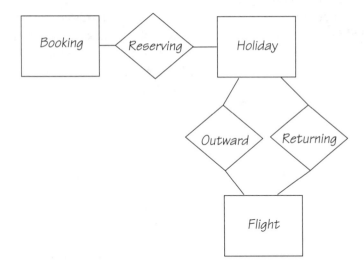

Figure 2.5.19: The beginning of the data model.

The relationships between HOLIDAY and FLIGHT are interesting. It is correct to model this as a single relationship with a one-to-many cardinality. One instance of the FLIGHT entity represents the outward flight, and another the returning flight. However, we feel that having an outward and a returning relationship demonstrates the system's policy far more effectively.

Next, let's look at the attributes of each of these entities. You know already that BOOKING REFERENCE GIVEN ON TELEPHONE is a unique identifier for

BOOKING. The booking form also collects other information about the passenger. A reasonable definition of the booking entity's attributes is

Booking = <u>Booking Reference</u> + Passenger Title + Passenger Initials
            + Passenger Name
            + (Passenger Age)
            + Passenger Address + Day Telephone + Night Telephone

The underscoring indicates the unique identifier, or *key field,* of the entity. Note that the passenger details are attributed to BOOKING. Each passenger is required to make a separate booking. The company does not keep information about passengers except when they make a booking, so there is no requirement for a passenger entity.

Some of the data in the forms could be attributed to the other entities. Take a moment to consider what attribution you would make.

There are more entities to add to the model. The summer flights brochure gives details concerning renting a car should one be required. Since the form has data that apply only to renting a car, it is feasible to make CAR RENTAL a potential entity. Note that the person making the reservation on the booking form can specify dates for the car rental that are different from the flight dates. As they are separate, you must say that these are attributes of the entity CAR RENTAL, and are not reused attributes from FLIGHT. The updated model is shown in Figure 2.5.20.

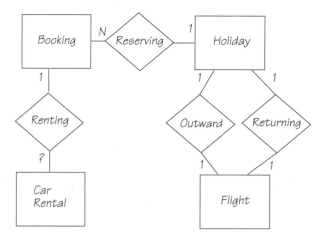

Figure 2.5.20: The data model to date. Cardinality has been added to provide additional information. We are uncertain whether a booking can be involved in renting more than one car. The form indicates only one, but that seems to be unreasonable business policy.

Passengers may pay by credit card or check. If there are two types of payment, each with its own data requirement, the model should show a supertype called PAYMENT with two subtypes: CHECK PAYMENT and CREDIT CARD PAYMENT. PAYMENT relates to BOOKING, since that is what is being paid for. The PAYMENT entities are shown in Figure 2.5.21.

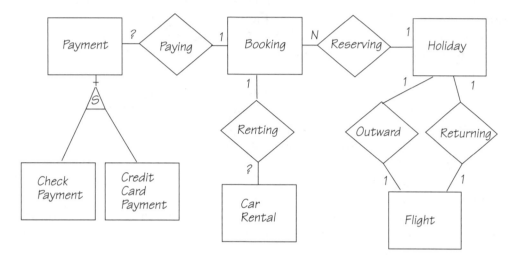

Figure 2.5.21: The data model showing all its entities and relationships.

Before going too far, you should write down whatever information you have about the known attributes. We suggest these sample definitions:

Booking = <u>Booking Reference</u> + Passenger Title + Passenger Initials
* + Passenger Name*
* + (Passenger Age)*
* + Passenger Address + Day Telephone + Night Telephone*

Car Rental = <u>Car Group + Date From + Date To</u> + Car Type + Price

Check Payment = Check Number

Credit Card Payment = Credit Card Number + Expiration Date

Flight = <u>Flight Number + Departure Date</u> + Departure Time + Arrival Date?
* + Arrival Time*

Holiday = <u>Holiday Code</u> + Duration + Destination + Price

Payment = <u>Payment Number</u> + Payment Date + Payment Amount

The data model is still only a *preliminary model*. We constructed it using the information from the brochure and the booking form. There must be more business policy for the system. For example, most charter flight operators book blocks of seats for certain flights. They get a discount from the airlines and make their profit by selling the seats to their own customers. So there is probably an entity called BLOCK BOOKING OUTWARD (or something similar) to hold details of how many seats are available. This would relate to the FLIGHT entity. There is a similar entity for the block booking for return flights. Naturally, as you talk to the users and model their business, you can expect to find more and more policy to add to the data model.

This approach to data modeling results in a rather fuzzy first-cut data model. Note that we referred to most of the entities and relationships in the model as "potential" entities and relationships, because you still have to confirm their existence with the users. At this stage, the model helps you to understand the nature of the problem. Later, when you model the event responses for Piccadilly, we will show you a way to confirm your model. Event-response modeling, which we will discuss later, also provides an alternative way to build data models.

The Mallard Travel model is intended as a demonstration of how you can build a data model from the data flows and the policy that you can determine from looking at existing brochures and forms.

Participation

We want to explain another requirement that you will eventually need to specify. As you have learned, building a data model leads you to a more and more detailed understanding of the data. Before finishing your Project, you will check the entities and relationships against the processes that create, reference, update, and delete them. At that stage, you will specify the rules for *participation* for each connection between an entity and a relationship. We have postponed mentioning this until now because the subject of data modeling is complex and you have had enough to think about. But now that you have built some data models, just spend a few minutes to think about how you will eventually define answers to the questions that may occur to you.

Looking back at the West Ashling Duck Race (Figure 2.5.8), you see the cardinal operators specify that any one sponsor may have more than one duck in the race. Now we discover that the ducks are owned by the Richmond Arms and kept in the back parlor when they are not being sponsored in a race. Stated formally, the DUCK entity has optional participation in the SPONSORING relationship. When we ask questions about the sponsor's participation in the relationship, we discover that the Richmond Arms is only interested in sponsors if they participate in a sponsoring relationship. The SPONSOR entity has mandatory participation in the SPONSOR-ING relationship.

Participation is specified by including the participation rules in the relationship definition in the data dictionary:

*Sponsoring = * Relationship. Keeps track of which ducks are being sponsored by which sponsor. Cardinality: for each Sponsor, there are many Ducks; for each Duck, there is one Sponsor. Participation: Sponsor mandatory, Duck optional. **

When you have learned more about the Piccadilly Project, you will be concerned with specifying the participation rules.

Summary

Well, that's it for building data models from policy statements and existing data. If you are not feeling slightly confused and uncertain, either you are indeed remarkable, or you have done this before. Building data models from policy statements is difficult, and by no means precise. Our intention in taking you through this chapter is to expose you to the thinking that lies behind this model, the information that the model shows you, and its contribution to the analysis of the system.

We mentioned several times in the chapter that you first have to understand why you need the data model. Then, when you build event-response models, you can make a more precise definition of the stored data.

What to Do

Build a data model for the system described in the exercise below. Use the sample program and notes in Figures 2.5.22 and 2.5.23. Compare yours with the sample answer and read the discussion in Chapter 4.3.

Exercise: The Barbican Data Model

The Barbican Centre is a large theater and arts complex in London. Figures 2.5.22 and 2.5.23 are two extracts from a monthly program of events. From these samples and the following user's description, derive the data model for the policy being used at the Barbican.

Here's the user statement by Molly Aire about the Barbican's policy: "People can book for performances at the Barbican by coming to the box office in person, by telephoning and using a credit card, or by mailing in the form with either a check or a credit card number and signature.

"We have a number of venues or performance areas in the complex. People can book in advance for any of them, except of course for the free events we hold in the lobby. The pricing policy is a little different for each of the venues. In the Hall, where we have orchestral music, the price of the seat depends upon the area of the Hall. I have given you a seating plan (Figure 2.5.22). Each performance has its own prices depending on the chosen seating area. We print these prices in the program. (See Figure 2.5.22.) There is never more than one performance a day, so there are no matinée prices.

Figure 2.5.22: Extract from the Barbican Centre program (reprinted by permission).

FEBRUARY WHAT'S ON WHERE AND WHEN

	BARBICAN HALL	BARBICAN THEATRE	THE PIT	BARBICAN CINEMA	FREE FOYER EVENTS	GUILDHALL SCHOOL OF MUSIC & DRAMA
WED 1		7.30 THREE SISTERS	7.30 ELECTRA	6.00/8.25 A WORLD APART (15)		2.30 TOUCHED / 7.00 TOUCHED
THU 2		2.00 THREE SISTERS / 7.30 THREE SISTERS	2.00 ELECTRA / 7.30 ELECTRA	6.00/8.25 A WORLD APART (15)		7.00 TOUCHED
FRI 3	7.45 CITY OF LONDON SINFONIA / RICHARD HICKOX / DAME JANET BAKER	7.30 THE CHURCHILL PLAY	7.30 A QUESTION OF GEOGRAPHY	7.20 BIRD (15)	5.30 PHILIP DYSON (CLASSICAL/LIGHT)	7.00 TOUCHED
SAT 4	8.00 ENGLISH CHAMBER ORCHESTRA / PHILIP LEDGER	2.00 THE CHURCHILL PLAY / 7.30 THE CHURCHILL PLAY	2.30 A QUESTION OF GEOGRAPHY / 7.30 A QUESTION OF GEOGRAPHY	11.00 OCC THE DARK CRYSTAL (PG) / 3.30 OCC THE DARK CRYSTAL (PG) / 7.20 BIRD (15)	12.15 TILBURY BAND (BANDSTAND)	
SUN 5	3.30 ERNEST READ SYMPHONY ORCHESTRA / RONALD CORP / 7.30 LONDON SYMPHONY ORCHESTRA / MYUNG WHUN CHUNG		3.00 A QUESTION OF GEOGRAPHY	3.30/7.20 BIRD (15)	12.30 LEIGH ETHERINGTON & MARK BENDOTT BAND (JAZZ)	
MON 6	7.45 HITS FROM THE SHOWS / NATIONAL SYMPHONY ORCHESTRA / JOHN OWEN EDWARDS	7.30 THE TAMING OF THE SHREW	7.00 THE BITE OF THE NIGHT	7.30 BIRD (15)		
TUE 7	7.45 LONDON ORIANA CHOIR / ENGLISH BAROQUE ORCHESTRA / LEON LOVETT	7.30 THE TAMING OF THE SHREW	7.00 THE BITE OF THE NIGHT	7.30 BIRD (15)	5.30 JANNA HENOLD / NEIL BLACKMORE (CLASSICAL/LIGHT)	1.05 CAPITAL WIND QUINTET
WED 8	7.45 SYMPHONY ORCHESTRA OF THE ROYAL ACADEMY OF MUSIC / JAMES LOUGHRAN / JOHN LILL	7.30 MEASURE FOR MEASURE	7.30 CYMBELINE	7.20 BIRD (15)		2.30 OPEN REHEARSAL VASARI
THU 9	7.45 LONDON SYMPHONY ORCHESTRA / RAFAEL FRÜHBECK DE BURGOS	2.00 MEASURE FOR MEASURE / 7.30 MEASURE FOR MEASURE	2.00 CYMBELINE / 7.30 CYMBELINE	7.20 BIRD (15)	1.30 TERRY SMITH DUO (JAZZ)	1.05 GUILDHALL BRASS BAND
FRI 10	7.45 CITY OF LONDON SINFONIA / JAN LATHAM	7.30 THREE SISTERS	7.30 ELECTRA	6.00/8.25 DEAD RINGERS (18)	5.30 OMEGA BRASS QUINTET (CLASSICAL/LIGHT)	
SAT 11	7.45 DETROIT SYMPHONY ORCHESTRA / GÜNTHER HERBIG	2.00 THREE SISTERS / 7.30 THREE SISTERS	2.00 ELECTRA / 7.30 ELECTRA	11.00 OCC BUSTER MALONE (PG) / 3.30 OCC BUSTER MALONE (PG) / 6.00/8.25 DEAD RINGERS (18)	12.15 MUSIC OF THE WORLD MARIANY	
SUN 12	7.30 LONDON SYMPHONY ORCHESTRA / RAFAEL FRÜHBECK DE BURGOS			3.30/6.00 DEAD RINGERS (18)	12.30 EL DORADO JAZZ	
MON 13	7.45 ENGLISH CHAMBER ORCHESTRA / JEFFREY TATE / YO YO MA	7.30 MEASURE FOR MEASURE		6.00/8.25 DEAD RINGERS (18)		1.05 LEDER RECITAL
TUE 14	7.45 VALENTINE'S DAY LOVE CLASSICS / LONDON CONCERT ORCHESTRA / BARRY WORDSWORTH	7.30 MEASURE FOR MEASURE		6.00/8.25 DEAD RINGERS (18)	5.30 ROY YARAHAN TRIO / DELIA MARTIN (JAZZ)	1.05 LUNCH RECITAL

Figure 2.5.23: Events at the Barbican Centre.

"The Theatre is where plays are presented. The prices here are the same regardless of the play, but they vary depending on the area inside the Theatre. We also have different prices for matinées and evenings. They are

Evenings		Matinées	
Stalls	£15.00	Stalls	£13.00
Circle 1	£15.00	Circle 1	£13.00
Circle 2	£11.00	Circle 2	£9.50
Circle 3	£7.50	Circle 3	£6.00

"In the Pit, where we have smaller plays, all seats are £10 in the evenings and £8.50 for matinées. School children and senior citizens can get a £5 ticket for matinées only.

"Cinema prices are £3.50 for adults. Senior citizens and children pay £2.50.

"We accept bookings from our mailing list subscribers as soon as they receive their program through the mail. Our program covers one calendar month of events, and we mail it about three months in advance. Other people can book as soon as we release the program one month after mailing."

Trail Guide

● Easiest: Now that you know how to build data models, it's time to do one for Piccadilly. The task is in Chapter 1.3 *What About the Business Data?*

■ More Difficult and ◆ Most Difficult: Although this was not officially part of your trail, we're glad you made it through the chapter. Your next task is to build a data model for the Piccadilly Project. It's waiting for you in Chapter 1.3 *What About the Business Data?*

❊ Promenade: By now you have seen different types of system models, and read about viewpoints that are useful when building system models. Your trail to this point may have been a little disjointed, but there is method to this madness. We wanted to present the views and models in a way that we felt would be most comfortable for you. We also wanted you to have the correct preparation for the next important topic: the essential viewpoint of the system's process. This is in Chapter 2.10 *Essential Viewpoint.*

2.6 MORE ON DATA FLOW DIAGRAMS

Before You Reached Here ...

● Easiest: You have come here from Chapter 1.3 *What About the Business Data?* In this chapter, we leave the data model temporarily to return to the data flow diagram. Previously, in Chapter 2.2 *Data Flow Diagrams,* we have considered the data flow diagram as a model of the system's function. This chapter provides additional information that you'll need in the Piccadilly Project and in your regular systems analysis work.

■ More Difficult: You are presumed to know this material, but it won't hurt to glance through this chapter in case there is anything new to you. If you want to stay on your trail, go to Chapter 2.8 *Current Physical Viewpoint.*

◆ Most Difficult: We don't know how you got here, as this chapter is way off your trail. If you don't want to stay here, turn to Chapter 1.4 *The Piccadilly Organization.*

❊ Promenade: We've spared you this chapter as it deals with the finer points of data flow diagrams, and we don't think that you have to know about them. You will probably be more interested in Chapter 2.8 *Current Physical Viewpoint.*

A Working Model of the System

The data flow diagram is a model of the system, built in the first place to help you understand the system. When you and the users have achieved an identical understanding of the system, the model becomes the specification of the system. If you wish the designer to design and implement a working system, the specification must be a *working model* of that system.

Look over the model of Miss Tweedy's Dating Service in Figure 2.6.1. Don't worry now about any new notation, because we'll get to it in this chapter. Instead, consider how this model explains what the system does and whether it is a working model of the system.

174

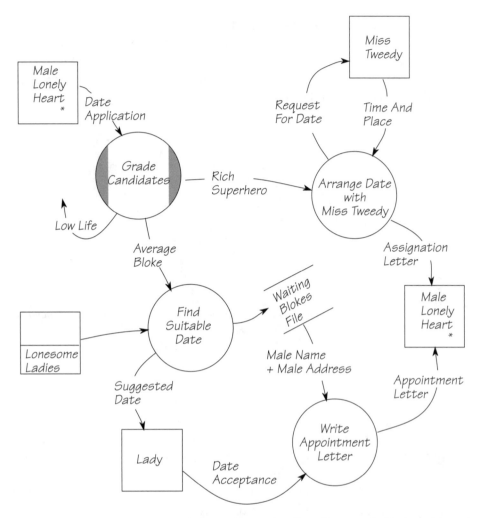

Figure 2.6.1: A data flow diagram of Miss Tweedy's Dating Service. Miss Tweedy runs a dating service, but she'd rather marry a rich man and retire. She has left instructions that if any eligible men apply, they are to be referred directly to her.

This system exists to provide data to the outside world. The lonely ladies expect that the system will provide them with information concerning suitable men who wish to meet them. We can see this expectation represented by the data flow SUGGESTED DATE and the terminator LADY. Miss Tweedy also has expectations that the system will supply her with a husband.

You can say this is a working model of the system provided you can demonstrate that the model produces the correct output when it is given the expected input. This correctness convention is called the *Rule of Data Conservation*.

The Rule of Data Conservation

In Figure 2.6.2, for example, the process FIND SUITABLE DATE needs the data flow AVERAGE BLOKE to contain enough data to find a suitable pairing in the LONESOME LADIES data store. The combination of data from the flow and the store is used to generate SUGGESTED DATE and an entry in the WAITING BLOKES FILE.

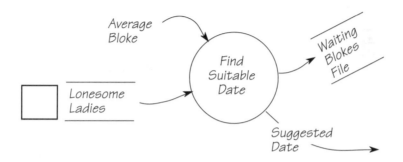

Figure 2.6.2: *The Rule of Data Conservation states that each process in the data flow diagram must be able to produce the output data flows from its input.*

Naturally, you would find the rule easier to demonstrate if you had first defined the data content of the flows and stores. Chapter 2.9 *Data Dictionary* discusses how to define the data, but for the moment, just look at the outputs and think about what data is needed in the incoming flows to make this process work. Along with the incoming data, a process may also use internal calculations and any constant information contained within the process to construct its outputs. Additionally, a process often needs to have access to the current time, day, and date. Rather than repeating clocks and calendars all over the model, you may assume that each process knows the current time, day, and date.

> Each process must receive data that are both sufficient and necessary to construct its output.

In Figure 2.6.2, the output flow to the WAITING BLOKES FILE would contain almost identical information to the AVERAGE BLOKE flow. The SUGGESTED DATE flow would contain the lady's name and address (from the LONESOME LADIES data store), and the name of the man (from AVERAGE BLOKE). There is nothing to suggest that this process breaks the Rule of Data Conservation.

When you can prove that all the processes in the model can construct their output from their input, you can say your model is a working model.

Triggering Processes

A process can be active only when it has data to work with, so it must wait for the data flow to deliver the necessary data. So we say the data flow triggers the process. In Figure 2.6.3, DATE ACCEPTANCE triggers the process WRITE APPOINTMENT LETTER. At some stage, the process reads the data store WAITING BLOKES FILE for additional information to construct the outgoing flow.

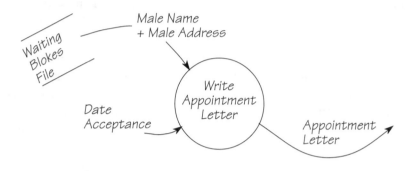

Figure 2.6.3: A process becomes active when the incoming data flow arrives.

Since a data flow makes a process active, it follows that the production of data controls the system's activity. That is, the system's operation is regulated by the processes themselves as they manufacture data for each other. There is no external entity directing which processes are to be active and which should remain inactive.

This makes the data flow diagram a natural way of representing the system. After all, in an office, you don't see the office staffers standing around waiting their turn to become active. They start work as soon as they have data to process. (At least, this is how the office staff should behave.)

Also, consider what you are trying to do. Your task is to define the requirements of the system. These requirements must not contain elements that are part of the current control or management mechanism. You do not know during the analysis stage if these mechanisms are appropriate for any future implementation of the system. If you don't yet know *what* the system has to do, you have no way of knowing *how* it is going to do it.

> The data flow diagram is a model of the system that contains only the data and functional requirements. It excludes all forms of control mechanisms.

Naming Data Flows

Every data flow must have a name. Choosing a name is important, and the name should be self-explanatory so that it contributes to the reader's understanding of the system. Because the data flow diagram is a model of some reality, the data flows exist in the real world. For this reason, data flows are usually easy to name—you just use their real-world name. You want names that communicate the meaning of the data. Test your data flow names by asking the question, "Does this name relate to the subject matter of the system, or is it concerned with how the data flow is implemented?" Sometimes, depending on the answer, you will need to rename the data flow.

For example, we worked on a system in which the users referred to a data flow called GREEN REPORT. When we asked what about the report made it green, we found out that the report was printed on green paper, but it actually contained information about debtors. Since the name did not communicate the subject matter of the data flow, we changed the name, with the users' agreement, to DEBTORS REPORT.

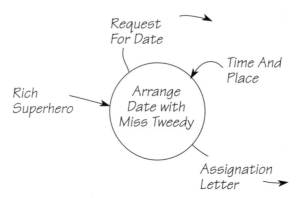

Figure 2.6.4: Each data flow has a name that lets the reader easily comprehend that part of the system. The name should be one that is well-established in the users' business or, failing that, is a suitable invention by the analyst.

Always name the flows in your models *before* naming the processes. Try this as a test: Take any data flow diagram and put coins or something over the processes to hide their names. Does the model still make sense? It should, provided the data flows have been well-named.

The processes are there to transform the data flows. If all you can see of your system model are the data flows, the processes should be fairly self-evident. Now uncover the processes and this time conceal the

> Give all data flows a name that is both self-explanatory and recognizable.

names of the data flows and stores. Does the model still make sense? Not as much. The reason is that the processes are too general without the data flows to give them context. The point of this exercise is that if you do not understand the data, the processes make little sense.

Composite Data Flows

Use composite data flows when you want to have your cake and eat it, too. Let us explain. Suppose you have a data flow containing several data elements, or groups of elements. The most common way of dealing with this situation is to name the data flow and then define its composition in the data dictionary. This is a perfectly reasonable way to control complexity, especially in dealing with complex flows.

Another solution, which avoids the entry in the data dictionary, puts the names of the data elements directly onto a single, composite data flow.

Figure 2.6.5 demonstrates a composite data flow, consisting of two data elements. Each of the names attached to the flow is a recognizable element in its own right, and has its own definition in the data dictionary. By using this idea, you reduce the number of entries in the data dictionary.

Male Name + Male Address

Figure 2.6.5: *A composite data flow showing its components.*

Data Flows Are Always Named, Except …

A data flow without a name is like a millionaire without money. However, there is an exception. You may omit the name when the flow is to or from a data store and the data flow has exactly the same content as the store. In other words, the flow carries a hundred percent of the store's data elements. Since the store must be defined in the dictionary, it would be redundant to name and define the flow. Also, omitting the name makes the model more readable.

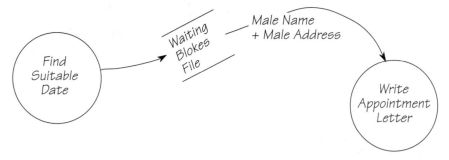

Figure 2.6.6: *The process* FIND SUITABLE DATE *writes all the data in the store* WAITING BLOKES FILE. *In this case, there is no need to label the data flow.*

However, sometimes the flow and the store are not the same. For example, in Figure 2.6.6, the flow coming from the store carries a name because it is a unique collection of data elements from the store. Later on, when you build essential requirements models, you will see examples of an exception to the rule for naming data flows.

Data Flows and Data Stores

A data flow represents data moving from one place to another. A data store is a static container of data. Data flows are deposited in the store for later retrieval.

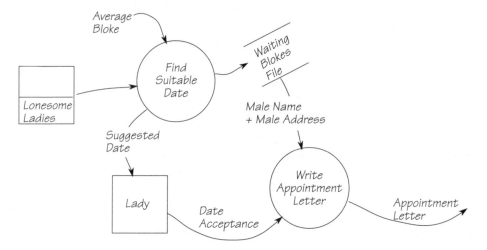

Figure 2.6.7: The system remembers the data in the store WAIT-ING BLOKES FILE and the system uses the data when it is time to produce an APPOINTMENT LETTER.

The data store indicates a time delay between the storage and the retrieval of the data. Before showing a data store on your model, be certain that the time delay is necessary for the system to carry out the business policy. You'll often come across data stores that have no policy-related reason to exist. An example of this is shown in Figure 2.6.8.

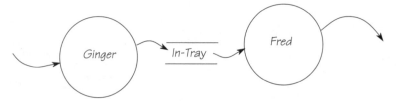

Figure 2.6.8: The IN-TRAY is not a real data store. It exists because one processor (FRED) is not fast enough to keep up with the incoming data from GINGER. There is no business policy reason for this data store.

Normally, data stores that are part of the business policy will show an important difference between the order of storage and the order of retrieval. For example, male candidates are stored in the WAITING BLOKES FILE in a random manner. They are extracted in a sequence that corresponds to the responses from the ladies, which may be a very different order. On the other hand, FRED probably just takes whatever is on top of the IN-TRAY, and there is no policy connected to the order of storage and retrieval. In fact, this applies to almost all first-in/first-out or first-in/last-out stores.

> A data store is valid if, between storage and retrieval, there is a delay related to the policy and the order of storage and retrieval is determined by the policy.

When a data store exists for a business policy reason, it must be used to both store and retrieve data. A read-only or a write-only data store cannot be correct because such a store is not normally part of a working system. However, there are some exceptions. A write-only data store may be there for archival purposes, and another system may create a read-only store to be read by your system. In both cases, the data store is part of another system and so acts as a terminator. The appropriate notation for these cases is a terminator symbol that incorporates a data store. This is called a *terminating data store.*

A terminating data store can also be represented as a regular data store, but with a terminator symbol beside it. Either notation is acceptable. Just remember to define the terminating data store in the data dictionary.

Lonesome Ladies

Note that in Figure 2.6.7, the process that updates a data store is shown with a data flow pointing to the data store. Suppose that the system files waiting blokes in alphabetical order. Why is there no data flow from the data store into the process? Isn't it necessary to locate the correct position?

Lonesome Ladies

The data flow diagram shows only the data needed for the processing. The process in this case is adding data to the store. Is it necessary for the process to read the data store before adding to it? Only if there is some business policy attached to the position that the new data occupy in the store. The current system may happen to store the candidates in alphabetical order, but that is probably for the convenience of the filing clerk; you would be hard-pressed to find any business policy about the order of filing.

Also, do not confuse data stores with databases. Records are normally stored in the database in a particular order, but that is because of a technological requirement of the particular database management system, not because of the business policy. Imagine that the data store has no technology, so there is no technological reason to store data in any order. It can be dumped into the store as it happens to appear. Update processes do not need a flow from the store to the process unless the data from the store are necessary for the processing policy, or unless the store's data leave the process by means of another data flow.

A Common Error

Many analysts attempt to build models without a process to store or retrieve data (see Figure 2.6.9). The storage and retrieval of data within the context of the system must be controlled by processes, not terminators, within the context; there must be a process that says, "I'm going to remember this data" and performs the actual storage. Remember you have declared the terminator to be outside your system.

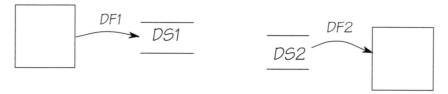

Figure 2.6.9: *This is wrong. Data cannot flow directly from a terminator to a data store, nor from a store to a terminator.*

Similarly, there must be at least one process to retrieve the stored data and assemble the data in the manner expected by the terminator. The corrected version of the above model is shown in Figure 2.6.10.

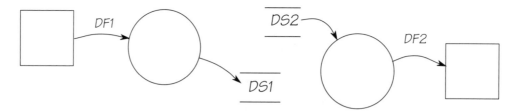

Figure 2.6.10: *The model is corrected by including processes to store and retrieve data.*

Trivial Rejects

A *trivial reject* is not usually subjected to detailed analysis, and occurs when the rejected data simply leave the system. Take, for example, the case when a male applicant is not considered suitable to become a client at Miss Tweedy's (Figure 2.6.11).

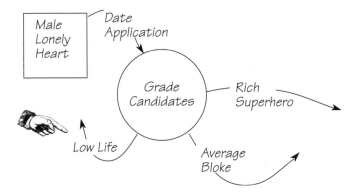

Figure 2.6.11: The trivial reject, LOW LIFE, is shown as a hooked arrow. There is no need to connect this flow back to the terminator.

The rejection of a candidate does not affect anything else in the system. The system has no policy for writing a rejection letter; Miss Tweedy's staff simply throws the application away. In this case, the system does not have to undo any processing, nor does it do any extra processing because of the rejection. When you write a description of this process in the form of a mini specification (the topic of Chapter 2.12 *Mini Specifications*), you will simply say that low-life candidates are ignored. The trivial reject notation is a convenient way to highlight this policy until you are ready to write your specification.

Well-Defined Processes

Each bubble in the data flow diagram must represent a well-defined process. "Well-defined" means that the bubble contains either a single function, or a group of functions, closely connected by data. A well-defined process is easily recognized by users because they can give it a sensible name that clearly describes its functionality.

A well-defined process has *narrow interfaces* with other processes, that is, there is only a small amount of data connecting it to the others. If the process is truly functional, it does not need large quantities of data from outside. Well-defined processes are self-contained, giving and getting only a minimal amount of data to each other.

Now suppose that instead of partitioning the system by its functionality, you broke it into tasks done by groups of people. Now a system model could look something like the bizarre example in Figure 2.6.12.

Functional processes have narrow interfaces. Our colleague Tom DeMarco[*] puts it: *The best partitioning is that which results in the narrowest interfaces.*

---

*To read Tom DeMarco's explanation of this and other concepts, we refer you to his classic *Structured Analysis and System Specification.* Englewood Cliffs, N.J.: Prentice-Hall, 1978.

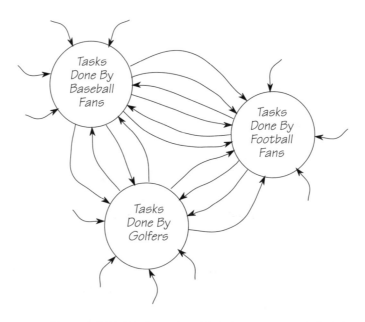

Figure 2.6.12: We hope that this is the worst-partitioned data flow diagram you'll ever see. The functional part of the system has been ignored in favor of the sporting predilections of the employees. Note the relatively large amount of data flowing between the bubbles.* This is because you could normally expect to find several people combining to complete a function. However, if a football enthusiast does part of the function and a golfer does the rest, a data flow has to pass on the half-completed result.

Indefinable Processes

One of the problems confronting most analysts is the seemingly indefinable process, the one that appears impossible to determine exactly the rules used by the process. Sometimes, the users are so familiar with a process that they find it hard to tell you exactly what data they use. Information such as telephone calls and facts the user remembers are so routine they become invisible. The users don't think of their information in terms of data flows and data stores. But you must. Similarly, the data elements do not have to be stored in a computer system for them to be necessary to the process. If they exist in any form, they must be a part of your specification.

Sometimes, despite knowing all the data, the process still appears to be indefinable. It is difficult to write a sensible mini specification when a user says, "I don't know how I do it. I've been doing it a long time. I just use my experience." While

*The data flows have not been named. Your authors are of the opinion that data flow names for the half-completed functions from such a crazy partitioning are beyond their imagination. We suggest that if these unnamed flows cause you offense, you name them after your pets or your relatives. The names would not be any sillier than the partitioning.

writing a deterministic mini specification that includes experience or judgment is impossible, it can be shown in the data flow diagram.

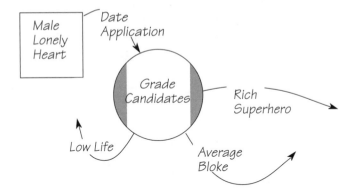

Figure 2.6.13: The process GRADE CANDIDATES is done by Zsa Zsa, a veteran of many successful marriages, who considers herself experienced in this area. The shading of the bubble indicates judgment being used in the process. It is impossible to write a consistent policy for this process. Zsa Zsa merely has to say, "I like this one" for the candidate to qualify.

This shading of the bubble in Figure 2.6.13 signifies that it may be impossible to write a deterministic policy for the process. Because the policy for this process requires indefinable human experience and judgment, it cannot be implemented on a computer in its current form. The annotation alerts the systems designers of a process in the system that cannot be automated and cannot be improved, but must remain as part of the system.

Drawing Data Flow Diagrams

Finally, we come to the task that many people find the most daunting: putting pencil to paper and actually *drawing* a data flow diagram. They're afraid that their diagrams will look more like the webs of hallucinating spiders than a serious model of a business system.

As ever, your authors are here to help. We have three prescriptions: One is automated in nature, another is advice, and the third is your attitude.

The automated prescription is a CASE tool. If you have not used CASE (computer-aided software engineering), the basic idea is that you build your models using software designed to help with diagramming and checking the consistency of your models. Most CASE tools are capable of producing models that are legible, which solves the problem for people who worry about their artistic ability. You will undoubtedly use CASE tools during your analysis career. There are situations in which they are almost indispensable. We would never choose to do systems analysis without some form of automated help. However, do not be in too much of a hurry. There is a learning curve. If you already have a CASE tool and you are experienced in using it, then use it for this project if you like. If you don't already have one, you will add considerably to the time needed to complete this book by starting to use one now. Remember that your aim is to understand the *thinking behind systems analysis*. Once you have done that, you will be in a position to make effective use of a CASE tool.

Here's some advice on making your drawings more presentable. First, keep in mind that no points are awarded in systems analysis for artistic merit. *The important thing is to record your analysis of the system.* However, there are a few things you can do to make your diagrams more attractive to readers.

Minimize crossing or extremely long data flows whenever possible. This is easier if you judiciously repeat terminators and data stores to put them closer to the processes that need them. When you do repeat something, highlight it by putting an asterisk in both copies of the repeated item.

> Male
> Lonely
> Heart
> ✻

Aim the arrow of your data flow at the center of the bubble, data store, or terminator. This makes a big difference in the appearance and readability of the diagram. The flow looks as if it is meant to go somewhere. Also, make it appear to come from somewhere. Draw the arrow's tail leaving from the center of the object.

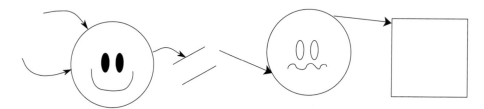

Figure 2.6.14: Aim the data flows at the center of their target. Do not let them emerge at strange angles. Draw data flows as curves with narrow arrowheads.

A few more tips on drawing legibly:

- Draw the data flows as curves; straight ones look too formal (though some CASE tools force straight lines, as you'll see in some of our own diagrams that we produced using a CASE tool). Curved flows are more friendly.
- Use narrow arrowheads.

- Use a thicker pencil for the flows than for the labeling.
- Make the bubbles bigger than you think you will need (that way you can fit the process name inside).
- Get a drawing template to make the circles and squares.
- Always be prepared to redraw the model.

Last, adopt the proper attitude: Consider what you are trying to do with these drawings. The goal here is *communication*. You are not trying to impress, but you are trying to show your understanding of a system to the users. Our experience has been that the users are not interested in great art, but they are interested in their system. We urge you to draw most of your models at the users' desks. This way you are saying, "This is not intended to be art, it's a model of your system. Tell me where I'm wrong. Tell me where it's incomplete."

Whether you draw well or not, an important part of your attitude is *fluency*. You have to become very familiar with these models, so familiar that you can express your ideas without thinking about the models, their conventions, or your own drawing skills. Drawing and redrawing makes you think about how you are using the model, and it makes you think about the system you are drawing. Each time you redraw the model, you find more questions, and each question gets you closer to a complete and accurate requirements model. Fluency leads you to build models that tell the story of the system, to draw models that reveal the patterns of the system. However, fluency is something that we cannot teach; you must learn it by yourself. The best way we know is by drawing, and redrawing, as many systems as possible.

What to Do

Make an honest attempt at the following exercises. They will give you practice in building data flow models. As with any other tool, the more practice you have, the better you become.

When you are happy with your models, go to Chapter 4.4, which contains some sample answers and a discussion of our approach. The Trail Guide at the end of the chapter gives your next destination.

Exercise 1: Any Defects?

Study the data flow diagram in Figure 2.6.15. List all the errors you can find and compare your list to the one in Chapter 4.4.

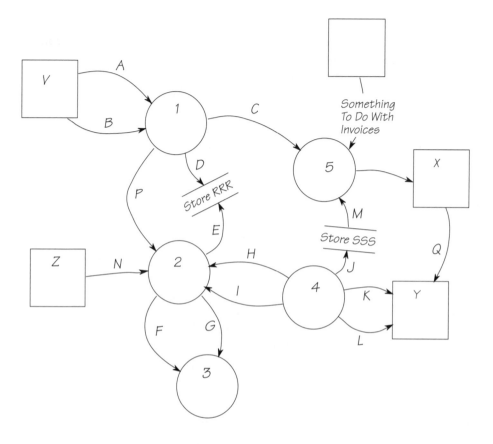

Figure 2.6.15: This data flow diagram has some problems. List all those you find.

Exercise 2: Can You Improve This?

Figure 2.6.16 shows a data flow model for an airline passenger check-in system. It is not a good example of a data flow diagram and needs improvement. Redraw it so that you eliminate all its problems.

Most of the faults in this model concern the Rule of Data Conservation. Study each process, and determine if it has enough incoming data to produce its outputs. If not, what new data does it need? Then, redraw the model to show your improvements. Check your revised model with the one in Chapter 4.4.

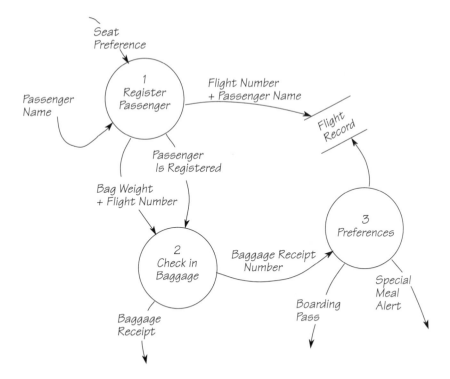

Figure 2.6.16: We're sure you can do better than this.

Exercise 3: The Clearing House Revisited

In Chapter 2.2 *Data Flow Diagrams,* we asked you to draw the context diagram for the Government Research Paper Clearing House. Use the sample context diagram in Figure 2.6.17 and the reprinted description to draw a data flow diagram for the Clearing House. Your diagram is the first breakdown of the context, so you have to select the right level of detail. This means your diagram should demonstrate the functionality of the system without having so many bubbles that the model becomes unreadable.

Here's the description of the operation: The federal government maintains the Government Research Paper Clearing House for unclassified research and development papers. The Clearing House sends scientific research papers to any citizen who asks for them. So that citizens know what is available, indexes of recently published papers are sent out each month.

Citizens may request to be added to the subscribers list for no charge and, upon being added, will be sent all the indexes currently on file.

For a fee, a citizen can request to have the full paper sent to him. This request must first be checked to see that the citizen is a subscriber and the index is correct,

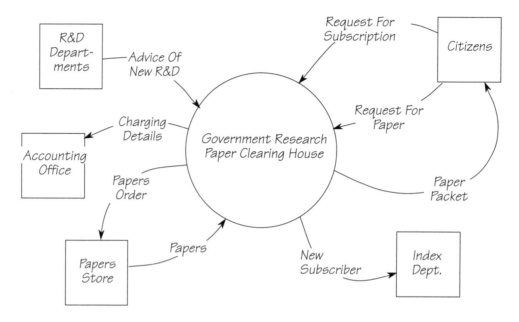

Figure 2.6.17: Context diagram for the Clearing House.

then the paper is sent. The invoicing for this service is handled by the Accounting Office.

The procedure for mailing has been in place for some years. When a paper is requested, the Clearing House generates an address label with a tear-off strip identifying the paper to be sent. Each weekday evening, an order for the required papers is sent to the Papers Store. The store sends the papers the following morning, and the Clearing House workers match the papers to the labels and ship them. The tear-off portion of the label is discarded before shipping.

The indexes are maintained and shipped by the Index Department. This department is run by two women who use personal computers to maintain their list of subscribers. They are also responsible for shipping the indexes to subscribers. The women do their jobs very well. In fact, they have won awards several times for their efficiency.

Draw a data flow diagram for the operation described above. Remember that all the data flows that enter or leave the context are part of your diagram, so use these boundary data flows as a way of getting started. Check your solution against the one given in Chapter 4.4.

Exercise 4: La Cave du Morey Saint-Denis

Morey Saint-Denis has collected some of the finest wines of Bordeaux and Burgundy. His *cave* (wine cellar) is the envy of his neighbors in Rue du Faubourg St. Honoré, and most of the gourmands of Paris. However, Morey Saint-Denis has to work hard at maintaining his collection.

After decanting the wine for the table, the *sommelier* keeps the empty bottles and uses their labels to update the cellar log once a week. This log is a record of the existing stock.

Morey Saint-Denis is offered wine for sale by some of the better merchants, and his favorite vineyards write to him if they think the vintage will be a good one. Morey doesn't buy every wine offered. He buys only if his stock of that wine needs replenishing, and if his research indicates that the wine will be *facile à vivre* (good-tempered).

Good-tempered wines are those that mature gracefully in the *cave*. They must not mature too quickly, nor reach their peak later than they should. Morey keeps an extensive file of the performance of each vineyard's wines over the years. He subscribes to several wine newsletters, and gets information from trusted friends as well as occasional reviews in wine magazines.

The Saint-Denis family is quite wealthy, but Morey maintains an annual budget for his wine purchases. He buys wine only when there are sufficient funds available. He occasionally changes the budget in mid-year, especially if his investments in the Paris *Bourse* (stock market) do better than expected.

Draw two data flow diagrams for the system Morey Saint-Denis uses to maintain his wine cellar. One is the context diagram, showing the inputs and outputs of the system. The other data flow diagram is the first breakdown, and shows the functionality. Use the Clearing House models as a guide for constructing these new ones. There are sample diagrams of this system in Chapter 4.4.

Trail Guide

If you returned here after Chapter 4.4, find your trail.

● Easiest: We will continue our study of data flow models. If you are building a model of a large system, you'll need to build a leveled model. You'll find out how in Chapter 2.7 *Leveled Data Flow Diagrams.*

■ More Difficult: This chapter is not on your trail. If you are reading it, that suggests you will rediscover your trail in either Chapter 2.8 *Current Physical*

Viewpoint; or, if you have already been there, see Chapter 1.4 *The Piccadilly Organization.*

◆ Most Difficult: This was not intended reading for you. Chapter 1.4 *The Piccadilly Organization* is the best destination for you.

❋ Promenade: It's hard to know why you have strayed this far off your path. If you found this chapter interesting, you'll most likely want to go to Chapter 2.7 *Leveled Data Flow Diagrams.* If you want to get back to your correct trail, go to Chapter 2.8 *Current Physical Viewpoint.*

LEVELED DATA FLOW DIAGRAMS 2.7

Before You Reached Here ...

● Easiest: In Chapter 2.6 *More on Data Flow Diagrams,* you have looked at some of the details of data flow diagrams. Before your education is complete in this regard, you need to know how to build a leveled set of data flow models. This chapter explains how leveling helps to overcome the difficulty of modeling large systems.

■ More Difficult: On this trail, you should already know about leveling. If not, read on. Otherwise, turn to Chapter 2.8 *Current Physical Viewpoint.*

◆ Most Difficult: This chapter is not required reading. We suggest that you pick up your trail in Chapter 1.4 *The Piccadilly Organization.*

❋ Promenade: Leveling is a way of building models of systems that are too large to show in a single diagram. That means virtually all systems. To see how it's done, stay with this chapter. However, leveling is not on the ❋ Promenade Trail, and the appropriate place to rejoin your trail is Chapter 2.8 *Current Physical Viewpoint.*

Most of Today's Systems Are Big

As computing equipment gets better and faster, more and more of our businesses become computerized. To remain competitive, companies must build software systems that include more and more features. The effect is that since most of the systems you must analyze these days are big, you'll be building models of large systems.

One reason to build a data flow model is to specify the system. Naturally, if the model is an accurate specification, it will include all the details of the system. If you are specifying a large system, then your completed model must contain a large number of bubbles.

When you need, say, several hundred bubbles to specify a system, it is not reasonable to fit them all onto a regular page. You could try to find a large enough sheet of paper, or you could draw very small bubbles in the hope of getting them all on one page, but having all those tiny bubbles on one huge sheet of paper is too complex for human comprehension.

To build a data flow model of a large system, you have to build a *leveled* model. The idea of a leveled model is not new; you may have already used one. For example, when you drive a long distance, you probably find the best route using a map that shows all the states or countries you'll drive through. As you drive along, you use more detailed maps that show the local roads. When you arrive at your destination city, you switch to an even more detailed map to find your way through the streets. The different maps represent different levels of detail that you need for navigation. The same principle applies to system models.

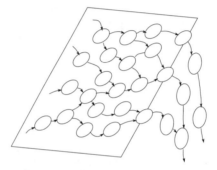

Figure 2.7.1: The problem is that most systems contain too many bubbles to fit on a normal sheet of paper.

We also understand things in a leveled way by starting with a high-level view that takes in the whole system, and then descending into the details. In systems analysis, you view the whole system by using the context diagram. (See Figure 2.7.2.)

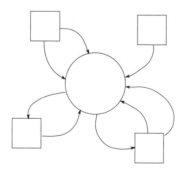

Figure 2.7.2: The context diagram shows the whole system in one bubble. The data flows around the context define the boundaries of the system.

Once you establish the context reasonably well, you reach the next stage of understanding by breaking the system into its major components. A context diagram and its breakdown are shown in Figure 2.7.3. However, the task is not always as simple as the diagram shows. Beware that the lower-level breakdown often reveals data flows that were missed at the context level. When this happens, you must then go back to make the appropriate corrections to the context diagram.

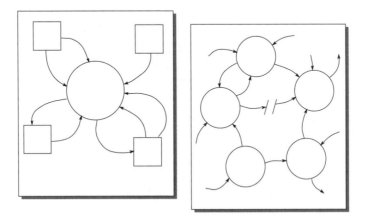

Figure 2.7.3: The first breakdown of the system shows how the major components interact.

After declaring the major components and interfaces, you investigate each of them in turn. Naturally, you do this by partitioning each component into its major components and declaring the interfaces between them. In Figure 2.7.4, the high-level diagram (Diagram 0) is called the *parent diagram* and each of Diagrams 1 through 5 is called a *child diagram*.

The diagrams at each level show a manageable number of processes. If any of these processes is too large or too complex, then draw a lower-level diagram to partition it into still smaller and simpler pieces. You continue decomposing until you arrive at processes that are small enough to be specified.

How Much Detail at Each Level?

We began this chapter by dismissing the notion of having several hundred bubbles on one page. It's beyond the ability of most humans to comprehend such a diagram. So just how much information can you get on one page before befuddling your reader (or yourself)?

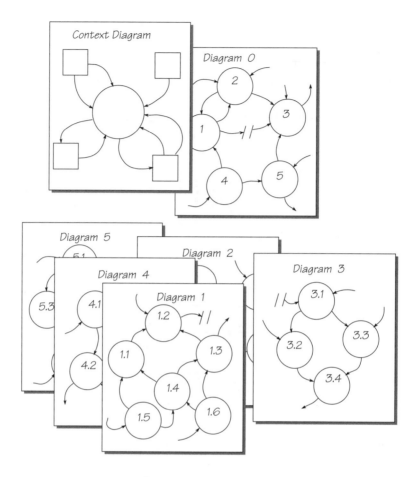

Figure 2.7.4: The top levels of the set. The context diagram declares the whole system, and Diagram 0 gives a manageable breakdown. The lower-level diagrams give a breakdown of each parent bubble at the higher level. This is called a top-down model.

The answer is connected to the number seven. Almost forty years ago, psychologist George Miller conducted research into the ability of humans to deal with simultaneous concepts. Miller established that at seven, plus or minus two, humans began making an intolerable number of errors.[*]

For completely unscientific reasons, we agree with Miller. Our experience has taught us that seven processes on one page seems to be the limit for users and analysts alike. As soon as there are seven bubbles with their associated data flows and data stores, they comfortably fill a page without being too complex. Similarly, if there are

[*]George A. Miller, "The Magical Number Seven, Plus or Minus Two: Some Limits on Our Capacity for Processing Information," *The Psychological Review,* Vol. 63, No. 2 (March 1956), pp. 81-97.

more than seven data flows into or out of a process, the process is probably too complex. We repeat, this is not scientific. We simply *pretend* there is a limit of approximately seven. If you also pretend the same thing, and resist building models with more than about seven bubbles per page, then your models will be easier to comprehend.

Numbering the Bubbles

Numbering the bubbles in a model is for identification purposes only. In Figure 2.7.4, note that

- The numbers *do not* indicate the processing sequence. They have no meaning other than identifying a specific bubble.
- Each of the child diagrams takes as its title the number of its parent. For example, bubble 1 begets Diagram 1, bubble 2 begets Diagram 2, and so on.
- The bubbles in the child diagram take the number of the parent, and each adds a decimal identifier. For example, the children of bubble 1 are numbered 1.1, 1.2, 1.3, and so on.

Identifying bubbles by a combination of the parent's number and a decimal makes it easier to navigate around the specification. It means that from any level, both the parent diagram and the lower-level child diagrams can be readily identified. For example, to see the breakdown of a bubble, look for the diagram with the same number as the bubble. Conversely, to find the parent diagram, look for the one that contains a bubble numbered the same as the child diagram.

Functional Primitives

The little guys at the lowest level of the set are called *functional primitives* because they have no component parts that can be thought of as functions. The function is so simple and has no internal component that itself can be thought of as a function, so you have no need to decompose it further.

There are no internal data flows in functional primitives. The functions in a diagram are separated by data flows. If you cannot find any flows within a bubble, it is likely there are no functional subcomponents for that bubble. Remember that a data flow is a genuine collection of data elements, not just the intermediate results of a calculation that have no interest outside the bubble.

Eventually, when you have decomposed so that you can completely describe a bubble in a page or less of text, then you can stop partitioning. These one-page specifications, known as mini specifications, are fully discussed in Chapter 2.12 *Mini Specifications*. For the moment, remember that this is the level of the system that describes the details of the decision making, calculation, and data handling.

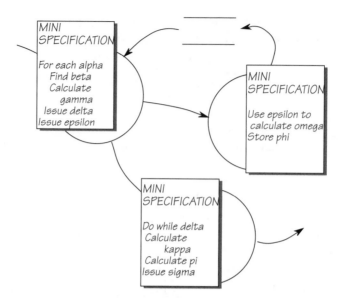

Figure 2.7.5: *Functional primitives are small enough to adequately describe their processing requirements with a one-page mini specification.*

Using the Imaginary Expanded Diagram

Imagine that you had the time and the patience to join all the functional primitives by their data flows. The model would then show only the processes from the lowest level of the set. This *expanded diagram* would look something like Figure 2.7.6.

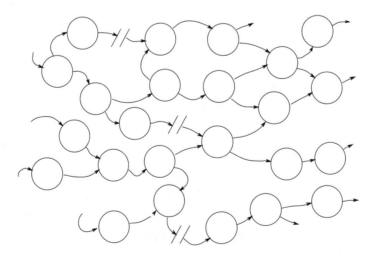

Figure 2.7.6: *This expanded diagram is an imaginary diagram to illustrate the idea of joining all the functionally primitive processes.*

We don't suggest that you build such an expanded model; it negates the idea of the top-down analysis already done. Besides, there are too many problems involved. You probably couldn't find a sheet of paper large enough to accommodate the hundreds, or possibly thousands, of bubbles. And even if you did, the resulting model would be almost impossible to read. If you could manage to read it, the effort needed to maintain this diagram would be definitely beyond your budget.

So just *imagine* such a diagram because you can learn something from it. The functional primitives represent the actual processes of the system. The higher levels of the model are simply convenient summaries to make the model readable. Let's see how we can use this imaginary expanded diagram.

Repartitioning to Suit Your Purpose

The reason for having the expanded view is that the bottom level represents the actual system, and you can regroup the system's processes to make a more logical partitioning at the higher levels. Regrouping is necessary if the original partitioning reflects the way the current system happens to be implemented. (*Partitioning* means breaking a system into manageable pieces, and the partitioning of many models reflects the processors and human organizations currently doing the work.) If this partitioning distorts your view of the system, you may reassemble the primitives to make a better partitioning for your model. Let's see how this works.

In the example (Figure 2.7.7), the partitioning in Diagram 0 was less than ideal. The large number of *interfaces* indicates that functions have been split between the bubbles. To correct the model, you must regroup so that functions, or closely related groups of functions, complete their processing within a bubble. When bubbles fulfill their functionality, they do not have to pass around data flows carrying incomplete results, and the interfaces between them are subsequently much narrower. To see the functionality within each bubble, you need to decompose it using a lower-level diagram. Figure 2.7.7 shows that these can be joined to make the expanded diagram.

Now look for groups with closely related functionality. These groups will have narrow interfaces to the other groups. The objective is to repartition the model so that the interfaces are as narrow as possible. Figure 2.7.8 shows the groupings, and the revised Diagram 0.

The regrouped model reflects a more *functional partitioning,* and so it is a better, more rational model on which to base the design of a future system. In addition, being able to repartition or regroup the bubbles gives you some needed flexibility:

- Sometimes, your users will give you a lot of detail before you can form the big picture of Diagram 0. In this case, it is more convenient for you to model some of the middle levels before repartitioning upward to produce the top levels.

- Sometimes, you will make a bad partitioning on your first attempt at modeling the high levels. When this happens, model down to the lowest level, and then regroup upward to make a better partitioning at the top.

Remember that if you are able to decompose a bubble, you are also able to recompose one from lower-level bubbles. Decomposing and recomposing are only possible if you observe a strict convention of keeping the levels in balance with each other.

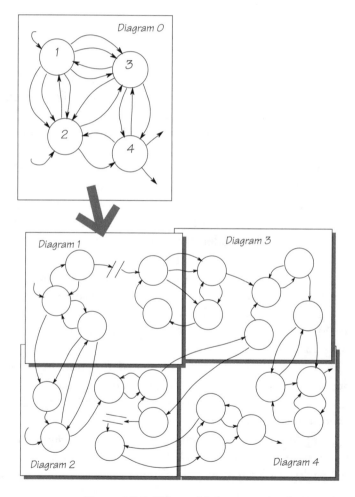

Figure 2.7.7: This model shows two levels. Note the poor partitioning in Diagram 0. The bubbles share too many interfaces, which indicates an inferior current implementation. Diagram 0 decomposes into four lower-level diagrams that are joined on the large plane at the bottom. This is the expanded diagram.

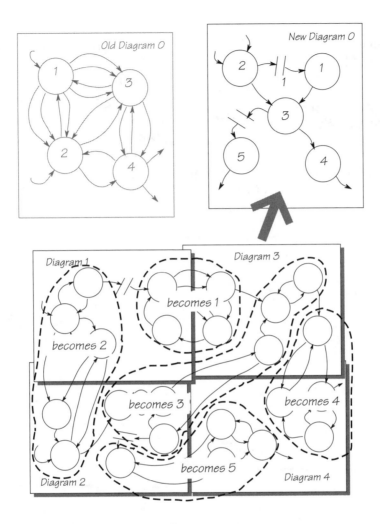

Figure 2.7.8: The analyst has regrouped the bubbles, as shown by the dotted lines. The expanded diagram is leveled upward to make the new Diagram 0.

Balancing

Two levels of a model are balanced when the child diagram is processing the same data that enter and leave the parent bubble. Although the idea of balancing may seem obvious, too many projects get into difficulties because they ignore the *balancing rule*.

The set of diagrams shown in Figure 2.7.9 balances between the levels. The data flows in the child diagram reflect those to and from the parent bubble. Now

Balancing means that all the data flows interfacing with the parent bubble are external data flows in the child diagram. Alternatively, you could say that all external flows in the child diagram connect with the parent bubble.

suppose that when the analyst drew the child diagram, he included another flow (let's call it к) leaving bubble 3.3. The diagrams are now out of balance. Why? Was the new flow overlooked at the higher level? Or is it just plain wrong and should not be part of the lower-level diagram? Whatever the reason, it must be corrected before proceeding. Similarly, if a data flow appears in the parent diagram, but not in the child diagram, this imbalance suggests that the analysis at the lower level is incomplete.

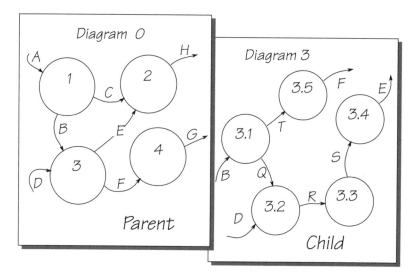

Figure 2.7.9: Two diagrams from a leveled set. Child Diagram 3 is in balance with its parent: All of the data flows into and out of the parent bubble appear in the child diagram. Similarly, the child's external data flows—B, D, E, and F—are shown entering or leaving the parent bubble.

If the diagrams do not balance, you cannot repartition them as we did in the example in Figure 2.7.8. Diagrams that do not balance are simply incorrect, and will not make an accurate specification for the new system.

Balancing Using the Data Dictionary

When you analyze large systems, it is sometimes convenient to bundle a number of data flows into a single flow to make the high-level diagrams easier to read. Then, when you draw the lower levels, split these bundled data flows into their components to show the functionally primitive processes.

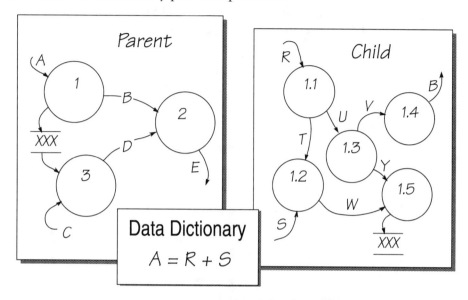

Figure 2.7.10: Parent bubble 1 shows an incoming data flow A. However, A is a bundled data flow containing the flows R and S. As the lower-level diagram shows the component flows, you must use the data dictionary to balance the model.

In the example shown in Figure 2.7.10, the child diagram at first appears to be out of balance with its parent because it doesn't show the data flow A. All data flows are defined in the data dictionary, and in this case it reveals that A is composed of R and S. As both of these flows appear in the child diagram, the model complies with the balancing rule.

Balancing the Data Stores

Data stores in leveled models have their own balancing rule: Show data stores at the highest level where they are used by more than one bubble and at every relevant lower level. When you decompose a process that accesses a data store, that store must appear in the lower levels if the diagrams are to remain in balance. When only one bubble in the diagram accesses a store, it looks a little odd; however, it is necessary if your reader is to make sense of the low-level diagrams.

Just as you can group data flows in higher-level diagrams, you can also have composite data stores. When there are many data stores in the system, some analysts prefer to collect them into a composite store, such as an "accounts database" or "operations file," and then break them into smaller data stores in the child diagrams. The advantage of this approach is that the decomposed stores show the precise data usage of the low-level processes.

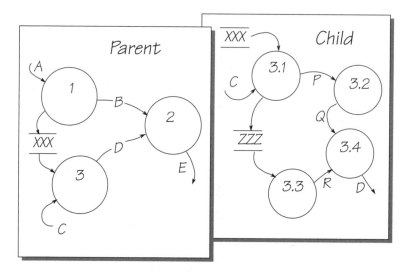

Figure 2.7.11: The parent diagram shows two bubbles sharing the data store xxx. It also appears in the child diagram as input to bubble 3.1. The child diagram reveals data store zzz, which was hidden inside bubble 3 at the higher level.

Summary

You can use leveling to control the number of bubbles you wish to show in any one diagram. You also use leveling to present a top-down model of the system. You would like your end result to be a specification that is organized in a top-down manner because it is easier to control and to check for consistency and completeness. Although you can think of analysis as a top-down activity, it is rarely done this way. The reason is that you typically get the system information in a random manner, as the users supply it. As you interview different people in an organization, they give you information at different levels. To cope with this situation, you must feel comfortable about building your models at any level, and decomposing and recomposing to make the most logical partitioning of the system.

It is perfectly acceptable for you to draw a data flow diagram without knowing precisely at which level the diagram fits into the system model, and you may decompose and recompose several times before you establish it at the correct location. There is nothing wrong with that; it is just part of the craft of systems analysis.

You must become comfortable with the idea of working top-down, bottom-up, middle-up, or middle-down.

The balancing rule ensures the model's integrity by proving that each child diagram is a faithful decomposition of the parent. The high-level models act as a guide to the details in the lower-level models.

What to Do
The four exercises that follow reinforce the lessons of this chapter. Try the exercises, and then check the answers and discussions in Chapter 4.5.

Then, consult the Trail Guide at the end of this chapter to find your next destination.

Exercise 1: Find the Leveling Problems
There are some problems with the leveling in the set of diagrams shown in Figure 2.7.12. Highlight all the errors you find. Ignore the single-letter names in this and subsequent models. They are used only to simplify the diagrams.

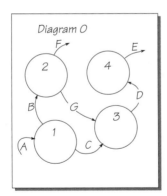
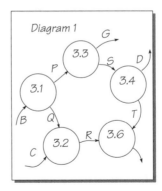

Figure 2.7.12: A set of sample parent-child diagrams to test your skill at balancing.

Exercise 2: Balancing Data Stores

The set of diagrams in Figure 2.7.13 has some problems with the data stores. Highlight the errors and tell why you think each one is wrong.

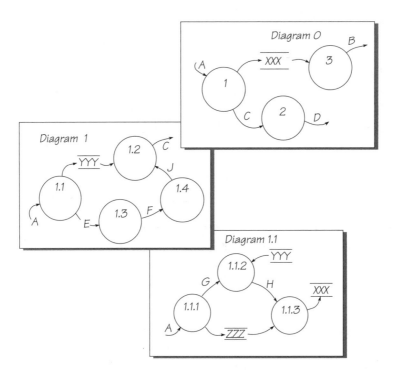

Figure 2.7.13: A set of parent-child diagrams with some data store balancing problems.

Exercise 3: Draw the Parent Bubble

Figure 2.7.14 shows a child diagram. Draw the parent bubble. Group the data flows to make the parent as readable as possible, and say what data make up any new flows introduced into your new diagram.

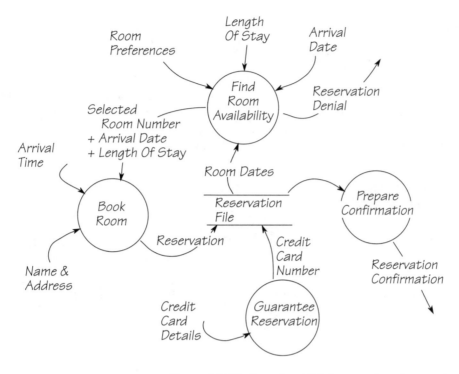

Figure 2.7.14: From the child diagram, draw its parent.

Exercise 4: Repartition the Model

Here is a leveled set of diagrams, but they are not very useful. Diagram 0 is badly partitioned. Can you make the model more useful by repartitioning in some way?

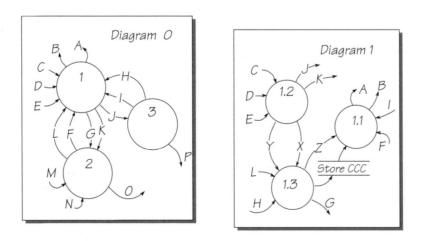

Figure 2.7.15: A badly partitioned set of diagrams.

COMPLETE SYSTEMS ANALYSIS

Trail Guide

This duplicates the Trail Guide in Chapter 4.5.

● Easiest: You now have all the information you need about data flow models. But when you build a model, it needs a viewpoint. One that you will find useful as you continue to explore the Piccadilly system is the current physical viewpoint. We discuss this in your next chapter, 2.8 *Current Physical Viewpoint*.

■ More Difficult: This chapter is not part of your trail; your next chapter is 2.8 *Current Physical Viewpoint*.

◆ Most Difficult: This was not intended reading for you, and Chapter 1.4 *The Piccadilly Organization* is the best destination for you.

❊ Promenade: We didn't intend you to be here, but hope that you found leveling interesting. Turn now to Chapter 2.8 *Current Physical Viewpoint*.

CURRENT PHYSICAL VIEWPOINT 2.8

Before You Reached Here ...

● Easiest: You've looked at further refinements of the data flow model in Chapters 2.6 *More on Data Flow Diagrams* and 2.7 *Leveled Data Flow Diagrams*. Now we add a viewpoint to those refinements. After you read this chapter, you will return to Piccadilly to build a current physical model of its operations.

■ More Difficult: You've arrived here from Chapter 1.3 *What About the Business Data?* This chapter will give you a viewpoint that you'll use when you continue to model Piccadilly's business.

◆ Most Difficult: The viewpoint chapters are not part of your trail as you have elected to do the Piccadilly Project without our help. However, you may decide to review this chapter. The next part of the project to be modeled is the whole organization, starting in Chapter 1.4 *The Piccadilly Organization*. The "Strategy" heading in that chapter explains our approach for analyzing the business, and it involves the current physical viewpoint. If you wish, read this chapter to see how we do it.

❈ Promenade: The correct route for you to get here is from Chapter 2.3 *A Variety of Viewpoints*. If you are reading the Textbook sequentially, this chapter will still make sense, but for your purposes, it's easier to follow the trail.

What Is a Current Physical Model?

A *current physical model* is a replica of the users' existing operation. We use the terms "current" because the model shows the system as it exists at the moment, and "physical" because the analyst portrays the actual procedures and devices used to implement the system. When you present a picture of the system that the users can easily recognize, you break down at least one barrier to your communication with them.

The current physical model documents the existing reality.

In Figure 2.8.1, the model intentionally shows implementation characteristics, as the names are meant to be recognized by the users. The names in the bubbles are the device or process names as the users know them. Data flow names are, in most cases, the users' names for the documents and packets of information that move around the office.

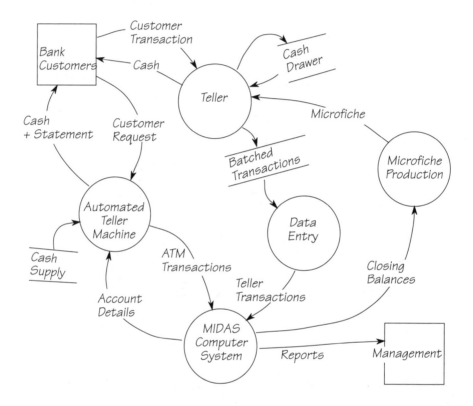

Figure 2.8.1: The current physical model for part of a bank's operations. This model depicts the system as you would see it if you visited the bank.

Why Build a Current Physical Model?

There is a paradox: The current physical model is a model of the system that you are about to replace. Why spend valuable time doing this? Why, for example, include microfiche production in a model when the intention is to replace it with an on-line terminal for the teller?

There is an old joke about a tourist who stops his car on a country road and asks a farmer how to get to a certain town. The farmer chews on it for a while, and eventually replies, "You can't get there from here."

The joke makes use of an absurdity: You can always drive from one town to any other provided they are on the same land mass. However, systems analysis is different. You are trying to find the "perfect" requirements for the system. At the beginning of an analysis project, without information about the users' business, you have very little chance of finding your way to the perfect requirements. This time, you may not get there from here.

Let's go back to the example in Figure 2.8.1. We mentioned above that the bank intends to replace microfiche and give the teller an on-line terminal. Why not specify the terminal right away? Well, how will you know what data the teller needs if you do not know what data he uses at the moment? By including the microfiche in your current physical model, you know where to look to see the data that the teller uses. Omit it, and finding a place to start the analysis becomes more difficult.

The current physical model is not the specification of the system. It is merely a way to start the analysis. At the beginning of analysis, the users describe their business to you in a way that typically includes the mechanisms they use to get the work done. While these mechanisms are not the basic policy of the system, you build the current physical model to help you understand the existing system and to begin to gather requirements for the new system.

When you capture the system in its current form, you collect many of the new system's requirements. Despite any bad reputation or glaring faults the current system may have, it is not totally worthless and may even have many functions that are making a positive contribution to the business. So there are some parts of the current system that must be included in any future system. You may implement them differently with new technology, but their underlying business policy, their *essence,* remains almost unchanged. Depicting the existing system with a current physical model is the first step toward ensuring that you include all the essential functions and essential data in your future system.

> The current system may be far from perfect, but you can learn a lot from it.

Your future system must be relevant to the users' business. That much is obvious. Your system will be relevant if it fits into the way the users do their business. Gone are the days when you could put together whatever system was most convenient to you or easiest to build, and tell the users to take it or leave it. Building a current physical model gives you the opportunity to study and understand the way they do things now, before trying to improve them.

211

History is littered with spectacular failures that happened by developers' failing to understand the current system before implementing "improvements." In the early days of Australia, English settlers brought with them their domestic cats and dogs. Although these animals were not native to the colony, the settlers reckoned the addition of a few household pets would improve the country somewhat. What they failed to understand was that most of the native Australian birds were ground dwellers, not tree dwellers as in their former home in England. In the absence of predators, the cats and dogs within only a few years of their arrival made the ground-dwelling birds extinct. Perhaps if the settlers had understood the existing situation, they wouldn't have introduced their "improvements."

Gaining the Users' Confidence

One of the most immediate benefits you notice when you build a current physical model is the positive response from the users. People always like it when others take an interest in their work. Your model, which reflects the way their business operates, confirms your interest and builds cooperation. It gives your users the opportunity to recognize and understand your models, and to verify the correctness of your analysis. Almost as importantly, during the brief physical modeling activity, the users become familiar with your analysis style, and it will be easier for them to contribute to your modeling effort.

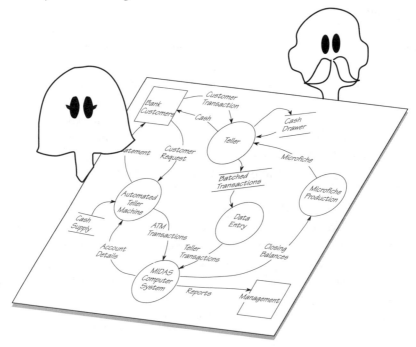

Figure 2.8.2: The user and the analyst work together to build an identical understanding of the business.

You are not trying to gain the users' confidence just to feel a warm glow. At some stage of the project, you will have to recommend ways to implement a new system. The users aren't likely to have any confidence in your recommendations if you haven't demonstrated your understanding of the existing one.

One other aspect of physical modeling is worth mentioning here. We always use data flow diagrams in our consulting business to record our interviews with users as they talk. We then show them our diagrams, talk them through the models, and confirm that we have captured exactly what they said. We solicit corrections for our misunderstandings, and continue modeling until both sides agree that the model is a faithful replica of the system under study. Then we hear the magic words, "Yessir! That's my business!"

Defining the Context of Analysis

The overriding reason for building a current physical model is to negotiate and verify the scope of the system you are about to study. It is critical for any analysis project to correctly define the boundaries of the system. Where does it start and where does it end? It is equally critical that the users and the analysts have the same, agreed understanding of those boundaries. The context diagram, in Figure 2.8.3, for example, defines the boundaries and shows how the system fits into the outside world. This should be the first of your analysis models and should be built as soon as possible after the start of the project.

The context diagram shows the boundary data flows that run to and from the terminators. It is these flows, and not the terminators, that define the boundaries of the system. How?

Your early interviews with the users will reveal some objects that appear to be outside the domain of study, so it is reasonable to show these as terminators. However, keep in mind that the terminator symbol really means "that part of the terminator that interacts with my system." To know precisely which part, you need to examine the data flow.

For example, PRODUCER and ACTOR are both shown as terminators in the context diagram (Figure 2.8.3). But shouldn't they be part of the system? Couldn't the producer auditioning an actor be considered a necessary component of a casting system? In this case, no. The diagram clearly shows you that the audition takes place outside the system. The system tells the producer about the audition with the data flow AUDITION SCHEDULE, and the PRODUCER DECISION data flow advises the system of the result.

On the other hand, actors come in to the Regional Theater Casting offices to apply for a part. In this case, their bodies are physically inside the system, but the data flow ACTOR'S REGISTRATION determines how much of a role they play in it (no pun intended).

213

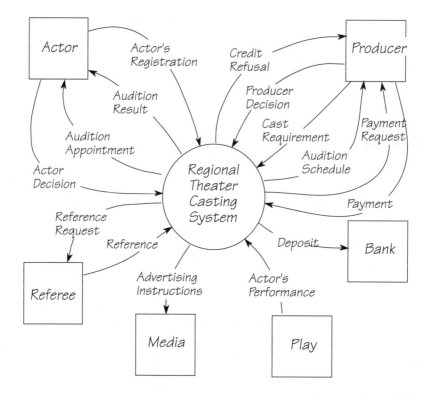

Figure 2.8.3: The context diagram for the Regional Theater Casting System. The system finds actors for parts in plays being produced for the small drama theaters around the country. The context diagram is used to negotiate the precise domain of the analysis effort.

Note the flow ADVERTISING INSTRUCTIONS going to the MEDIA. Everything to do with the advertising, such as writing the copy and designing the layout, happens inside the system; otherwise, this flow would not be in the context diagram. See what a precise definition of the context this is. Without this flow in the context diagram, you would have to rely on a statement such as, "The system includes advertising." What does that mean? Does it mean that the system is expected to print the advertisements? Or does it mean that someone decides what to advertise and what to send to a production house for the finished ad? Here, the context diagram is saying, "Every activity that contributes to the data in the flow ADVERTISING INSTRUCTIONS is included in the domain of analysis." By knowing what those data are, you know what the activity is. If the activity to be studied were different, the boundary flow would be different.

The boundary data flows also define the system's stored data. In the early part of the Piccadilly Project, you built a preliminary data model using the context diagram as input. Later in the project, you will see how you can partition this diagram in different ways depending on what you need to see. It will prove to be very useful.

It is vital that the users, management, and project team all agree on the context of the project, and agree as early as possible. Otherwise, you run the very real risk of analyzing the wrong part of the users' business. The area to be analyzed is the part of the business that the user has some need or desire to change, and any other areas in contact with the changes.

There are two types of functions: those to be changed, and those to be computerized. The new system may result in the computerization of manual processes, or the reimplementation of an existing computer system. For your analysis to work correctly, your context must be large enough to include all those functions that have a possibility of being computerized, together with the manual functions in contact with the computer system.

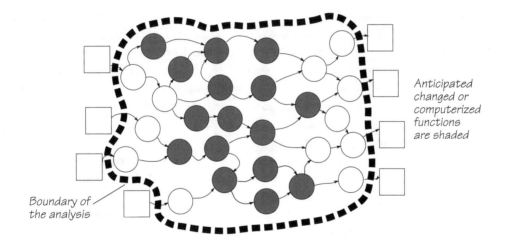

Anticipated changed or computerized functions are shaded

Boundary of the analysis

Figure 2.8.4: The context of the system. Note how the analysis covers a much larger area than the target of change. By doing this, you ensure that the changed functionality, probably a computer system, fits exactly into the users' business.

The context of the system in Figure 2.8.4 is everything within the dotted boundary line: Some of the analyzed functions will not be automated or changed in any way. However, analyzing them is not a waste of effort. To build an appropriate computer system, you must fully understand the surrounding processes: those processes that produce the input data so that your computer system always receives the data it expects; and those that use the computer's output. Otherwise, you cannot be sure that you're generating relevant information.

If the context of study is too small, you'll fail to take advantage of automation opportunities in the users' business. If the context is too large, you'll waste analysis effort. If the context is erroneously located, you'll deliver a system that doesn't provide the facilities needed by the end user. As any of these mistakes will result in an unsuitable future system, you need to spend the necessary effort to establish the right context.

Building a Current Physical Model

Once the context is agreed upon, then you make the first attempt to partition the system. The first breakdown, called Diagram 0, is a rough breakdown of the context. As we've said, the current physical model reflects the business as the users see it, so it is likely to include departments, people, devices, and so on. Consider Diagram 0 in Figure 2.8.5.

There are, admittedly, some problems with this very physical model and its messy interfaces. However, its objective is to emulate reality, not to be great art. Sometimes, the systems you study are not partitioned along purely functional lines, and your diagrams may look even more untidy. It doesn't matter. As the analysis progresses, you'll level downward from these partitions to find a more logical system. At the moment, the diagram serves you well as a starting point.

Diagram 0 verifies the context. As you build this diagram, you'll likely discover some details to make you alter your context diagram. This is normal and should not cause concern. Indeed, some analysts prefer to construct the context diagram and Diagram 0 in parallel, since both diagrams require similar information from the users, and one can't be finished without the other.

Whichever way you develop your Diagram 0, it now provides you with the basis for gathering the details.

Gathering the Details

Once you are reasonably confident that the interfaces are correct, you can treat each of the bubbles in Diagram 0 as a separate system. Each process and its data flows in the diagram form the context for a lower-level study. This time, however, restrict your study to whatever happens inside that process. Analyze the process by interviewing the appropriate users, and by drawing data flow diagrams from their answers.

As you descend into the details, you'll begin to see a more logical partitioning emerge. The partitioning at the top level is usually political, and therefore not subject to reason. At the lower level, the bubbles are more functional and reveal the true business policy. (We discussed this leveling process fully in Chapter 2.7 *Leveled Data Flow Diagrams.*)

216

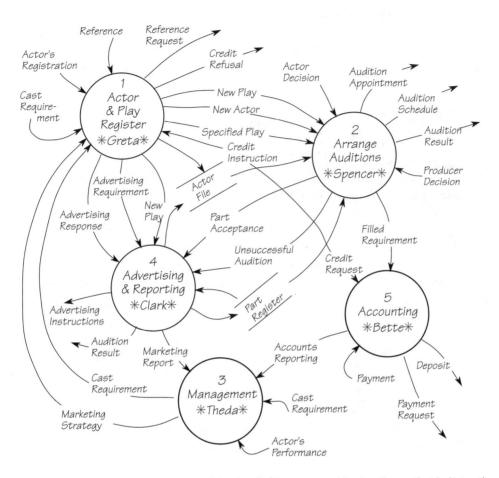

Figure 2.8.5: This Diagram 0 uses a partitioning theme that imitates the existing departments. The data flows mirror the documents and information that travel between the departments and between the company and the outside world. The users' names are added to the bubbles to make the model easier to recognize.

How Detailed Should It Be?

The current physical model provides a common language between the users and the analyst. Therefore, you build this model as long as you need a tool to help you communicate with the user.

Remember that the model is only a stepping stone to later stages of the analysis. It is a waste of time to build a perfect current physical model containing a complete set of lower-level diagrams, together with the supporting data dictionary and mini specifications for every process. There is no need for this elaboration. You'll be able to get what you want without all of the lower-level diagrams. We recommend that you build no more than two levels below Diagram 0, and only go that far for large systems. Collected documents can substitute for written entries in the data dictionary. Manuals of procedures and interview notes are a satisfactory substitute for mini specifications.

> The current physical model should contain only as much detail as you need to gain a clear enough understanding of the system to be able to build the essential requirements model.

The objective of your analysis is to discover the underlying policy: the essential requirements. We'll discuss this topic at some length in Chapter 2.10 *Essential Viewpoint*. For the moment, let's say the *essential requirements* are those that are necessary for the business and are completely independent of any technology. As desirable as it may seem, you'll have difficulty bypassing the current physical model altogether in your search for the essence. Most of the users you'll encounter are knowledgeable about their business, but not that many of them can completely describe the abstract, essential requirements. So you are likely to include physical characteristics in your early models to allow the users to recognize their system.

How many physical characteristics you include varies enormously. It depends on the users, your knowledge of the system, and your own skill as an abstract thinker. Some users are adept at realizing that physical features are only part of the implementation and happily accept models without them. Others have trouble recognizing the essence of the system.

The physical modeling activity always ends as soon as you are able to build the essential models. Sometimes, early in the project, you are able to gather all the information you need to build essential models. Sometimes, you have to do a certain amount of current physical modeling before you and the users can see the essentials; and sometimes, having started the essentials, you need to augment your knowledge by backtracking and modeling more of the current physical viewpoint.

There are occasions when you build current physical models for a different reason. Management may need to see what the organization is doing. Recently, we were asked to consult with a medical insurance company. The company had expanded so rapidly that its managers were having difficulty keeping in touch with their own infrastructure. We worked with the staff to build a physical model of their organization. This "warts-and-all" model was used by the staff and management to see what was going wrong with their way of doing business. They were able to identify the areas in most urgent need of attention, to find duplicate and redundant procedures, and to locate the most fertile areas for further study. The current physical model is still being used as a springboard for new analysis projects.

Summary

Whether you build a current physical model depends on your users and yourself. Sometimes, the users need to see recognizable physical details in the models before they can contribute to the analysis. Sometimes, you and others on the analysis team need to build this kind of model to fully understand the business. Sometimes, very little physical modeling is necessary before you can get on with building your essential models. Your responsibility is to build as much current physical model as you need, and no more.

Trail Guide

● Easiest: You now have some more data flow modeling skills and a way of looking at the current system. Return to the Piccadilly Project where you'll build models of the entire company. You'll find this assignment in Chapter 1.4 *The Piccadilly Organization*.

■ More Difficult: We are bringing you through the viewpoint chapters of the Textbook because systems analysis is easier and clearer if you adopt a single viewpoint for each of your models. We suggest you use the current physical viewpoint in the Piccadilly Project while you are becoming familiar with the television business. The next stage of the Project is found in Chapter 1.4 *The Piccadilly Organization*.

◆ Most Difficult: You are off your trail if you are reading this, but you can find it again in Chapter 1.4 *The Piccadilly Organization*.

❋ Promenade: You aren't reading this book to do the work of modeling the Piccadilly system, so don't follow the others to Section 1. Instead, we want to show you another very different viewpoint in Chapter 2.4 *Data Viewpoint*.

2.9 DATA DICTIONARY

Before You Reached Here ...

● Easiest: You have built a model of the current implementation of Piccadilly in Chapter 1.4 *The Piccadilly Organization*. Before you can consider that model to be a working representation of the system, you have to define its data. This chapter discusses how a data dictionary records the meaning of the data flows and stores in your model.

■ More Difficult and ◆ Most Difficult: This chapter is not required reading. You may rejoin your trail in Chapter 1.5 *Building the Data Dictionary*.

❋ Promenade: The model in Chapter 2.2 *Data Flow Diagrams* contains the processes, data flows, and data stores that make up the system. This chapter deals with specifying the data. (Later, you'll see how to specify the processes.) The data dictionary gives an exact meaning to all the flows and stores in the data flow model.

Working Models

We stress again and again that the best way to specify a system is to build a working model of it. A *working model* is one that demonstrates each process in the data flow diagram is capable of manufacturing its outputs from its inputs, and that each entity and relationship in the data model can supply or store the data needed by all the processes.

What exactly are the inputs and outputs and entities? So far, you have used meaningful names to indicate the content of the flows, but have trusted that each of the data flows in the model somehow contains whatever is needed. You have also believed that the data stores and entities contain whatever data needed. They probably do, but you must prove it.

The *data dictionary* is the part of the model that provides definitions of the data flows, data elements, stores, entities, and relationships. After you have built the dictionary, you know that you have a precise understanding of the data, and that your model really works.

The Meaning of Your Data

The definitions in the data dictionary serve the same purpose as the definitions in any dictionary. If you need to know the meaning of a word, you look it up in the dictionary to find the precise meaning of the word. It is exactly the same with the analysis data dictionary. You define all your flows and stores in the dictionary, and your readers use the dictionary to find the exact meaning of any data component of your model.

Let's look at why this is important. Consider the fragment of a data flow model in Figure 2.9.1. This is a common enough process, found in the many thousands of order processing systems around the world. Despite the everyday nature of the process, and although most people working with business systems have seen quite a few examples of order forms, what are the exact contents of the data flow ORDER FORM in this particular system? Experience tells us that an order form usually contains some customer information, something about the purchased goods, and probably some pricing information. This seems like a reasonable description of an order form, but it is not at all precise. A specification that quoted "some customer information, something about the purchased goods, and probably some pricing information" does not provide the system implementors with what they need. Different programmers would write different programs producing different order forms. If it is to be useful, a specification must be exact.

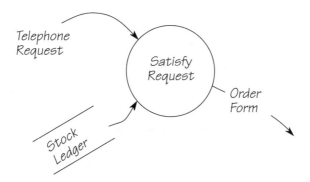

Figure 2.9.1: *This process receives a request by telephone from a customer. It checks the stock availability, and produces an order form that is sent to the warehouse for dispatch.*

Suppose that instead of the vague sentence quoted above, you had a definition that looked like this:

Order Form = Customer Ident + Date Of Order
+ Time Of Order + Telephone Clerk Ident
+ {Item Code + Quantity Ordered + Unit Price
+ (Discount Code)}
+ Requested Dispatch Date

With this data dictionary entry, there is only one possible interpretation of the data flow. You write it like this so that you, the users, and the implementor all share a precise and identical understanding of the data.

Defining Terms in the Context

When you read the above definition of the order form, did you think that it was different from order forms you've seen before? It probably was. There is no universal law that decrees the contents of order forms. In different systems, "order form" means different things. So you must define what an order form means within the context of this system.

Consider the following illustration of context from S.I. Hayakawa's classic book *Language in Thought and Action:*

> Let us see how dictionaries are made and how the editors arrive at definitions. What follows applies, incidentally, only to those dictionary offices where first-hand, original research goes on—not those in which editors simply copy existing dictionaries. The task of writing a dictionary begins with reading vast amounts of the literature of the period or subject that the dictionary is to cover. As the editors read, they copy on cards every interesting or rare word, a large number of common words in their ordinary uses, and also the sentences in which each of these words appears, thus:
>
> pail
> The dairy pails bring home increase of milk
> Keats, *Endymion*
> I, 44-45
>
> That is to say, the context of each word is collected, along with the word itself. For a really big job of dictionary writing, such as the *Oxford English Dictionary* (usually bound in about twenty-five volumes), millions of such cards are collected, and the task of editing occupies decades. As the cards are collect-

ed, they are alphabetized and sorted. When the sorting is completed, there will be for each word anywhere from two or three to several hundred illustrative quotations, each on its own card.

To define a word, then, the dictionary editor places before him the stack of cards illustrating that word; each of the cards represents an actual use of the word by a writer of some literary or historical importance. He reads the cards carefully, discards some, rereads the rest, and divides up the stack according to what he thinks are the several senses of the word. Finally, he writes his definitions, following the hard-and-fast rule that each definition must be based on what the quotations in front of him reveal about the meaning of the word. The editor cannot be influenced by what *he* thinks a given word *ought* to mean. He must work according to the cards or not at all.[*]

Hayakawa is talking about writing dictionaries using "first-hand, original research." This applies to you, too: The analysis data dictionary is certainly an original dictionary. Note what Hayakawa says about the editor working within the context of the period or subject of the dictionary and not attempting to influence it by his own opinion. This is precisely what you must do as well: You must not assign meanings that the term had in some previous system, nor give your opinion as to what the term ought to mean. The data dictionary entries define the meaning of data *within the context of the system being studied.*

Problems of Accuracy

Whenever a document appears in the model as a data flow, instead of writing data dictionary definitions, you might be tempted to include a sample of a document in the specification. This turns out to be a bad idea in the long term because the document focuses on the physical implementation of the data, rather than the meaning behind the implementation.

Since the document is designed for the current system, it may not be suitable for any new implementation. The document focuses on the solution to a problem and often hides the real problem. To understand the problem, you need to see the pure data. You must write definitions that ignore both the medium and any data that are dependent on the medium. In other words, you write a description of only the information content of the document.

There is always a strong possibility that a document can be misunderstood: Columns are often headed with cryptic labels; the document may contain redundant information; most documents contain at least one example of ambiguity; and often the design of a document imposes an unintended meaning. The data dictionary definition, however, is written in a manner that cannot be misunderstood. That means you need a suitable notation.

*From *Language in Thought and Action* by S.I. Hayakawa. Copyright © 1963, 1964 by Harcourt, Brace & World, Inc. Reproduced with permission by George Allen & Unwin Ltd.

Notation

To specify the data flows and stores, you define their data content by using these operators:

> = *anything after this is combined to make up a flow/store/entity*

> + *shows the combination of components*

For example,

> *Applicant Registration = Applicant Name + Applicant Address*
> *+ Applicant Date Of Birth + Applicant History*

This data dictionary entry means that each time the model refers to APPLICANT REGISTRATION, it means the combination of the applicant's name, address, date of birth, and history.

The dictionary also shows

> *Applicant Address = House Number + Street Name*
> *+ (Apartment Number)*
> *+ City + State + Zip*

The parentheses enclosing APARTMENT NUMBER indicate that this data item is optional, or that every APPLICANT ADDRESS doesn't have to include it. For example, most people living in New York City have an apartment number in their address, but residents of Pooletown, Oklahoma (population 420), do not use apartment numbers, as everybody lives in a house.

Some data items need to occur a number of times:

> *{Anything enclosed in braces repeats}*

For example,

> *Applicant History = (Current Job)*
> *+ {Previous Job}*

This entry says that an applicant may have a current job, and some number of previous ones. Let's say the system is only interested in the three most recent previous jobs. Additionally, let's say that the company policy is that it doesn't consider applicants who have never worked. You could then write this definition:

Applicant History = (Current Job)
           + 1:3 {Previous Job}

The cardinal operators (1:3 preceding the braces) add information to the definition. It now says that every applicant registration must have had at least one previous job, and not more than three. However, the restriction of three previous jobs is there for policy reasons. When the company interviews applicants, it doesn't take into account any but the last three jobs. The cardinal operators have no bearing on the physical capacity of the current form. Maybe the current form can only accommodate two jobs, so the applicants can overcome this by writing the third job on the back of the form. The logical limit is still three, even though the physical limit of the form is two.

When cardinal operators are not shown (which is the usual case), the lower limit is zero, and the upper limit is some undetermined number. For example, consider this entry:

Applicant Name = {Forename}
           + Surname

The business policy allows the applicant to have no forename at all or as many as he wants. While this may at first appear to be silly, in the context of this system, it is reasonable. Applicants will have a surname and probably, but not necessarily, one or more forenames. There is no upper limit to the number of forenames, as members of the English and Spanish aristocracies have a considerable and unpredictable number of forenames.

The current policy yields the definition *1:3 {Previous Job}*. You would write *1: {Previous Job}* if the system considers *all* the applicant's previous jobs; or *:3 {Previous Job}* if the system relaxes the rule prohibiting first-time workers, but retains the upper limit.

Use cardinal operators only when you have something special to say. Usually, the context of the definition makes the operators apparent to your readers.

On some occasions, you need to make a selection between data items. Let's say you are analyzing a system that sends a newsletter to its customers. The customers may elect to have their mail sent to either their home address or the office. The notation for selection is

[|] The square brackets enclose a number of choices. The choices are separated by a |. It says [pick this | or this | or this].

The data dictionary entry for the address reads

Mailing Address = [Home Address | Office Address]

Enlightening notes in the data dictionary are occasionally useful.

> *✶ Comments are enclosed by asterisks ✶*

For example, the analyst may add

Mailing Address = [Home Address | Office Address]
✶ Home addresses are preferred ✶

Strictly speaking, this comment is not necessary as it is not defining data. However, it probably saves more time than it wastes, so you may think of it as a beneficial surplus. Please don't make indiscriminate use of comments. They should only add information that is of interest to the analysis.

Further Decomposition

Some definitions include items that need further breakdown. For instance, consider this entry:

Applicant Registration = Applicant Name + Applicant Address
+ Applicant Date Of Birth + Applicant History

Any component that is itself made up of components must be further decomposed. Thus,

Applicant Name = {Forename}
+ Surname

Applicant Address = House Number + Street Name
+ (Apartment Number)
+ City Name + State Name + Zip Code

Applicant History = (Current Job)
+ 1:3 {Previous Job}

Keep breaking these components down until you reach a primitive level. Such a primitive is called a *data element*.

Data Elements

A data item is primitive if, within your context of study, you can give it a value. Therefore, it is something you can't reasonably break down any further. For example, the definition given above for the applicant's address includes a house number. From the analyst's point of view, the house number is a primitive item of data, and therefore has the definition:

*House Number = * Data element **

Of course, you could define house number as

House Number = 1:6 {Digit}

But this definition provides no real enlightenment. Any numerical piece of data can be defined as *{Digit}*. The values for this data element are called *continuous*. There is an almost infinite variety of numbers that identify houses. Since most people understand this, there is nothing more of interest to say about house numbers. Once you define it as * *Data element* *, you have nothing more to add. You and your users have no direct concern with the size of the item, and that implementation detail can be safely left until later.

There is an exception to the rule. Some data elements have a strictly limited number of possible values. These are called *discrete* elements, and their possible values are shown in your dictionary. For example, this book was written on three Macintosh® computers, which are linked by networking software that allows any of the three to be file servers. An appropriate definition looks like this:

*Server Id = * Data element **
    ["Ian" | "James" | "Suzanne"]

If the values of a data element are defined elsewhere, you can use the data dictionary entry to point to the defined values:

*Country Dialing Code = * Data element. See booklet supplied by telephone*
    *company. **

You can also use data dictionary comments to record a data element's unit of measure:

*Mountain Height = * Data element. Height above sea level measured in feet. **

Defining Calculations

Mathematical calculations are traditionally placed in the mini specifications. However, many analysts have discovered the convenience of defining algorithms in the data dictionary using the comment notation:

Total Floor Area = * Data element. Calculate as room length times room
width. Units: square meters. *

This definition tells us that TOTAL FLOOR AREA is a data element that can be calculated by the algorithm. If the calculation of TOTAL FLOOR AREA is referred to in more than one process, it makes sense to define the algorithm in the data dictionary rather than duplicating it in many mini specifications.

Defining Data Stores, Entities, and Relationships

Some of the most important data in the system are stored. Most of the data that enter the system through the incoming boundary data flows end up in a data store. Figure 2.9.2 illustrates a data model that was derived from a data store. The data dictionary should contain definitions of the data stores, as well as the data model's entities and relationships.

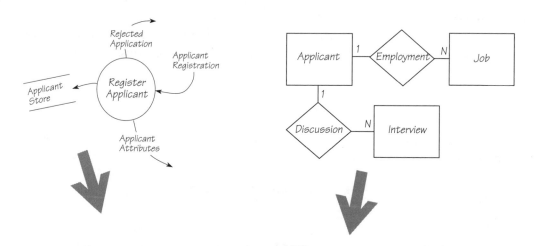

Figure 2.9.2: To complete the data dictionary that supports your data flow models, define all the stores and the flows. Similarly, to complete your data model, define all the entities and relationships.

Applicant = <u>Applicant Name + Applicant Date Of Birth</u> + Applicant Address
+ Date Registered + Salary Required

Applicant Attributes = Applicant Registration

Applicant Registration = Applicant Name + Applicant Address
+ Applicant Date Of Birth + Current Job + Salary Required
+ {Previous Job}

Applicant Store = {Applicant Name + Applicant Date Of Birth
+ Applicant Address + Date Registered + Salary Required
+ Current Job
+ {Previous Job}
+ Interview Comments}

Discussion = * Relationship. Cardinality: for each Applicant, there are many
Interviews; for each Interview, there is one Applicant. *

Employment = * Relationship. Cardinality: for each Applicant, there are many
Jobs; for each Job, there is one Applicant. *

Interview = <u>Interview Session Number</u> + Interview Comments

Interview Comments = * Free text comments on the applicant and his/her
attitudes and aspirations *

Job = <u>Type Of Work + Employer Name</u> + Date Started + Date Ended

Rejected Application = Applicant Name + Applicant Address
+ Reasons For Rejection

Definitions of data stores, entities, and relationships are not fundamentally different from definitions of flows. However, there are several special differences to note.

First, definitions of data stores are enclosed by braces { } because the data in a store repeat. In our example, the applicant file holds information about several hundred people.

Second, if an occurrence of the store, entity, or relationship has to be identified, its unique identifiers are underlined and, by convention, placed first in the list of attributes. For example, in Figure 2.9.2, each interview carries a unique INTERVIEW SESSION NUMBER. Note how a composite identifier (a combination of several data elements) identifies the APPLICANT and JOB entities.

Third, the relationships in Figure 2.9.2 do not contain data, so the data dictionary records them with only a comment, just like a data element. Naturally, if they held data, you would define the data in the same way as if the relationship were a regular data store. Relationships are formed whenever there is a business need to remember the association between entities. Some analysts use the dictionary entry to record the business role of the relationship, including why it is necessary, and what conditions must exist with the participating entities to form the relationship.

Note the correlation between the data store APPLICANT STORE and the data model. A rule of thumb is to form each repeating group into an entity that relates to whatever came before the group. Because of this convention, some analysts write data dictionary entries for all the stores before attempting their data models. However, this heuristic only works when the original data stores have already been logically partitioned.

What Do You Put in the Data Dictionary?

The answer to this question is

EVERYTHING

Every piece of data that the system uses must be recorded in the data dictionary. This means every data flow, data store, entity, relationship, and data element has an entry in the dictionary. Until the dictionary is complete, the model is not complete.

We have another answer to the same question:

NOT EVERYTHING

You must restrict your data dictionary to information that is useful to the analysis. This means the dictionary shows the composition of the data flows, data stores, entities, and relationships, and the meaning of each data element. Information about the physical implementation of the data is not important to the analysis. Remember that you are writing the *requirements* for the system—not designing its implementation. The requirement is for a data element to exist; it is not for the element to have seven numeric digits with two decimal places, nor is it a requirement to note that the element is currently held in packed binary format.

This information may well be needed later by the implementors and database designers. If you include all such information in your dictionary now, you will add unnecessary clutter. During analysis, concern yourself with what the system has to do. While you are still gathering the requirements for the system, you are hardly in the best position to decide the implementation. After analysis, when you know all the requirements, you can determine the best way to implement the data, and then it will be appropriate to add the design decisions to the data dictionary.

Aliases

Aliases are found in the dictionary when several names exist for the same item, or when several analysts inadvertently use different names for the same item. For example, suppose there is a collection of data elements that some people call "Buying instruction" and others call "Purchase requisition." Defining this item twice introduces redundancy into the dictionary and makes the specification misleading. How do you treat aliases?

You must first decide which of the names is more descriptive, and then write a pair of definitions. One definition is for the preferred name, like so:

Buying Instruction = * Alias Purchase Requisition *
*   Proposed Supplier Ident + Required Delivery Date*
*   + Ordering Authorization*
*   + {Item Required + ...*

The entry for the other term is as follows:

Purchase Requisition = * See Buying Instruction *

In the example, the ** Alias Purchase Requisition ** comment acknowledges the existence of another name, but anybody looking up that name is referred to BUYING INSTRUCTION for the full definition.

Avoid aliases whenever possible by encouraging people to adopt standard naming conventions. This is worthwhile, as the existence of synonyms makes the dictionary more difficult to manage. You must always check before writing a new definition that your intended entry does not already exist in the dictionary under another name. This is easier if you compare your data dictionary definitions with other analysts who are modeling the same data, or who have data flows or data stores that interface with your models. Since the data dictionary tightly connects the data model and the process models, this tool should be used to control data names and prevent aliases, especially at the data element level.

Summary

The data dictionary adds a strengthening element to your data flow models. Your dictionary defines all the data shown in your diagrams, and with this knowledge of the data, you should be able to prove that each process can produce its outputs from its inputs. The data dictionary also completes your data model. By defining its entities and relationships, you now have a complete understanding, as well as a written description, of the system's stored data.

It's a good idea to think of the data dictionary as something that you build in parallel with the other models. Don't wait until the other models are almost complete before starting the data dictionary. Waiting means that you must write several hundred, possibly several thousand, dictionary entries all at one time—a tedious task, to say the least. Also, if you postpone writing a definition, you risk forgetting its meaning and you lose the greatest benefit of the dictionary: forcing you to think about the data and to understand more precisely how each of the processes, entities, and relationships uses the data. When you understand all that, you understand the system.

Now let's write some definitions.

Exercises

Here is a reminder of the data dictionary operators:

    =   is made up of
    +   and

What to Do

Write data dictionary entries for the exercises that follow. They're designed to build your skill in writing data dictionary entries. Then, compare your answers with those in Chapter 4.6.

    [Select this | or this | or this]
    {This item is repeating}
    (This one is optional)
    * Here is a comment *

Write data dictionary entries for the following:

1. Define the title page of this book. (That's the one with "Dorset House Publishing" near the bottom of the page.)

2. Assume this statement is correct: "The selling price for a line in the order form can be derived in a number of ways. Sometimes, a salesman has negotiated a special price. If a discount rate has been written on the form, then a discount has been given to the customer. The discounted price is calculated by subtracting the discounted standard price from the standard price." Use the following definition for a line on the order form:

Order Line = Product Description + Product Code + Pack Size
+ Quantity Ordered + Quantity Sent
+ (Discount Rate)
*+ **Selling Price***

Write the data dictionary entry for SELLING PRICE.

3. Within the context of your system, a person's identity is made up of his name, address, and date of birth. If a social security number is available, that alone is sufficient as an identifier. Write the definition for PERSONAL IDENTITY.

4. The client identifier is the client's name, or an acronym. Sometimes, both are used. Define CLIENT IDENTIFIER.

5. One line of an invoice shows a description of the product, the quantity sold, and the selling price that is an undiscounted price or a discounted price or a special price plus handling. Define INVOICE LINE and SELLING PRICE.

6. An application shows the applicant's name, address, and telephone number. The telephone number is written as an area code, the number, and an extension if he has one. Not all applicants have a telephone number. Define APPLICATION.

7. When traditional breweries make deliveries of ale to English pubs, they often use container sizes that are part of the folklore of British drinking. The containers are called pins, firkins, hogsheads, barrels, and puncheons. Assume that each pub on the delivery route gets several containers that are listed on a BREWERY DELIVERY NOTE. The containers delivered to one pub are not necessarily all the same size. For background information, a pin holds 4 gallons, a firkin 8, a hogshead 16, a barrel 32, and a puncheon 64. Define BREWERY DELIVERY NOTE.

All data stores shown in the data flow diagram must be defined in the data dictionary. In exercises 8 and 9 are some descriptions of data stores. Write the data dictionary entries for each data store.

8. A consulting company maintains a file of client information. Each client has a record in the file that is made up of the client acronym and the client name. Following these is a number of jobs that are active for that client. Each job must have a unique identity for the job, the type of job, and the start date. The

name of the main contact appears after all the jobs for that client. Define CLIENT FILE.

9.   The Regional Theater Casting System that you saw in Chapter 2.8 *Current Physical Viewpoint* keeps information about actors in a card file. There is a green card for each actor showing his name, address, and date of birth. Following each green card are some more cards. There are white ones that relate to the parts the actor has applied for, and yellow ones for the parts that the actor has actually played during the past ten years. Both cards carry a part classification, and every actor has at least one white and one yellow card.

Each yellow card has the date started and the date ended for each part, the salary, the producer, and any reviews by the critics. Each white card shows the producer who offered the part, the date on which the successful actor is expected to start (the "start date"), and the salary offered, together with a short description of the role ("determined youngest daughter suppressed by tyrannical family," "bashful young man secretly in love with the boss's daughter," and so on). The casting agency keeps actors on file even when they are not in the process of applying for a part. Define this ACTOR FILE.

Trail Guide
This duplicates the one in Chapter 4.6.

● Easiest: The next step is to return to the Piccadilly Project to begin its data dictionary. Go to Chapter 1.5 *Building the Data Dictionary*.

■ More Difficult and ◆ Most Difficult: You didn't have to come here. Your next task for the Piccadilly Project is in Chapter 1.5 *Building the Data Dictionary*.

❊ Promenade: You've now seen the models that systems analysts use when they try to understand and specify systems. But there are more questions: What is the point of modeling the current system if that is to be replaced by something new? How can an analyst build a model of a new system without first understanding the true policy of the current one? To solve this dilemma, we introduce the concept of *viewpoints*. When applied to system modeling, using a certain viewpoint means that any unwanted information is filtered out and ... Wait. Turn to Chapter 2.3 *A Variety of Viewpoints*, and all will be explained.

ESSENTIAL VIEWPOINT 2.10

Before You Reached Here ...

● Easiest and ■ More Difficult: You have read in Chapter 1.7 *Strategy: Focusing on the Essentials* why the Project needed to take a new viewpoint. This chapter describes the essential viewpoint that you'll use for most of your systems analysis work.

◆ Most Difficult: You didn't have to come here, but we're happy you did. The essential viewpoint is among the most important aspects of systems analysis. We urge you to stay for this and the next chapter. If you feel you really must move on, then Chapter 1.8 *Identifying Events* is next for your trail.

※ Promenade: Now that you've seen the data modeling chapters (2.4 *Data Viewpoint* and 2.5 *Data Models*), you are ready to look at another abstract viewpoint: the essential requirements. This and the next chapter (read them sequentially) discuss how the system's logical policy is modeled.

Systems That Go Wrong

Unfortunately, too many computer systems are built that are unsuitable for the users' needs or that require an unusual amount of maintenance. Additionally, there are too many systems that are actively disliked by the users.

These systems go wrong because they were built to the wrong requirements. When the requirements are wrong, the implementors cannot fail to build the wrong system. In order to get it right, the analysts must find the *real requirements*. The real requirements are not merely a rehashed version of the current system, nor are they the educated guesses of the analysts about the system they'd like to see. The real requirements describe the business being done by the system without any bias toward any implementation. These are the *essential requirements*. Essential requirements are also known as "logical requirements" or "business policy requirements."

235

Creating a New System from What Exists

The point of most software development is to replace a system, which may or may not be a computer system. The system may be completely manual, or a collection of manual processes, some machines, and some inspired human brainpower. Whatever the implementation, some kind of business system currently does the job.

Figure 2.10.1: Even software that appears to be completely new and different is usually a reimplementation of some existing system. Take word processing, for example. This seemingly recent phenomenon is really a computerized replacement of what typists have been doing with typewriters, scissors, and glue since the nineteenth century. The implementation may look different, but the basic requirements are the same.

For example, since VisiCalc was introduced in 1978 by Dan Bricklin and Robert Frankston, spreadsheets running on personal computers have spread at an almost plague-like rate. But the spreadsheet does not signify a new requirement. It only automates a task that accountants and business people have been doing with pencil, paper, and calculators for many years. The growth in popularity of automated spreadsheets has come about because they make the task inexpensive and accessible.

The point is we build systems for which there is a need. By and large, most of our software development is concerned with the reimplementation of some system that already exists in some other form. How do you construct a new system to replace the existing one?

It would be very convenient at this stage for a miracle to occur; and after a blinding flash of light, you'd find the old system replaced by a shiny new system that meets all of the requirements. The convenience of miracles is offset by their improbability.

Perhaps you could just rewrite the existing computer programs. This would give you the opportunity to exchange the old code for new, with the hope that the old bugs will be buried and no new ones brought to life. But a simple rewrite means that you would spend a lot of time implementing a system that carries along with it all of the old technology and idiosyncrasies. Rewriting does not guarantee that the new system will be more appropriate or that it will satisfy the users any more than the old one did.

Simply reimplementing a manual system on a computer means that the new system would be no more than a simulation of the manual processes, and that it would contain all the restrictions of manual technology. For example, some time ago, a government department was building a computer system to automate a manual one that had existed for many years. The users (they also had existed for many years) specified the system. The manual system contained a form that had to be filled out with the details of each client. The form was to be automated and subsequently become a screen in the computer system. The implementation of this screen carried a strange requirement. If the operator made a mistake while filling out the screen, the screen had to be cleared and filled out again. This may seem absurd when you consider the capabilities of moving cursors and replacing text on a screen, but consider what this screen was replacing. It replaced a form that clearly stated, "This form must be completed without erasure or alteration." The computer system had obliged. It was a faithful reimplementation of the manual system.

What was wrong with the new system? It was not that the developers failed to take advantage of the available technology. They used computers and terminals that gave everybody convenient access to the data. But they did fail to find the true requirements. Instead of finding the requirements, they assumed the implementation was the requirement. Not so.

Identifying the Essence of the System

> The requirements are called essential because they represent the reason that the enterprise exists.

The essential requirements are independent of any implementation. They exist because of the meaning, or purpose, of the system. The essential requirements represent the fundamental business of the enterprise.

Systems are built to pay bills, invoice customers, control production, guide aircraft, and so on. Systems are not built to use snazzy new workstations, or to program the database, or to construct a local area network for the intellectual challenge. While these technologies may be used to implement systems, they are not the reasons for the system's existence. (At least, we sincerely hope not.)

237

To find the essentials, imagine the system as if it had been implemented without any technology—no computers, no humans, no databases, no paper files, nothing. Then, imagine all the processes that are there to support the technology can be put aside for the moment; set aside the computer edits that check that the input data have been input correctly (if there are no computers, there is no reason to keypunch or to scan the data). There is no need for audits to unearth the mistakes made by the fallible humans (no humans, remember), or for backups of the databases to protect against disk failure, or for check numbers and duplicate reference numbers to repair a file that's out of order. In fact, anything to do with the implementation of the system is gone.

The illustration in Figure 2.10.2 shows banking as it was a hundred years ago. Not many of you readers have firsthand memory of banking as it was then, but you are bound to have read about it or seen movies depicting the era. Now consider the twentieth-century equivalent (Figure 2.10.3):

Figure 2.10.2: Banking in the nineteenth century. Bankers wrote in their ledgers using quill pens. They wore frock coats and conducted their business with a hushed, genteel air.

The Bank of Scotland is prepared to make home banking available to anyone with a television set connected to the Prestel network. Prestel is an information exchange network that can display data on specially equipped television sets. Via the home banking terminal, a television set, and the Prestel adaptor, account holders can access the Bank of Scotland's central computer and view their accounts, transfer money, pay bills, balance checkbooks, and do most of the other things related to a bank account.

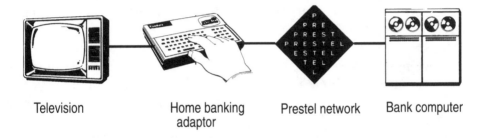

Television Home banking adaptor Prestel network Bank computer

Figure 2.10.3: Home banking from the Bank of Scotland. The keyboard is plugged into a Prestel-equipped television set. The bank customer uses the Prestel network to communicate with the bank's computer.

How much has banking changed in the hundred years between the illustrations? Not much. Both of the implementations shown above satisfy the same essential requirements; they are both involved in the same banking business. If you had built an essential requirements model of a bank a hundred years ago and did the same today, you would see very little difference between the models. The banks still take deposits and lend depositors' money. What has changed is the technology. The essential requirement to provide the customer with information about an account is satisfied by either a handwritten statement or a display on a home television screen.

Why Model the Essential Requirements?

Modeling the essential viewpoint may seem like a lot of work. Is it worth it? There are some compelling reasons for investing the effort. For example, we must avoid the habitual reimplementation of the current system's technology. While that technology may have been the right stuff yesterday, repeating the implementation prevents you from taking advantage of tomorrow's technology. When your essential view is independent of any implementation, you can select the most suitable technology for your new system. Besides, you cannot begin to select the most appropriate technology until you know the essential requirements.

The system that you are about to build should reflect its business policy. By building the essential

> The reason for thinking of the system as independent of its implementation is to discover the essence of the system. In this way, you avoid mistaking the implementation for the requirements. The essential viewpoint sees the system as an implementation-free set of business requirements.

model, you get a perfect understanding of the business. Then, by understanding the business, you can understand the implementation.

The most expensive component of today's software is maintenance. Maintenance programmers seem to be constantly buried under a pile of change requests from the users. Some requests are made because the business policy has changed, or sometimes because federal, state, or local governments have introduced new legislation that affects the business (essentially, it's the same thing). Businesses and the laws they operate under are constantly being altered. However, if these were the only modifications you needed to make to the installed system, the maintenance burden would be relatively tolerable.

Apart from the above changes or alterations to the hardware or system software, all other maintenance changes are caused by the failure of the systems analysts to specify the right system. Either the analysts have specified the wrong system, or they have failed to discover all the requirements. Change requests in this case are really belated attempts to build the correct system.

When you are able to capture and implement the correct requirements in the first instance, the installed computer system accurately reflects the users' business. In other words, the system does what the users really need it to do, and there is no need to correct the system through maintenance.

Essential Stored Data

In Chapter 2.5 *Data Models,* we discussed building a model of the system's stored data. In that model, you assembled entities and relationships in a manner that excluded any redundancy, duplication, or data that existed only to support the implementation. When you eliminated all that, you were left with the essential stored data. In other words, the data model shows you the data independent of any storage technology (Figure 2.10.4).

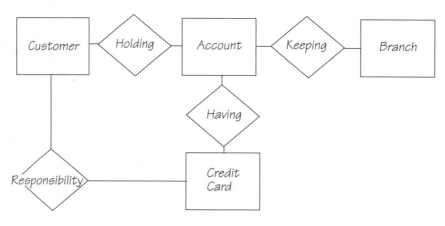

Figure 2.10.4: The data model represents the essential stored data for the system.

Stored data are considered essential only if there is a business reason for the system to remember the data. Nonessential data are eliminated from the data model because that data (such as foreign keys and pointers) exist only to support the storage technology. Nonessential data are not part of the essential system. Just as a process is not essential if it uses nonessential stored data, data cannot be essential if the data exist because of a nonessential process.

How do the essential stored data relate to the essential processing requirements? The most fundamental role of essential stored data is to provide the raw materials to the processes that manufacture the system's information outputs. The essential purpose of the system is to provide data and services to the outside world. Most of the data that flow out of the system through the boundary data flows come from the system's store of data, so we can say that the essential processes that provide the system's outputs are utilizing essential stored data.

> Essential processes and essential data are two parts of the same system.

How are the data stored? By essential processes on the input side of the system. Any system must have essential processes that transform the input boundary data flows into essential stored data.

Most essential models follow the pattern that is illustrated in Figure 2.10.5.

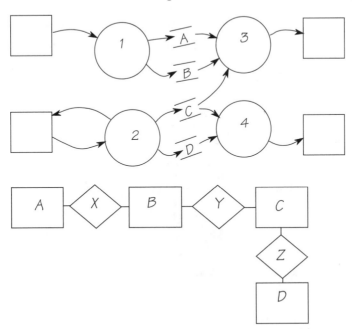

Figure 2.10.5: Some essential processes store the essential data; others retrieve it. The essential data stores in the process model are actually the entities in the stored data model, shown below it.

Note how the data flows that enter the system from the outside world are stored (by processes 1 and 2), to be later retrieved by essential output processes (3 and 4). When such essential data flows are stored, you should break them into their entity components. The data flow model, therefore, shows each entity type as a store. Whenever a process accesses two or more stores (which are really stores of entities), then it establishes or uses a relationship between the entities. The data flow model in Figure 2.10.5 thus leads to the data model shown below it. This aspect of essential data and processes is covered in more detail in Chapter 2.11 *Event-Response Models.*

Where Do You Find the Essential Requirements?

As we said, most systems are developed to replace an existing system, and while it may be far from perfect, the existing system has supported the business enterprise for some time. Therefore, some of the essential requirements for any new system must exist in the current system.

The current physical model, being a view of the existing implementation, includes (somehow) many of the essential requirements. Unfortunately, this model also includes many other things; some are concerned with the physical implementation of the essential requirements, and some are redundant, obsolete, or just plain wrong. So the task is to separate the wheat from the chaff: to find the essential requirements and discard everything else.

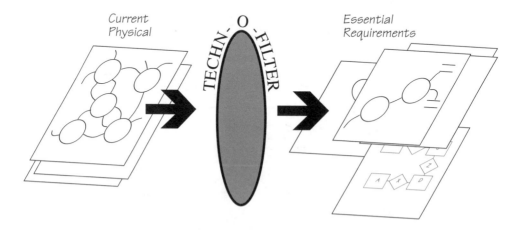

Figure 2.10.6: The current physical model must be run through a technology filter to remove the current implementation and thus reveal the essential requirements.

Summary

Building a current physical model is one way of finding the essential requirements. However, it is sometimes a more roundabout way than is really necessary.

You can often build an essential requirements model without first having a complete current physical model. When you move on to Chapter 2.11 *Event-Response Models,* you will see examples of using the context diagram to determine the events, and from there, go directly to building the essential event-response models. (All of this terminology is explained in Chapter 2.11.) What this comes down to is that after building the context diagram, some or all of the work associated with the current physical model can be avoided.

Each project is different, so we cannot give you fixed rules as to how much current physical modeling is necessary. You must judge for yourself how much physical modeling is useful and how much is simply wasted effort. The two most important factors governing how much physical modeling you need to do are the accessibility and the certainty of the business policy.

If you have detailed knowledge yourself of the subject matter of the system, you should be able to establish the context and go straight to the essential viewpoint. If other people have the business knowledge but you have immediate access to them, again you can avoid physical modeling. However, if you do not have immediate access to people who know the business policy, or if the people are geographically distributed, or if the context or aim of the project is unclear, then you would be wise to start by building a context diagram, Diagram 0, and one or two lower-level diagrams, with a reasonably complete data dictionary.

We do recommend that you not build a complete physical model with all the processes broken down to the functionally primitive level because it has always proved to be uneconomical. The best way to decide if you have done enough current physical modeling is to try building the essential model. If you get stuck, you can always backtrack and do some more current physical modeling.

Trail Guide

● Easiest: You'll continue to examine the essential viewpoint. In Chapter 2.11 *Event-Response Models,* we'll discuss the most efficient method of modeling the essence of the system.

■ More Difficult: Your next destination is Chapter 1.8 *Identifying Events,* where you'll start to build the essential model for the Piccadilly Project. However, you could go easy on yourself by following the ● Easiest Trail for the discussion in Chapter 2.11 *Event-Response Models.*

◆   Most Difficult: This was not intended reading for you. Chapter 1.8 *Identifying Events* is the place to rejoin the Piccadilly Project.

❋   Promenade: The most successful method of modeling the essence is by building event-response models. Your next chapter is 2.11 *Event-Response Models.*

EVENT-RESPONSE MODELS 2.11

Before You Reached Here ...

● Easiest: In Chapter 2.10 *Essential Viewpoint,* you have read about the essential view. Now we'll discuss how to build event-response models that make use of this viewpoint. The idea behind event-response modeling is simple, but to use it effectively, you must understand it correctly. Read on.

■ More Difficult: Strictly speaking, this chapter is not on your trail. However, event-response modeling is very important and may not be as straightforward as it seems. As you are already here, read through the chapter. If you find that you already know about these models, then skip to Chapter 1.8 *Identifying Events.*

◆ Most Difficult: You are traveling on a trail where you presumably know about these models, so Chapter 1.8 *Identifying Events* is the most logical (no pun intended) destination.

❋ Promenade: You're here because of the importance of event-response modeling, a strategy for building the essential requirements model. Most systems analysts use it because it is the simplest, easiest, and most straightforward method of deriving the real requirements of the system. Additionally, because events are small units of work easily recognized by the users, an event-partitioned essential model has advantages when it comes time to implement the system. Later, we'll talk about using event-response models as the basis for estimating project effort.

Building an Essential Requirements Model

A current physical model uses the physical processors, such as departments, people, machines, computers, and so on, as its partitioning theme. A rather exaggerated current physical model might look like the one shown in Figure 2.11.1.

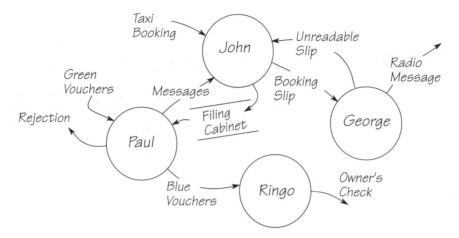

Figure 2.11.1: *Current physical models use the current implementation as their partitioning theme.*

As we've said many times before, basing the requirements models for a system on the current implementation is not a great idea. The way the business is currently operating conceals the essential requirements for the system. So if the internal mechanisms of the system aren't the best place to look for these requirements, you must look elsewhere. Looking outside the system may be rewarding.

Systems exist to provide services to the outside world. The services have one thing in common: They are provided as a response to some event that takes place outside the system. For example, in Figure 2.11.2, when customers decide they want a taxi, they do it outside the context of the taxi system. When they decide to make a payment for their account, again they make their decision outside the system's context. Such a happening outside the system is called an *event*.

Events

We are concerned with events because they provide us with a method for making a logical breakdown of the system. This method is called *event partitioning*. To make it, you must focus on what happens in the world outside your analysis study. Specifically, you are looking at the happenings, or events, that take place in the terminators. Your system knows that an event has occurred because it receives a data flow from a terminator. The system has a unique response to each event.

Each event produces a unique data flow. It must be unique because the system can determine which event has happened by the data content of the flow. Only by recognizing the event can the system produce the correct response.

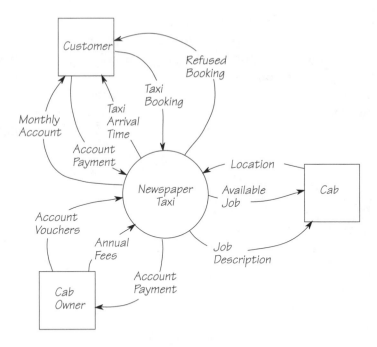

Figure 2.11.2: The context diagram shows the system's connections to the outside world. By inspecting the connections, you can discover the system's services. The outgoing data flows indicate the taxi company's services to its customers and cabs. The system uses the data in the incoming data flows to provide these services.

Events are given a unique name. For example, the event shown in Figure 2.11.3 is called *Customer wants a taxi*. The name of the event provides you, the analyst, with an indication of what the response should be. When the name is descriptive, it provides you with a relevancy check when you must search through a system to discover the appropriate response.

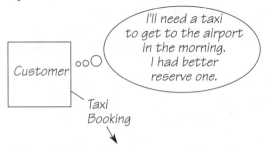

Figure 2.11.3: An event happens outside the system. The system has no control over when or why it happens, but it knows that the event has taken place because the event produces a data flow.

Identifying the Event Response

An *event response* is a system's predetermined reaction to an event. It is a collection of all the actions that respond to the event, regardless of where they happen in the current implementation. Ignore processor boundaries. If a process or data store anywhere within the system is part of the reaction, include it in your event-response model. In the example in Figure 2.11.4, the system has predetermined actions that are triggered by the arrival of the data flow TAXI BOOKING.

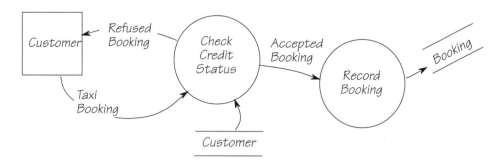

Figure 2.11.4: The event-response model for the event **Customer wants a taxi**. *The system identifies its response from the incoming data flow* TAXI BOOKING. *The response includes all the processes and data needed to react to this data flow.*

Note that in the event-response model in Figure 2.11.5 (or in any other event-response model), the response is bounded by terminators and data stores. A terminator supplies data to the system, but the behavior of the terminator is outside the control of the system; you cannot predict when it will send its data. Similarly, if the system sends a data flow to a terminator, you cannot predict how long the terminator will take before it reacts. In the same way, when a data item is stored, you probably cannot predict how long it will remain there until it is retrieved. Both the terminator and the data store are time-delaying mechanisms.

The significance of time-delaying mechanisms surrounding the event response is that the response itself takes place in its own discrete and continuous period. When the system is responding to an event, processing continues until the data flows reach the final data stores or terminators. Just as you ignore processor boundaries, you also ignore delays normally associated with crossing processor boundaries (waiting for an available operator to key in data, the time taken to print information, and so on).

Simply put, when a data flow enters the system, the event-response processing continues until the system has accessed all of the data stores concerned with the response and, whenever necessary, has sent the appropriate data flows to the terminators.

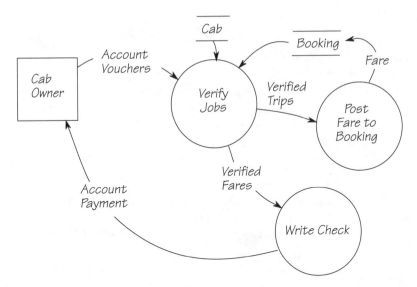

Figure 2.11.5: The event-response model for the event **Cab owner submits account vouchers**. *The response begins when the account vouchers arrive, and ends when the system has recorded the data and has sent the account payment to the cab owner.*

Identifying Temporal Events and the Time Trigger

So far, the events discussed have all been external. *External* events start outside the system at a terminator and trigger the system's response with a data flow. There is another type of event response called a *temporal* event because it is triggered by the arrival of a predetermined time. For example, let's say that on the fifteenth of each month, the system produces its monthly accounts. The arrival of the fifteenth day is the event. Its response is shown in Figure 2.11.6.

Figure 2.11.6: The response to the temporal event **Time to produce monthly accounts**. *The system gathers stored data, processes the data, and then communicates with the outside world.*

Another way for a temporal event to be triggered is by the passage of time in relation to stored data. For example, the dispatcher always calls for a taxi ten minutes before the booking time to give the selected taxi time to arrive at the customer's pickup point. So an event happens whenever it is ten minutes before a recorded booking. The system's response to this event is shown in Figure 2.11.7.

Temporal event responses begin when it is time to provide a service. The processing for the response starts inside the system, usually with a process accessing stored data. The event response finishes when the system sends one or more data flows to the outside world.

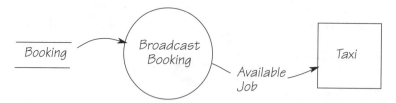

Figure 2.11.7: *The model for the temporal event* **Taxi booking becomes due**. *There will be another external event happening shortly when nearby taxis radio their locations.*

Finding Events

The context diagram is the place to start identifying events, since it shows all the data flows crossing the system boundary and since all these flows are, one way or another, connected to an event response. The best strategy is to first identify all external events. When a data flow enters the system's context, it does so because an external event has occurred. Any input data flow is associated with just one external event. This means that by inspecting the incoming data flows, you can determine all of the external events that affect your system.

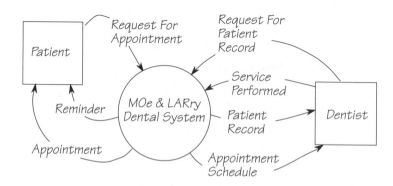

Figure 2.11.8: *The context diagram for the MOe & LARry Dental system. Each of the data flows is a part of an event response. How many events can you determine from this diagram?*

Some external events simply store the incoming data. Some, as part of their response, send data to the outside world. In other cases, responses to external events generate more than one outgoing flow. So each of your external events may be linked to one or more of the outgoing flows.

What about the output flows that are not linked to an external event? These are the products of temporal event responses. Any flow whose name suggests a periodic production (for example, REMINDER and APPOINTMENT SCHEDULE in Figure 2.11.8) originates with a temporal event response. By a process of inspection and elimination, you should be able to attribute the boundary data flows to their respective events.

> All of the data flows in the context diagram must be accounted for. They either trigger a response or are the result of an event response.

Naming Events

After you have found all the events, you need to give each event a name. The name must reveal the nature of the event, and thus give a clue as to the expected response. For example, *Patient gets a toothache* describes an event, but it gives no indication of what it has to do with our system. *Patient needs appointment* not only tells you and your reader what the event is, but also what kind of response the system must make.

A guideline for naming external events is to use this template:

- *External event name = Terminator name + reason for the data flow*

For example, using this template, the event that produces the data flow REQUEST FOR PATIENT RECORD is called *Dentist needs patient record*. A template for naming temporal events is

- *Temporal event name = Time to + action that must be taken*

In the dental system, this template gives you a temporal event called *Time to produce appointment schedule*.

Compiling the Event List

Once you have named the events from the context diagram, the most convenient way of keeping track of them is to use an *event list*. This list is simply an inventory of all the events to which the system responds. As the analysis progresses, you will find such a list useful for recording which events have been modeled, and which remain to be done. You will also find that the list helps you to categorize the events into groups by those that have a similar response. This can save you much work later if you need to copy the pattern of the response from one event-response model to another.

Here are the events that were derived from the context diagram in Figure 2.11.8 for the MOe & LARry Dental system:

Event Name	Associated Data Flows
1. Patient needs appointment	REQUEST FOR APPOINTMENT (IN)
	APPOINTMENT (OUT)
2. Time to send appointment reminders	REMINDER (OUT)
3. Dentist needs patient record	REQUEST FOR PATIENT RECORD (IN)
	PATIENT RECORD (OUT)
4. Dentist performs service	SERVICE PERFORMED (IN)
5. Time to produce appointment schedule	APPOINTMENT SCHEDULE (OUT)

The numbers assigned to the events are arbitrary and do not indicate any kind of priority or time sequence. The associated data flows are listed to help you to account for all the events. The (in) or (out) following each data flow name indicates the obvious: whether the flow is into, or out of, the system. This notation makes it easier to account for all of the flows.

The purpose of the list is to help you keep control of your model. Some systems have several hundred event responses, and controlling the complexity of the models for these systems can be difficult.

For each event in the event list, build an event-response model by linking all the processes that respond to the event. This task is simpler if you keep in mind that an event is either an external or a temporal event. The event response follows the pattern for that type of event. Let's look at these patterns.

Seeing the External Event-Response Pattern

The external event is some happening in the outside world. The event response starts when the system receives a flow from a terminator. The response ends with a combination of flows to and from data stores and/or flows to terminators (Figure 2.11.9).

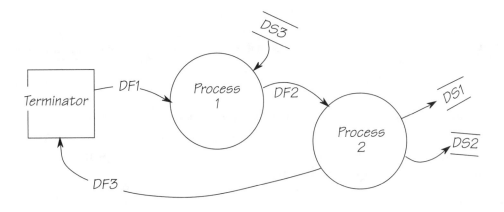

Figure 2.11.9: An external event-response pattern. The incoming data are processed, then stored. The data flow back to the terminator (DF3) is a receipt for the original message (DF1). This receipt flow is not always present in external event responses. Even if there is an outgoing flow from an external event response, it does not necessarily go back to the original terminator.

Typically, the incoming data go through a series of checks and validations, and are then added to a data store. For example, in the taxi system, the customer was checked to see that he was not a known bad payer, and then the request was added to the BOOKING store (Figure 2.11.4). Sometimes, the system acknowledges the initial message by sending some data back to the terminator (for example, in Figure 2.11.5, the ACCOUNT PAYMENT is sent in response to the ACCOUNT VOUCHERS).

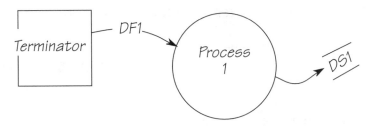

Figure 2.11.10: Sometimes, the external event response can be as simple as this. The data flow is stored without any access to existing stored data, and no outgoing flow is generated.

The external event response usually includes some combination of storing data, and/or referencing previously stored data. Use the two patterns shown in Figures 2.11.9 and 2.11.10 as a guide. All external event responses are variations on these themes.

Seeing the Temporal Event-Response Pattern

A temporal event is triggered when it is time to do something, or when a predetermined time occurs. Temporal events have stored data at one end of their chain of bubbles, and a terminator at the other.

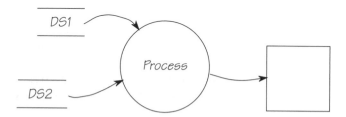

Figure 2.11.11: *The usual pattern for a temporal event response is to collect data from the system's stored data, then process and send the data to the outside world. Sometimes, temporal event responses store data.*

Temporal events tend to be relatively simple. In this example (Figure 2.11.11), there is only one bubble, which is quite common, as most temporal events are simply reporting on stored information.

Now that you know what the models look like, let's build an event-response model using the dental system example.

Using the Current Physical Model to Build the Essential Model

In Chapter 2.10 *Essential Viewpoint,* we discussed the need to model the system from the point of view of the business policy in order to find the essential requirements. Breaking the system into event responses is only the first step toward finding these requirements. Now we have to find the essential event response and build the model.

The information needed to build an essential model may come from several sources: the current physical model (if you built one), interviews with the users, sample reports, observations of the work as it is done, job descriptions, sample input documents, existing computer systems, and so on. Let's start by looking at the current physical model as a source of event-response information.

The way to use this model is to link the processes that make up the system's response to each event. This usually involves following the trail of processes through several diagrams. For example, Figure 2.11.12 shows the model that results from joining the processes that respond to event 3 *Dentist needs patient record* in the MOe & LARry Dental system.

Whenever you use the current physical model as input to building your event-response models, you'll probably include processes extracted from several diagrams. Then you'll notice a major advantage of event-response modeling: You have a cohesive unit that is small enough to examine in detail. It now becomes far easier to assess whether a process is a duplication, redundant, dependent on the current implementation, or part of the essential policy of the system. If any processes or data are not part of this policy, you must eliminate them.

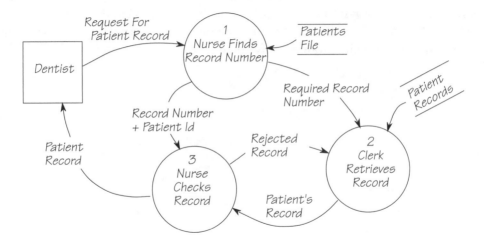

Figure 2.11.12: This event response is a faithful recording of the way the current implementation responds to the event **Dentist** **needs patient record**. *The dentist knows only the patient's name and address, so those items are part of the request data flow. The nurse looks up the patient's record to find the* RECORD NUMBER *because the filing clerk, who works in a different department, can only find the records by number. Since the nurse doesn't trust the clerk, the nurse checks the record to ensure the clerk has located the correct patient record.*

Modeling the Essential Processes and Stored Data

The essential stored data and the essential processes exist to support the system's business policy, and they are so closely linked that we can consider them together.

If the system stores data, naturally enough, the data must come into the system via an incoming boundary data flow. In other words, there is an external event response that stores the data in essential memory. On the right–hand side of Figure 2.11.13, you see a temporal event response retrieving the essential stored data. The essential rules apply: If the data are stored, the data must be essential and needed for the services provided to the outside world. Data are essential only if referenced by at least one other event.

The data flows to and from the terminators (the boundary flows) are initially considered to be essential flows. The reason for this is that when the context of the system is agreed, the boundary data flows are the data provided by or needed by the system's services; that is, the terminators provide or expect these flows.

You've seen how the event responses store or retrieve essential stored data. A problem can arise if an external event response stores data from a boundary data flow, and no other event response references that data. If this happens, you may not be able to change the actual physical data (the terminator will probably continue to send the data as always), but you must adjust the definition of the incoming bound-ary data flow to reflect the data that the system really needs. The new implementa-tion of the system will ignore the unwanted items.

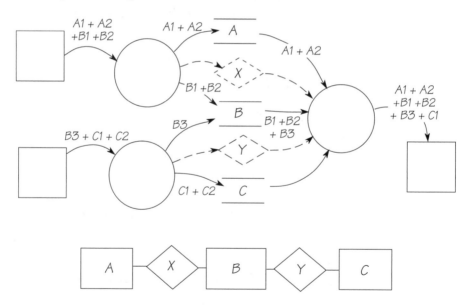

Figure 2.11.13: An event occurs and the resultant data flow enters the system. The incoming data flow contains A1, A2, B1, and B2. These are named to suggest the enti-ties that would hold that data. The event response stores the A data in the A entity, stores the B data in the B entity, and creates the relationship X. (The relationship is shown dotted only to suggest its creation and use. We do not advocate showing rela-tionships in event-response process models unless the relationship contains data.) The second external event response stores B3, C1, and C2, and creates relationship Y.   The temporal event response retrieves the stored data, processes the data, and sends the resultant data flow to the outside world. The data model to support this fragment of the system is shown below the event responses.

Developing the Event–Response Data Model

While considering the data being stored and retrieved by event responses, you can also consider the data associated with each event response. You have already isolated the essential processes necessary for one event response. Now do the same for the essential data. To illustrate, let's look at a typical example of how an event response uses stored data. The event-response model for event 1 of the MOe & LARry Dental system, *Patient needs appointment,* is shown in Figure 2.11.14.

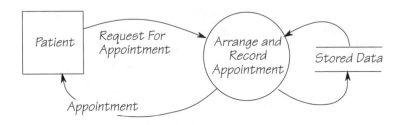

Figure 2.11.14: An intermediate version of the event-response model for the MOe & LARry Dental system. The stored data are undetermined as yet, and so are shown simply as STORED DATA.

In the figure, there are some things to consider:

- The data flows REQUEST FOR APPOINTMENT and APPOINTMENT are boundary data flows; they enter and leave, respectively, the context of the system.
- The data added to STORED DATA must be part of REQUEST FOR APPOINTMENT or be calculated by the process; the data can't come from anywhere else.
- The data content of APPOINTMENT must come from STORED DATA, or from REQUEST FOR APPOINTMENT.

As REQUEST FOR APPOINTMENT initiates this event response, let's examine its data content:

Request For Appointment = Patient Name
+ (Required Date)
+ (Required Time)
+ Dentist Name
+ {Required Service}

257

The date and time are optional, as some patients will take whatever appointment is available. The patient designates a dentist—either Moe, Larry, or Curly—and indicates if he requires certain serious services.

What happens to the data? The patient's name is used to reference an existing patient or, if this is the first visit, to record a new name. This means there must be an entity in essential memory called PATIENT. There is no other information in the data flow about a patient, so the only thing we can say now is that there is an entity called PATIENT with an attribute PATIENT NAME.

The next two items in the data flow are used to negotiate an appointment with the designated dentist. If the patient requests any services, they must be related to the appointment. Using the information from the data flow, you can construct a model of the data used to support this event response. The result is shown in Figure 2.11.15.

Now let's examine the outgoing data flow:

Appointment = Patient Name + Dentist Name + Appointment Date
+ Appointment Time

All of the data in this outgoing flow can be retrieved from the data model (Figure 2.11.15), so it confirms our view of the stored data.

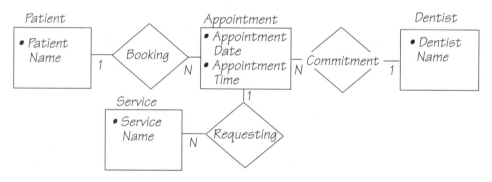

Figure 2.11.15: This event-response data model is built from the data flows connected to the event Patient needs appointment. The data elements are all attributed to the appropriate entities.

In the intermediate version of this event response (Figure 2.11.14), we used the data store STORED DATA temporarily until we could identify the real data. Now that you know what the essential data are for this event response, you can replace the temporary STORED DATA with stores that have the same name as the entities in the later version of the event-response data model. If you had already built a first-cut data model, you could use it to help you build the detailed event-response data model.

The event-response model shown in Figure 2.11.16 uses the stored data that you have determined to be essential. When your event-response model uses only essential data and shows only essential processes, then you have captured the real requirements. However, it is sometimes difficult to know if all the processes and data are essential. The following discussion gives some hints on building the essential viewpoint.

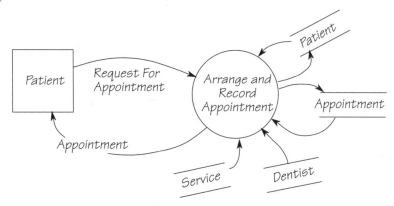

Figure 2.11.16: *The final version of the essential event-response process model. The data stores are the entities from the data model. All the data model attributes are available to the process.*

Refining Event–Response Models

When you build an event-response model directly from the current physical model, it may contain some implementation-dependent processes and data. The following types of processes are suspicious, and you can usually eliminate them from your essential requirements:

1. *Audits* verify that a process produces the correct result. Audits are built into systems because the processors used to implement the system are imperfect; they make mistakes, are dishonest, or whatever. However, the essential requirement eliminates all processor technology, and so eliminates all errors that can be made by the processors. Thus, there is no need for audits to check the processors.

2. *Edits* check the validity of data flows that travel from one processor to another. As you are ignoring processor boundaries, you can leave this kind of edit out of your essential event-response model. However, it may be necessary for the system to edit data it receives from the outside world, because you cannot influence the terminators, which are outside your context of study. So it is

259

necessary to include "border guard" edits as the first line of defense. (Just as a border guard may check your passport as you leave a country, you may need to install a border guard edit to check a flow's validity.)

3. *Transporters* are processes that move data from one processor to another, for example keypunching, scanning, or bar-code reading. These processes exist only because of the implementation; there is no essential policy involved.

4. *Translators* are processes that make the output from one process understandable to another; for example, an analog to digital conversion, or the printing of a report that is used inside the system's context. These processes are not part of the policy; they exist merely because of the way the system happens to be implemented.

5. *Backups* are not needed. By removing the technology, you remove the potential for failure, and hence the need for a recovery method.

If you are in doubt about a process, try decomposing it into its primitive components. You'll find it a lot easier to separate essential and implementation processes when they are at their lowest level.

It may seem radical to throw out processes that normally make up a large portion of most systems. However, keep in mind you are trying to do two things: You are trying to find the essential requirements for the system, and you are eliminating anything that could influence the future implementation of the system. Besides, you aren't changing the current physical—that view still exists. You are merely taking a different view of the same system. Later, once you have a complete picture of the requirements, it will be time to build a view of the new implementation.

Eliminating False Data Stores

A *false data store* is a data store that is part of the existing system for some reason connected to the implementation. Because they are invented for nonpolicy reasons, you should eliminate false data stores from your event-response models. Take for example, the event response to *Cab owner submits account vouchers* shown earlier in Figure 2.11.5. When the processes were brought together, we intentionally ignored the processor boundaries. Now let's redraw the model to show the two different people who are involved in processing this response (Figure 2.11.17).

Note that in this example, the false data store (Ringo's in-tray) doesn't represent the end of the response to an event. The in-tray owes its existence to the current implementation that uses people, and the person in this case has a problem working fast enough to keep up with demand throughout the day. This false data store does not contribute to the policy of the system and should not be included in the essential event-response model. Similarly, files in which the input is batched before further processing are false. They are there for efficiency reasons only. As you do not yet know the future implementation, you don't need to include these batch files.

However, be careful that you do not eliminate files when there is a genuine reason for batching data. For example, there are many event responses for which it is necessary to collect data for a time before processing it further. For example, if company policy required the account payments in Figure 2.11.17 to be made once a month, the system would need to store the vouchers until it was time to process the payments. While that situation would make the in-tray necessary, it should be renamed to reflect the essential purpose of the data, rather than the implementation.

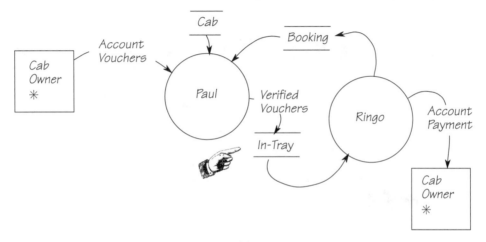

Figure 2.11.17: Ringo, who writes the checks and updates the bookings, is out to lunch. So Paul puts his output, the verified vouchers, into Ringo's in-tray.

Refining an Event–Response Model: An Example

Let's continue with the dental system example to show how to refine an essential event-response model. Consider the event-response model for the event *Dentist needs patient record*, repeated in Figure 2.11.18. Keep in mind that it was built from the current physical model by bringing together all the processes that respond to an event. Which processes do you think are there only because of the way the system happens to be implemented?

Let's start with the PATIENT RECORDS data store. This store is created in another part of the system and is kept in record-number sequence. Record numbers are controlled by the filing clerk who tells the nurse whenever a new number has been assigned. The nurse then adds the record number to the PATIENTS FILE in the dentist's office.

Whenever a dentist requires a patient's record, he asks for it using the patient's name (and, if necessary, address). The nurse has to look up the PATIENTS FILE, find the RECORD NUMBER, and give it to the filing clerk for retrieval of the patient record. The patients' records are stored in record-number sequence because that happens to be convenient for the filing clerk. The nurse's translation of name to record number is only necessary *because of the implementation*. If the data were stored in a manner that allowed it to be accessed by patient name and address, the translation step would be eliminated.

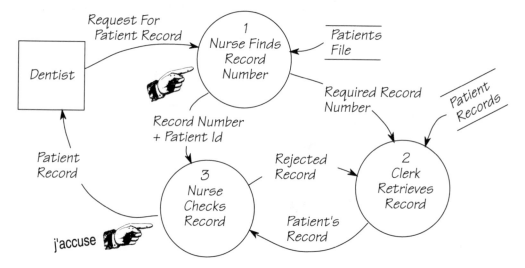

Figure 2.11.18: The event response for **Dentist needs patient record.** *These three processes were taken from the current physical model, and some of them are implementation dependent. What are the essential processes and data?*

The filing clerk employed in this office is paid the minimum wage, and is prone to making errors. The nurse always checks that the clerk has retrieved the correct PATIENT'S RECORD. However, if you take the essential view, every process must work correctly (there is no technology to go wrong), and it cannot produce the wrong record.

So it appears that the two processes done by the nurse are there only because of the existing implementation, and can probably be eliminated. However, before you make them disappear, consider whether a process is doing any storage or retrieval of essential stored data. Let's look at the definitions:

Request For Patient Record = * The address is supplied when there is a
　　　　　　　　possibility of several patients with the same name *
　　　　　　　　Patient Name
　　　　　　　　+ (Patient Address)

Patient Record = Patient Name + Patient Address
　　　　　　　　+ {Visit Date + Visit Time
　　　　　　　　+ {Service Identification}}

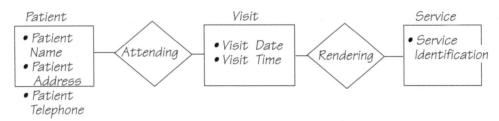

Figure 2.11.19: The event-response data model for the event
Dentist needs patient record.

By attributing the data elements from these two data flows, you can derive the
event-response data model shown in Figure 2.11.19.

The essential purpose of the event response is to use the incoming data flow
to retrieve information about the patient. There is no record number in the
incoming data flow, nor in the essential data model. The record number exists for
the convenience of people who are not trusted to read names. In other words,
someone in the past has chosen to use these people as the implementation, and
process 1 exists only to support that choice; there is no policy attached to that bubble. As you are modeling the essential processes and not the current implementation, you can eliminate process 1 from your essential model.

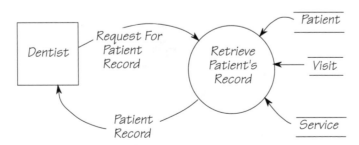

Figure 2.11.20: This is the essential event-response process
model for the event **Dentist needs patient record.** *The physical
data have been replaced by the essential equivalent, and all implementation-dependent processes have been removed.*

You can eliminate the audit process (bubble 3) because it checks what a previous process has done. However, there is no essential requirement for a process to make errors, so there can be no essential requirement for a process to check for errors. Process 3 is also banished. You are left with the model shown in Figure 2.11.20. Note that you could have started with the essential data model and worked backward from there to build this essential event-response process model.

Mid-Point Summary

Before going on to practice building some event-response models, let's review what we have. A summary is presented in Figure 2.11.21.

By now, you have the idea of essential event-response models, and it's time for you to build some for yourself.

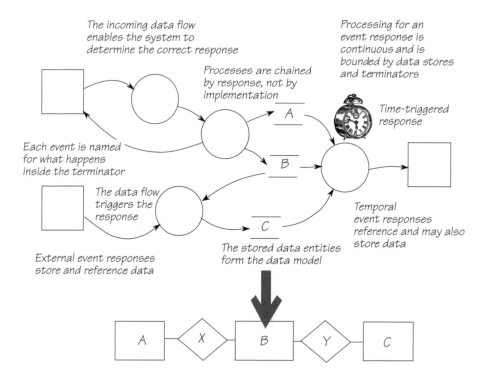

Figure 2.11.21: Review of building an event-response model.

What to Do

Following are two exercises that will give you practice building event-response models. Figure 2.11.22 shows the current physical model for the MOe & LARry Dental system. Answers to these exercises are in Chapter 4.7. Be careful to avoid looking at the answer to exercise 3, which comes up later.

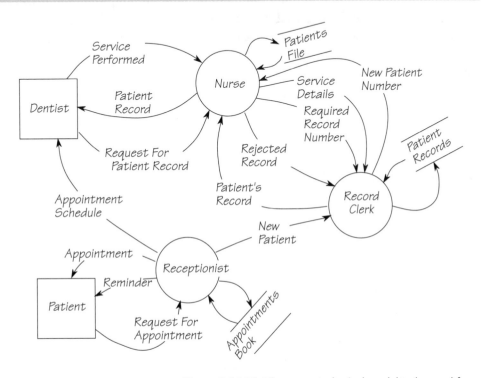

Figure 2.11.22: The current physical model to be used for these exercises.

Exercise 1: Dentist Performs Service

The event list shows event 4 *Dentist performs service*. Build the essential event-response data and process models for this event. The definition of the input data flow is

Service Performed = Patient Name
+ {Service Identification}
+ Visit Date

Exercise 2: Time to Produce Appointment Schedule

Fifteen minutes before a dentist is due to arrive at the office, the receptionist goes through the appointments and makes up the dentist's appointment schedule. The schedule lists what a dentist is doing that day. Build the essential event-response process and data models for event 5 *Time to produce appointment schedule*. Write a data dictionary entry for the data flow APPOINTMENT SCHEDULE.

Discussion: The Essential Data Model

By now, you have several event-response data models, with the same entities appearing in several of the models. You can consolidate the various event-response data models to form a *system data model* by connecting the models via their matching entities. The partial data dictionary for this model is

Appointment = Appointment Date + Appointment Time

Dentist = Dentist Name

Patient = Patient Name + Patient Address + Patient Telephone

Service = Service Identification

Visit = Visit Date + Visit Time

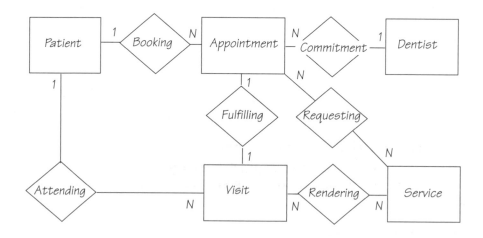

Figure 2.11.23: The system data model for the MOe & LARry Dental system. All of the entities and relationships were originally part of the event-response data models that you've studied.

Our knowledge of the event responses means we can add the correct cardinality to the data model. Naturally, if you had had this data model in the beginning (or if you peeked ahead and saw it), the event-response process models would have been easier to construct. Figure 2.11.24 shows how each event response uses part of the system data model.

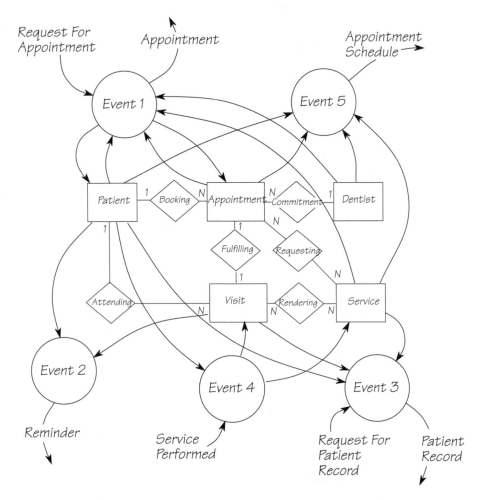

Figure 2.11.24: The event-response processes usually only access part of the system data.

There is no one best approach to building the essential model. Our preferred way of working is to construct a series of event-response process models, together with their associated event-response data models. The reason we take this approach is that building a system data model to cater for all of the details of the events is too confusing because the scope is too large to focus one's attention on the details. The event-response data model provides a way of focusing on the minute details of the data for one event. However, we can have the best of both worlds by amalgamating the event-response data models into one system data model (ideally, with a CASE tool to manage it). Another way to get started is to build at least a high-level version of the system data model first, then confirm it with the essential process models. This data model is useful as an inter- or intra-project communication tool. Either way works, so we have included both approaches in this book.

Discussion: Life in the Fast Lane

Now you know the complete (and laborious) way of arriving at the essential event-response process and data models. You have seen how an event response is modeled as it exists in the current system and then refined to leave only the essentials. However, it doesn't always have to be done that way. If you have enough knowledge about the system, you can shift up a gear or two.

Once you are comfortable with the event-response modeling approach, you can build essential models with a very abbreviated current physical model phase. The most minimal approach is to start the essential event-response models without any current physical modeling at all. If at any stage you have trouble determining the essential response, then you simply build some of the current physical model before resuming your essential approach.

The suggested fast-lane approach* uses these steps:

1.   Build the context diagram during initial interviews with the users.

2.   Use the data flows in the context diagram to determine the events. There must be an event response to account for every data flow. Add each event, together with its associated data flows, to the event list.

3.   Alternatively, if your users find it easier to identify events without the context diagram, list the event, then add the associated flows to the context diagram.

4.   For each event, ask, "What event happens if this event does *not* happen?" For example, *Customer pays* is one event. *Customer does not pay within the allotted time* is another. Add any resulting events to the list and new flows to the context.

5.   For each event,

   •   Interview the users about the responses to the event and build a one-bubble essential event-response process model (also known as an *essential activity model)*. Include only the essential processes and data.

* McMenamin and Palmer refer to a similar approach called "blitzing" in their book, *Essential Systems Analysis* (see the Bibliography).

- Draw the event-response data model.
- Write the mini specification for your essential activity model.
- If the process is too complex for one mini specification, use leveling to build a low-level model. Then, write a mini specification for each process in the low-level model.
- If you have trouble building the essential models, build a physical model of the current system's response to the event. Do only as much physical modeling as you need to understand the event response well enough to abstract the essentials.

6.   Do a CRUD check (more about this below) to test your context diagram and to discover missing events.

Identifying Custodial and Fundamental Processes

Our colleagues Steve McMenamin and John Palmer (see the Bibliography) have classified activities carried out by a system into two types: custodial and fundamental. A *custodial activity* maintains the system's essential memory. For example, the response to an event such as *Customer changes address* is part of the system only to keep the system's stored data current. In other words, this activity is one of the custodians of the system's data.

A *fundamental activity,* on the other hand, is one of the key reasons for having the system. For example, we might have constructed the system to send invoices to our customers and to accept their payments. These fundamental activities don't merely maintain the stored data, they are the reason for having it in the first place.

Sometimes it is difficult to differentiate between fundamental and custodial activities, and most event responses are a mixture of fundamental and custodial activities. The reason for introducing this terminology is that certain events, which are mainly concerned with custodial activities, are sometimes hard to discover because users don't think to tell you about them. Your CRUD check, as we'll discuss in a moment, will help you to discover these missing events.

Performing the CRUD Check

CRUD is a mnemonic for create, reference, update, and delete. The CRUD check verifies that you have identified every data element needed by every process within the context of study, and consequently all the events, and that all your event-response models are functioning correctly.

In other words, there must be enough event responses so that each attribute of stored data is CRUDed. At least, that is the principle. However, beware that some attributes will not be updated, and some may never be deleted, but all are created and retrieved. Missing CRUD actions for an attribute means that you may not have

> The CRUD convention says that every attribute in the data model must be created and referenced, and some will be updated or deleted.

captured all the event responses or that you may have some redundant data. Take, for example, the attributes of the APPOINTMENT entity in the MOe & LARry Dental system. There is an event to create them, *Patient needs appointment,* and an event to reference them, *Time to produce appointment schedule.* However, they are not updated, nor are they deleted. This means that we have forgotten two events: *Patient changes appointment* and *Patient cancels appointment.* Typically, you will add these custodial events to the list, build the event-response models, and inspect them carefully to see if other essential data are needed.

As well as there being enough events to install the system's stored data, there must be enough data to support the events that retrieve data. Any data element that is part of an output boundary data flow should be an essential stored attribute or should be derivable from essential data. Any data element that is used by an event response must either be an attribute of an entity or relationship, or it must travel into the system through a boundary flow.

Beware that some attributes may be CRUDed by more than one event response, so your audit should ask, "Are there any other conceivable uses of each attribute that have not been covered by one of the event-response models?" For example, in the MOe & LARry Dental system, the events *Time to send appointment reminders* and *Dentist needs patient record* both reference VISIT DATE. However, the references are for different reasons. The former is making a decision about whether it is now time to send a reminder to the patient, the latter is supplying the dentist with historical information about the patient.

> Use the attributes of the data model to check that all the essential processes have been modeled. Use the elements of the boundary data flows to check that all the entities have been included in the data model.

Exercise 3: Sid Edison's Radio Repairs

Try this exercise using the fast-lane approach. The following is a statement from Sid Edison, who owns a radio repair company. Although Sid is no relation of Thomas Edison, he likes his customers to make the association. He feels it helps his image. Read through Sid's statement on how radios are repaired at his company:

"Hi, I'm Sid Edison and I fix radios. My customers bring me their broken radios. You know the kind of thing. They play their rock music at full blast until the speaker collapses. Or they don't like something they hear and throw the radio across the room. Whichever way it's broken, they ask me to fix it.

"I always tell them how much I think the repair's going to cost. I either give them a firm quote for the work, or sometimes I charge them time and materials. In that case, I give them a note that explains the cost. I never start a repair job until I get an authorization from the customer. Sometimes, the cost of the repair is more than the radio is worth. I don't want to do the work and then get stuck with a radio the customer won't claim.

"Most of the repair jobs are pretty straightforward, but I get some tricky ones. I always look in my file for the manufacturer's specification just to make sure. Sometimes, I have to send off to the manufacturer for the full specification for the model before I can finish the job. The manufacturers don't charge for the specs, which is lucky. However, the parts suppliers charge for parts. If the repair needs some parts I don't have in stock, then my technician has to order some new parts from the supplier. These usually take a day or two to arrive. I send a pick-up notice to the customer when the repair is completed.

"The parts suppliers bill me at the end of the month, and I pay at the end of the following month. Customers pay cash when they pick up their radios"

Draw the context diagram for Sid's system, and make a list of all the events that Sid has described. Now add to your event list missing events that are concerned with custodial activities. It may help to sketch a preliminary data model while figuring out these events.

Build event-response process and event-response data models for at least three events. (Don't pick the easy ones.)

Compare your answers with those in Chapter 4.7. You may prefer to do this exercise one part at a time. In that case, do the context diagram and the event list first, and confirm your answer with Chapter 4.7's before listing the events for the custodial activities. Again, check your answer before building the event-response models.

Make sure that you are comfortable with event-response modeling before proceeding.

Some Notation Issues

For those of you following the ❋ Promenade Trail, the following will be of little interest. Skim through this section quickly to the Trail Guide.

You've seen event-response process and data models that store and/or reference data. Sample models are shown in Figure 2.11.25.

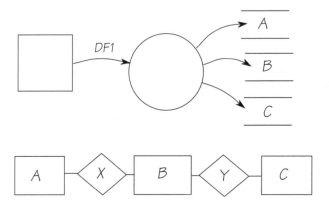

Figure 2.11.25: Sample event-response process model (above) and its corresponding data model (below).

When it stores the data, the process must create the relationships X and Y. Why don't we show the relationships in the event–response process model? A suggested model appears in Figure 2.11.26.

This model is more comprehensive when it shows the relationships. However, if the relationships X and Y do not contain data, the flows to them are not real data flows (they contain no data), so we prefer to leave them out of our event–response process models and to use the mini specification to specify the creation of the relationships and the accesses to essential data. If a relationship contains data that are stored by the event response, you should indicate the data flow and the relationship in your model.

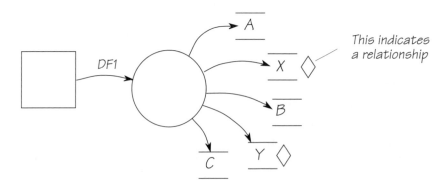

Figure 2.11.26: Event-response model that shows relationships.

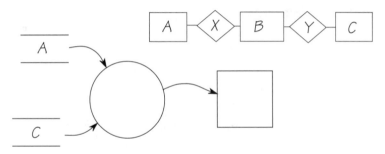

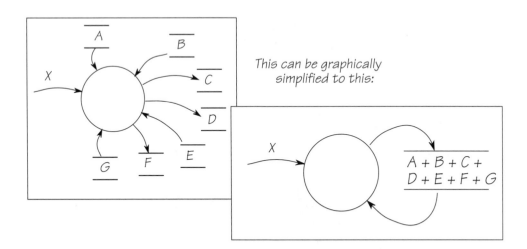

Figure 2.11.27: Data are retrieved from entities A and C. Entity B and its two relationships are used only to locate the correct instances of A and C. Therefore, it's not shown in the process model.

The example in Figure 2.11.27 follows the rule that the only flows shown in the model are those containing data. In this case, entity B does not provide data for the process, nor do the two relationships. Because there is no data involved, you can omit the entities, flows, and relationships.

Sometimes, a process needs too many stores to fit comfortably in the model, as shown in Figure 2.11.28. The solution is shown on the right.

This can be graphically simplified to this:

Figure 2.11.28: The many data stores are grouped into one store, which is named and its contents defined in the data dictionary.

Making Event Responses Unique

Each unique data flow entering the system has a unique response. For example, the system responds differently to an incoming data flow called POLICY CANCELLATION than it does to a data flow called AUTO POLICY CANCELLATION.

For example, in Figure 2.11.29, the incoming flow POLICY CANCELLATION is used when the customer wishes to cancel all policies. There is no data in POLICY CANCELLATION to tell the system which policies to cancel. So there must be a process to look at all possible policies to find those held by a particular customer. Compare that event response to the one shown in Figure 2.11.30. In this case, the system knows from the data content of AUTO POLICY CANCELLATION that the customer wants to cancel a car insurance policy.

> The content of the arriving data flow tells the system which event has occurred.

The response to each event is unique. That is, the collection of processes and data that make up the response are not exactly duplicated anywhere else in the system. However, a process that is part of one event response may be duplicated as part of others. Similarly with the entities: The same entities are used by many different event responses. Because they are part of one event response does not mean they cannot be part of another.

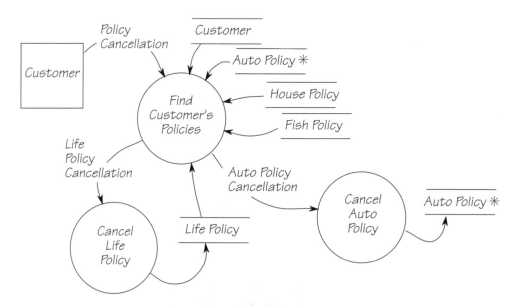

*Figure 2.11.29: The response to the event **Customer cancels policies.** Part of the system's response is to find all the policies held by the customer.*

The best way to deal with duplicate processes is to draw the process wherever it occurs. There is no point in amalgamating all the event responses that include a duplicate process. If you do this, you lose the control advantages of event partitioning and end up with a large, unmanageable, incomprehensible model. Later, when you write mini specifications, you will avoid duplication by writing one specification and cross-referencing it to all the processes that it defines.

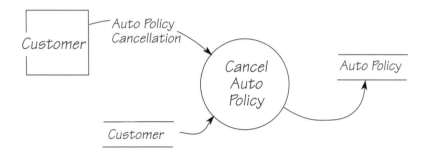

Figure 2.11.30: The response to the event **Customer cancels auto policy.**

Joining the Event Responses

Each event response has a separate event-response process model. For large-scale systems, this may mean building several hundred models. There are advantages to keeping them separate; but for management reasons, you may choose to combine them into a single model and see the whole system, rather than scattered fragments.

To connect the models, join those event responses that share stored data. An example of a connected model is shown in Figure 2.11.31.

In most systems, you'll find that the processes that use the same stored data have a strong functional relationship, so the joined event responses usually fall into coherent groups. Using these groupings, you can reduce the number of bubbles in the model by leveling upward. This leveling effort must seek to minimize the interfaces. In fact, the processes shown at the top level should have no data flowing between them. They should communicate only through the stored data.

Instead of building a leveled model, you can choose to use the CRUD check to manage the inter-event interfaces. You will see more of this approach in Chapter 1.15 *CRUD Check.*

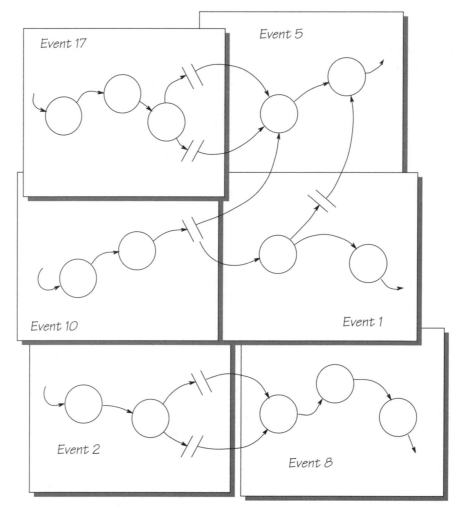

Figure 2.11.31: The event-response models are joined by their common usage of stored data.

Summary

The event response is the most convenient partitioning theme for the essential model. Event-response models are built by linking all of the processes that make up the system's reaction to an event.

There are two types of events: External events happen outside the system, and temporal events happen when the system must do something at a specific time.

Every data flow in the context diagram is associated with one of the events in the event list. Each input data flow in the context diagram indicates an external event. An external event can also be associated with anything from zero to many output data flows. Output data flows that are not associated with external events are in response to temporal events.

The essential event-response model ignores any processing that is there because of the current (or future) implementation, and shows only the essential processing.

An event response processes essential stored data and, in some cases, data derived from the essential stores. Each event response has its own event-response data model. Alternatively, if you prefer to model the event-response processes without an event-response data model, you need to use a system data model.

All of the essential attributes should be CRUD checked (to assure they're created and referenced, and possibly updated or deleted). This checks that all the events have been discovered and that all the data are relevant.

Event-response models can be built using the current physical model, or directly from the context diagram and user interviews. These models can be leveled upward to make a convenient system-level overview.

What to Do

If you're happy with the answers you got for the exercises, then go on to the Piccadilly Project. If you'd like more practice in modeling events, try completing the essential model for Sid Edison's system. Then check your answer against ours in Chapter 4.7.

Trail Guide

This duplicates the one at the end of Chapter 4.7.

● Easiest: Once you are comfortable with event-response models, return to the Piccadilly Project. Your first destination is Chapter 1.8 *Identifying Events,* where you will build the event list. In the chapters that follow, you will model and subsequently refine one of those events.

■ More Difficult and ◆ Most Difficult: This chapter was not part of your trail, and your Piccadilly Project work resumes in Chapter 1.8 *Identifying Events.*

❋ Promenade: On this trail, you avoid the hard work. While the others are off to build event-response models for Piccadilly, you'll go to Chapter 2.12 *Mini Specifications,* where you'll read about developing a specification for each bubble in the event-response model.

2.12 MINI SPECIFICATIONS

Before You Reached Here ...

● Easiest: You have built an event-response model for Piccadilly in Chapter 1.10 *Refining an Event Response*. Before the event-response model is complete, however, you must write a description of the processes in your model. These process descriptions are known as mini specifications; they're also called process specifications, P-specs, or transformation specifications.

■ More Difficult and ◆ Most Difficult: Your trails bypass this chapter. To rejoin your trail, go to Chapter 1.11 *Writing Mini Specifications*.

❄ Promenade: So far, you've seen the data dictionary used to specify the data flows and entities found in the event–response models. (You just came from Chapter 2.11 *Event-Response Models*.) This chapter completes the event-response modeling topic, as it discusses how mini specifications describe the system's processes.

Working Models

Let us reiterate: The goal of systems analysis is to build a working model of the system. The working model is based on the Rule of Data Conservation, which states that each process in the model must receive data that are both necessary and sufficient to produce its output. The data dictionary defines both the inputs and outputs of each process, and it can be used to prove that each process is capable of producing its outputs from its inputs.

However, something's missing. You need a specification that says what the process does with its inputs to produce its outputs. That is the role of the *mini specification*. Mini specifications, as the name implies, are small. They specify a small part of the system, the functional primitive, which can usually be described in a page or less. The mini specification describes a single, discrete functional component of the system.

Specifying the Functional Primitives

Your event partitioning breaks the system into functional pieces. Each event-response process model declares the essential processes and data concerned with that event. Now you need to specify the detailed processing rules.

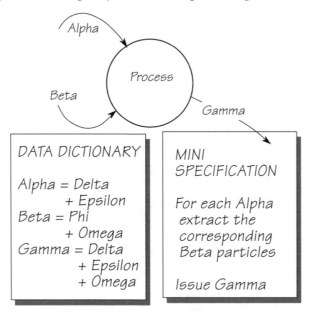

Figure 2.12.1: The complete and rigorous model. The data dictionary defines the data, and the mini specification describes the processing of that data.

You can model the system's essential processing at several levels. At the highest level, you group all of the system's responses to an event into one bubble. This one bubble is referred to as an *essential activity*. If the essential activity is small enough, you can specify it by writing a mini specification. Otherwise, you'll need to decompose the essential activity into some number of functional primitives using leveling (see Chapter 2.7 *Leveled Data Flow Diagrams*). Each of these primitives is then specified in the form of a mini specification. When you have written all the mini specifications, you have something that is quite remarkable: a specification of a system that is complete. Everything is specified, but only once.

Isn't It Late to Be Specifying?

We intentionally left this chapter about mini specifications until you had a chance to develop your process models. This is to make our point that you delay specifying

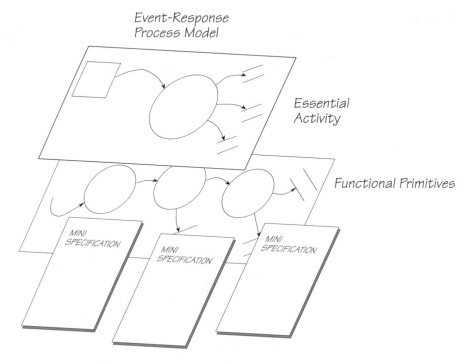

Event-Response
Process Model

Essential
Activity

Functional Primitives

MINI
SPECIFICATION

MINI
SPECIFICATION

MINI
SPECIFICATION

Figure 2.12.2: An event-response process model may have several levels. A complex essential activity will be broken down to its functional primitives, each of which is described by a mini specification.

the processes for as long as possible. By building the data flow diagrams, data models, and data dictionary first, you are refining your knowledge of the system. In the process of eliminating the implementation-dependent processes and data, you have reduced the requirements to their essential minimum, and there are fewer mini specifications to write.

However, postponing writing the mini specifications doesn't mean that you must wait until you've finished all the event-response modeling before you write your first specification. Sometimes, when you discover a piece of essential policy while doing current physical modeling, it makes sense to write the specification right then. Above all, don't write mini specifications for processes that are dependent on the implementation. Those processes are adequately specified with a current physical data flow diagram.

Sometimes, especially if you are asked to produce a high-level overview, it is useful to write a specification for a high-level process that is not a functional primitive. We refer to such a high-level specification as a *maxi specification*. Our advice is to avoid writing maxi specifications because they are a duplication of policy that will eventually be in several mini specifications. The high-level requirements are better specified by a combination of data flow diagrams, data model, and data dictionary. Normally, because you are using the models to talk about the system and raise questions, users accept them very easily. Put the strategy of writing a maxi specification into your analyst's toolkit only for extreme situations when you are having trouble communicating with someone who insists on a text specification.

Specification Techniques

Each of the system's processes is different, with one making many decisions, another doing extensive calculations, a third following many rules, and so on. Because of the differences, processes need different kinds of specification. When you write mini specifications, choose a technique that is appropriate for the process being described.

With an example, we'll demonstrate some of the most useful methods for specifying processes. As we do, evaluate them for their suitability for other types of processes. Look at the following fragment of a model and the accompanying user's statement. Neither gives you enough information to complete the specification. However, our attempt to write a mini specification for this process should reveal the missing parts.

Here is the user's statement: "This process is part of a system for paying contract workers at a construction site. The workers submit their time cards at the end of the week. Another part of the system edits the cards, so this process receives only accurate data. When the workers fill in their time cards, the hours are complete hours and not fractions.

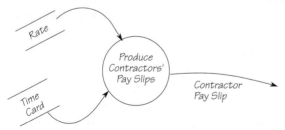

"Different work categories have different hourly rates of pay. Overtime worked under categories 4, 5, or 6 is not paid at overtime rates."

As we write a mini specification for the process PRODUCE CONTRACTORS' PAY SLIPS, we'll illustrate several specification techniques. First, we'll use a variant of our natural written language called structured language.

Figure 2.12.3: This process is part of a system to pay contract workers. The data dictionary gives some of the definitions for this model.

Contractor Pay Slip = Contractor Id
+ {Work Category + Normal Hours Worked + Loaded Rate
+ $ Amount + Overtime Hours Worked + Overtime Rate
+ Overtime $ Amount}
+ Total Hours Worked + Total $ Amount

Penalty Loading = * Data element. Awarded for dangerous or dirty work. *

Rate = {Work Category + Hourly Rate
+ (Penalty Loading)}

Time Card = Contractor Id
+ {Work Category + Normal Hours Worked
+ Overtime Hours Worked}

Structured Language

Structured language uses a subset of our natural language. The subset is restricted to a few verbs that manipulate the data items already established in the data dictionary.

The term "structured" is used because this language follows the same rules for combining statements as does structured programming. The structured programming rules were developed to make programs more concise, more readable, and provably correct. In 1966, the Italians Böhm and Jacopini in their landmark paper argued that programs written using only certain structures could be mathematically proved to be logically correct (see the Bibliography). Similarly, if you stick to the rules of structured language, it will be much easier to find the logical errors in your specifications.

Keep in mind that while structured language resembles some programming languages, the mini specification is *not* the specification of a computer program. You are still doing analysis and trying to state the essential requirements for a process. Therefore, avoid using any procedural terms that could influence the eventual implementation.

Let's consider structured language. Böhm and Jacopini reasoned that programs have a *sequence*. Sequence means the statements are read from top to bottom with each statement following the previous one. Be careful here that you not inject arbitrary procedures into the specification, and that any sequence in the specification is necessary for policy reasons, not because the current implementation uses it.

In our example, all the available time cards must be processed. This means that the actions to process each of them must be repeated until all have been done. Hence, the first Böhm and Jacopini construct: *repetition*. Its general form looks like this:

For each of these
      Do this

That means you would write

For each TIME CARD

As you need to look at all the hours for each work category on one card, add

      For each WORK CATEGORY

This second line is indented to show that it is subordinate to the first and also to make the specification more readable.

You were told that different work categories have different rates of pay. The data dictionary reveals that these rates are stored in the data store RATE. To look up the appropriate rate, use the general form

Find occurrence of the data **corresponding to** identifier

As you are using the category of work to locate the required entry in the RATE data store, you would write

Find the entry in RATE *corresponding to* WORK CATEGORY

Alternatively, a more relaxed version is

Find the entry in RATE *with the same* WORK CATEGORY

You may vary the verbs if you wish. Just be consistent so that your readers can understand your specification.

So far, our mini specification looks like this:

For each TIME CARD
      For each WORK CATEGORY
            Find the entry in RATE *with the same* WORK CATEGORY

Now you have to make some computations. These are usually written as an algebraic equation, thus:

LOADED RATE = HOURLY RATE *multiplied by any present* PENALTY LOADING
NORMAL $ AMOUNT = NORMAL HOURS WORKED *times* LOADED RATE

COMPLETE SYSTEMS ANALYSIS

Some analysts prefer to use symbols such as "$*$" instead of "times" or "multiplied by," and "/" rather than "divided by." Thus,

NORMAL $ AMOUNT = NORMAL HOURS WORKED $*$ LOADED RATE

The results of some of your calculations will be defined in the data dictionary as derivable or calculable data elements. If the algorithm for deriving a data element has been defined in the dictionary, your mini specification can reference the derivable element without having to repeat the algorithm.

Now you have to write a decision into your specification. This brings in the second Böhm and Jacopini construct: *selection*. The general form is

If this is true
    Do this
Otherwise
    Do this

To calculate the overtime, write

If the WORK CATEGORY is not 4, 5, 6
    OVERTIME $ AMOUNT = OVERTIME HOURS WORKED $*$ (HOURLY RATE $*$ 1.5)

If the condition is not true, there is nothing to be done. (Never write meaningless lines like "Else do nothing.")

Note that the user's statement did not give any rules for calculating overtime. We assumed that overtime pays fifty percent more than normal, and we would need to go back to the user to confirm this rate. However, it is only when you try to write a mini specification that you discover this kind of shortfall in the requirements and can then get the missing information.

Writing this statement also triggers the question, "What happens if contractors submit overtime hours in category 4, 5, or 6? Are they paid at normal rates for all the hours they work? Or can I disregard any overtime in these categories?" Let's suppose the user says that normal rates apply to all hours in any of these three categories. You could amend the above to read

If the WORK CATEGORY is not 4, 5, 6
    NORMAL $ AMOUNT = NORMAL HOURS WORKED $*$ LOADED RATE
    OVERTIME $ AMOUNT = OVERTIME HOURS WORKED $*$ (HOURLY RATE $*$ 1.5)
Otherwise
    NORMAL $ AMOUNT = (NORMAL HOURS WORKED
    + OVERTIME HOURS WORKED) $*$ LOADED RATE

This assumes that category 4, 5, or 6 overtime is paid at the loaded rate. (Check with the user.) Now to use the product of your calculations

Accumulate TOTAL HOURS WORKED
Accumulate TOTAL $ AMOUNT

Again, the verb "accumulate" is rather informal, but the meaning is clear. With this, we finish specifying WORK CATEGORY. Some analysts like to write "End For" at this stage to indicate the end of the actions to be repeated. (We don't because there is no need with indented subordinate clauses.)

The next statement applies to TIME CARD, and so its indentation must indicate that it is to be done after every WORK CATEGORY has been processed, but before a TIME CARD has been completed. The complete mini specification looks like this:

For each TIME CARD
    For each WORK CATEGORY
        Find the entry in RATE *with the same* WORK CATEGORY
        LOADED RATE = HOURLY RATE *multiplied by any present* PENALTY LOADING
    If the WORK CATEGORY *is not 4, 5, 6*
        NORMAL $ AMOUNT = NORMAL HOURS WORKED * LOADED RATE
        OVERTIME $ AMOUNT = OVERTIME HOURS WORKED * (HOURLY RATE * 1.5)
    Otherwise
        NORMAL $ AMOUNT = (NORMAL HOURS WORKED
        + OVERTIME HOURS WORKED) * LOADED RATE
    Accumulate TOTAL HOURS WORKED
      Accumulate TOTAL $ AMOUNT
Issue CONTRACTOR PAY SLIP

The verb "issue" means that all the necessary data for one CONTRACTOR PAY SLIP are gathered and sent along the data flow channel.

Some analysts prefer to number each line to further highlight subordinate lines. A numbered specification looks like this:

For each TIME CARD
    1 *For each* WORK CATEGORY
        1.1 *Find the entry in* RATE *with the same* WORK CATEGORY
        1.2 LOADED RATE = HOURLY RATE *multiplied by any present* PENALTY
        LOADING
        1.3 *If the* WORK CATEGORY *is not 4, 5, 6*
            NORMAL $ AMOUNT = NORMAL HOURS WORKED * LOADED RATE
            OVERTIME $ AMOUNT = OVERTIME HOURS WORKED * (HOURLY RATE * 1.5)
        Otherwise
            NORMAL $ AMOUNT = (NORMAL HOURS WORKED
            + OVERTIME HOURS WORKED) * LOADED RATE

1.4 Accumulate TOTAL HOURS WORKED
1.5 Accumulate TOTAL $ AMOUNT
2 Issue CONTRACTOR PAY SLIP

While numbering adds some formality to the structure of the specification and makes the subordination clear, we've found that the majority of users prefer to read a "more English" or less mathematical version of the specification without numbers. Some users have stated their preferences in such strong terms that we had to relax the language to a point like this:

For each work category on each time card, the following has to be done

> *First, find the entry in rate with the same work category as the time card*

> *Second, calculate the loaded rate by multiplying the hourly rate by the penalty loading if one is present*

> *If the work category is not 4, 5, 6 (these categories do not attract overtime)*

>> *Then the normal $ amount is equal to the hours worked multiplied by the loaded rate*

>> *and the overtime rate is the hourly rate multiplied by 1.5*

>> *and the overtime $ amount is overtime hours worked multiplied by overtime rate*

> *Otherwise (this is for categories 4, 5, 6)*

>> *Add the normal hours worked and the overtime hours worked and multiply that by the loaded rate*

> *Accumulate the hours worked and the overtime hours worked into total hours worked*

> *Accumulate the $ amount and the overtime $ amount into total $ amount*

When all that has been done for all the work categories on one time card, transfer the appropriate data items to the contractor pay slip and issue the contractor pay slip

This relaxed approach introduces a lot of redundancy, but its folksy manner may be helpful in some circumstances.

Recall that the mini specification is not intended to be procedural and should not influence any future implementation. Structured language infringes on this rule somewhat in that the statements must be written in some sort of order. For example, consider these two lines of the specification:

Accumulate total hours worked
Accumulate total $ amount

There is no reason why the hours have to be accumulated before the dollar amount; we just chose to write them that way. This case is quite innocent. Most reasonable implementors will realize that the order in which the two statements appear comes about by chance, and they won't feel bound to implement them in that order. Of course, you could write

Do the following in any order
 Accumulate TOTAL HOURS WORKED
 Accumulate TOTAL $ AMOUNT

But this seems to be overkill when no harm is being done by the order of these statements. In some cases, implying a procedure could lead you into procedural thinking and thus divert you from analysis. There are less procedural ways of writing mini specifications, the most popular being the decision table.

Decision Tables

A *decision table* is used when the process makes decisions based on the answers to a variety of questions. Let's take the same piece of policy you specified above, and build a decision table for it. When you have the two specifications to compare, consider which method better describes the policy of the process.

To begin, first look at the questions to be asked. The data dictionary shows

Rate = {Work Category + Hourly Rate
 + (Penalty Loading)}

So there is a question about whether or not PENALTY LOADING is present. The user's statement told us, "Different work categories have different hourly rates of pay. Overtime worked under categories 4, 5, or 6 is not paid at overtime rates."

This raises two more questions: Is the system processing hours in category 4, 5, or 6? Are overtime hours being worked? To start the decision table, write the questions

Penalty loading present?
Work category 4, 5, 6?
Overtime hours?

The table works by providing answers to combinations of these questions or, as they are referred to in decision tables, conditions. This means the table must contain an answer for every possible combination of conditions. Use this pattern to fill in the table: To discover how many combinations of conditions are possible, take the product of the number of possible values. In this case, each condition has 2 possible values. So the calculation is $2 * 2 * 2$, which gives you 8 columns in your table. Now fill in the table to reflect each of the 8 combinations of conditions.

Penalty loading present?	Y	N	Y	N	Y	N	Y	N
Work category 4, 5, 6?	Y	Y	N	N	Y	Y	N	N
Overtime hours?	Y	Y	Y	Y	N	N	N	N

Note how this pattern is built. The top row begins by listing all possible answers to the question. In this case, there are only two: *yes* and *no*. Repeat this combination until you run out of columns in that row. The next row doubles the pattern: two *yes*'s and two *no*'s. This doubling continues until you reach the last row, which will always be half one answer, and half the other. (If there were three answers, the last row would be one third of each.) Note that sometimes the correct response to a question is *don't care*. For instance, suppose the user had told you that even when contractors worked overtime for categories 4, 5, or 6, they submitted all their time as normal hours. Now the system can disregard the answer to the overtime question for those categories. Indifferent answers are shown as a dash. The table looks like this:

Penalty loading present?	Y	N	Y	N	Y	N	Y	N
Work category 4, 5, 6?	Y	Y	N	N	Y	Y	N	N
Overtime hours?	-	-	Y	Y	-	-	N	N

When the table is completed, it contains all combinations of possible answers to the questions. Now you must deal with the actions to be taken in response to the answers. Look at the data dictionary entry for the output:

Contractor Pay Slip = Contractor Id
+ {Work Category + Normal Hours Worked + Loaded Rate
+ $ Amount + Overtime Hours Worked + Overtime Rate
+ Overtime $ Amount}
+ Total Hours Worked + Total $ Amount

This entry says that you have to calculate the loaded rate of pay for each work category. The loaded rate varies depending on whether a penalty loading is present or not. You also have to determine if overtime is being worked, and to calculate the overtime pay. The total hours and total pay have to be accumulated.

Adding these answers to the table, you get

Penalty loading present?	Y	N	Y	N	Y	N	Y	N
Work category 4, 5, 6?	Y	Y	N	N	Y	Y	N	N
Overtime hours?	Y	Y	Y	Y	N	N	N	N
Calculate rate with penalty								
Calculate rate without penalty								
Calculate overtime								
Accumulate total hours								
Accumulate total pay								

Now all that remains is to mark which combinations of answers provoke which actions. You can do this by going back over the user's statement and consulting the data dictionary. Naturally, you would need to ask questions when the policy is not apparent. The final table looks like this:

Penalty loading present?	Y	N	Y	N	Y	N	Y	N
Work category 4, 5, 6?	Y	Y	N	N	Y	Y	N	N
Overtime hours?	Y	Y	Y	Y	N	N	N	N
Calculate rate with penalty	✓		✓		✓		✓	
Calculate rate without penalty	✓		✓		✓		✓	
Calculate overtime			✓	✓				
Accumulate total hours	✓	✓	✓	✓	✓	✓	✓	✓
Accumulate total pay	✓	✓	✓	✓	✓	✓	✓	✓

The algorithms for the calculations can be packaged with the decision table to complete this mini specification. Another alternative is to define the algorithms as calculable data elements in the data dictionary. For example,

*Overtime = * Data element. Calculable. **
        Overtime Hours Worked multiplied by Overtime Rate

This technique is particularly effective if the calculation is used in more than one mini specification.

To repeat, the decision table has the advantage over structured language of being nonprocedural. There is no reason why one question would be asked before another when it is obvious that all of them are going to be asked anyway. Nor is there any reason to think that one action would be taken before another. The table thus solves one problem, but raises another. Some users don't like decision tables. They can be seen as unfriendly, "technical" specifications. To retain the decision logic and yet make the specification more friendly, try a decision tree.

Decision Trees

A *decision tree* does the same job as tables, and some people find them easier to read. We leave it to you to decide. The decision tree for paying the contractors is shown in Figure 2.12.4.

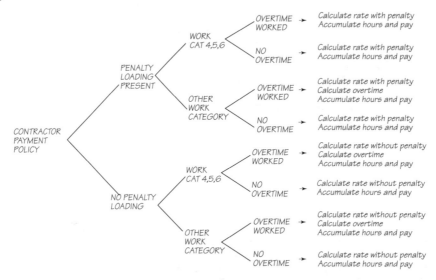

Figure 2.12.4: Decision tree for calculating the contract workers' pay slips.

Specifying Judgmental Bubbles

A judgmental bubble is one with a process that does not appear to be definable. Although bubbles that make judgments are rare, you have to know what to do about them when you find one. Such bubbles are annotated with shading down the sides.

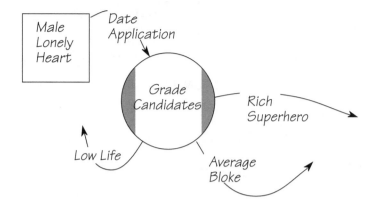

Figure 2.12.5: *An example of a judgmental process from Miss Tweedy's Dating Service. Zsa Zsa makes a decision on what kind of male is applying for a date.*

The process shown in Figure 2.12.5 is entirely subjective: "This vun I like," "This vun no fun at all," and so on. For truly subjective processes, the mini specification should state as many as possible of the criteria for making the decisions, and leave it at that. Later, when you start thinking about the implementation, the shading will highlight the processes that cannot be automated because they cannot be specified.

Be forewarned: Many of the seemingly judgmental processes are not subjective at all. There are well-defined rules; they just can't be seen, as when you hear, "I don't know how I do it. I've been doing it for so many years." You must investigate further to find the underlying policy for the process. Ask for all the information used by the process, ask for any existing written documentation about the process, and try to define all the algorithms, rules, guidelines, rules of thumb, telephone calls, and notes on the back of envelopes that you can find. The object of the exercise is to determine if the process is really a judgmental one, or if the process is a regular process just enveloped by the users' mystique and can be specified by one of the regular specification methods.

Specifying Data Storage and Retrieval

Any system has many processes that manipulate essential stored data, so it is appropriate to discuss how to write mini specifications for them. We'll use the data model in Figure 2.12.6 for our examples. Remember: An entity in a data model represents all the occurrences of that entity, and we show the entity as a data store in the process model.

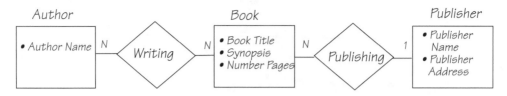

Figure 2.12.6: A sample data model.

In Figure 2.12.7, a process searches its data store, BOOK, to match the incoming enquiry data flow. The structured language form for this is

Retrieve the occurrence of BOOK *corresponding to* BOOK TITLE *in* BOOK ENQUIRY

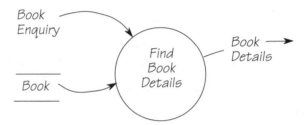

Figure 2.12.7: A process that retrieves one particular entity occurrence.

Keep in mind that you are not writing a computer program, and the data model is not a database. You are simply stating the requirement for this process to be able to access the stored data and to retrieve a particular book.

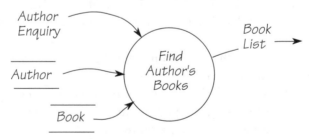

Figure 2.12.8: This process uses two entities: AUTHOR and BOOK. The AUTHOR ENQUIRY supplies the name of an AUTHOR, and the process lists every BOOK written by that AUTHOR.

To specify the process shown in Figure 2.12.8, write the following:

Find the AUTHOR *with the corresponding* AUTHOR NAME
Retrieve all associated occurrences of BOOK
Issue BOOK LIST

The word "associated" refers to those entities sharing a relationship with the first one. Association can refer to entities several relationships away from the original. Take, for example, the process shown in Figure 2.12.9. The mini specification reads

Find PUBLISHER with the corresponding PUBLISHER NAME
Retrieve all associated occurrences of AUTHOR

In Figure 2.12.9, the entities PUBLISHER and AUTHOR are separated by two relationships and an entity. If you look at the data model in Figure 2.12.6, you can see that it is necessary to access the entity BOOK to navigate the data model. However, there is no data in this entity and its relationships that are used by the essential process, so by convention we omit them from the process model to avoid unnecessary clutter.

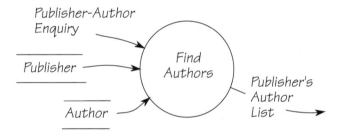

Figure 2.12.9: *In this event response, the task is to find all the authors that have written books for a designated publisher.*

Note that the cardinality in the data model must allow the necessary navigation. In this case, for any one publisher, the model allows the process to locate all the instances of book, and from each of those, all the authors. Sometimes, the entities are connected by more than one relationship. In these more unusual cases, you'll need to specify the relationships in the access path that interests you.

Now to store some data. Figure 2.12.10 shows a process that stores a book and its author. For the purposes of this example, assume that the publisher entity already exists in the system's stored data. (Another event will have previously stored it.)

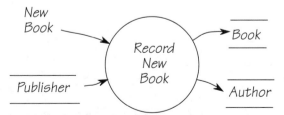

Figure 2.12.10: *The event-response process model for the event **New book** is received.*

The incoming data flow NEW BOOK contains the book details and the author's name. The mini specification looks like this:

Find the corresponding PUBLISHER
Establish an occurrence of BOOK and create a PUBLISHING relationship with
PUBLISHER
If the corresponding AUTHOR does not exist
> *Establish an occurrence*
> *Create a WRITING relationship between AUTHOR and BOOK*

Note that relationships are not shown in the process models unless they contain data that are stored or used by the process. Otherwise, the relationship is well enough specified by the data model. This minimalist approach is possible because the specification of every event response is composed of an event-response process model and an event-response data model.

Summary

When you decompose high-level bubbles to their functionally primitive level, you unearth the real processes of the system. A mini specification is written to describe each of these real processes. Its role is to describe what a process does to transform the incoming data flows and stores into outgoing data. The mini specification can take many forms depending on the policy being specified; some policies are best described in a natural language, some cry out for a decision table or a decision tree, and others can be described simply by showing the stored data affected by the process.

Mini specifications are normally written when the essential event-response models are built. If you specify the processes in a physical model, the chances are that you'll waste valuable time and effort specifying implementation-dependent processes.

Enough talk. Let's do some work.

What to Do

Here are some exercises intended to give you a gentle introduction to mini specifications before applying them in the Piccadilly Project. Write each mini specification using whichever method you think most appropriate. When you are happy with the result, check Chapter 4.8 for the sample answers.

Exercise 1: Hopper's Choppers

Larry Hopper runs a small helicopter service in the American Samoan Islands. The service ferries tourists, citizens of the islands, and employees of Hopper's Choppers from island to island. Larry has a scale of fares, each fare offering different levels of priority. He has also established some rules for booking passengers onto the helicopters. Larry has asked his daughter, Grace, to analyze his business system with the goal of improving service. One of her event-response models appears in Figure 2.12.11. The data store HELICOPTER BOOKINGS is modeled in Figure 2.12.12.

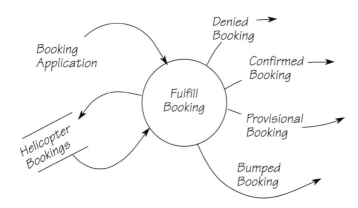

*Figure 2.12.11: The event-response models for **Passenger wants to make a booking** from the Hopper's Choppers system.*

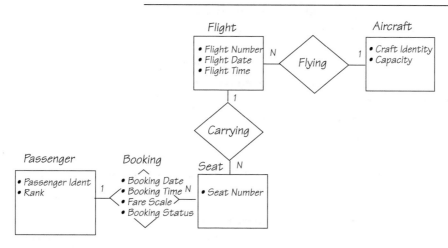

Figure 2.12.12: The data model derived from the store HELICOPTER BOOKINGS.

Grace's data dictionary shows

Booking Application = Passenger Ident + Flight Date + Flight Time
        + Number Of Seats Required + Fare Scale + Rank

Booking Status = ["Provisional" | "Confirmed"]

Bumped Booking = Passenger Ident + Flight Date + Flight Time

Confirmed Booking = Passenger Ident + Flight Date + Flight Time
        + Booking Status

Denied Booking = Flight Date + Flight Time

Fare Scale = ["Full Fare" | "Economy"]

Provisional Booking = Passenger Ident + Flight Date + Flight Time
        + Booking Status

Rank = ["Civilian" | "Emergency Employee" | "Employee"]

Here are Larry's rules: "Anyone who pays the full fare gets a confirmed booking if a seat is available. We bump provisional bookings if someone wants to pay full fare. People use provisional bookings because the fare is cheaper, and most of the time they don't get bumped.

"Employees fly free, so they don't get any priority except when there is some emergency and we have to get an employee to the problem fast. Then we will bump anybody to get the employee on a flight. We would only ever have one emergency employee per flight.

"The rules for bumping are simple: Cheap fares go first, the latest bookings go before the oldest."

Write a mini specification for the process FULFILL BOOKING.

Exercise 2: Terry's Ski Tuning Service

Ski tuning is the craft of repairing skis: filing the base, sharpening the edges, and removing all the nicks caused by skiing over rocks or other people's skis. Terry's Ski Tuning does an excellent job of making skis perform like new, but the part of the business that pays Terry is in real need of your help.

Write a mini specification for the process CALCULATE SKI TUNING BILL shown in Figure 2.12.13. You'll need to know that Terry and Joel are the master technicians, and that any service they perform is loaded by 15 percent. Services performed during the day attract a $20 surcharge. (The normal service is overnight, and Terry has to get someone to come in specially for any daytime work.) Tuesday and Wednesday nights are slow, so Terry offers a 20 percent discount for any work done on those nights.

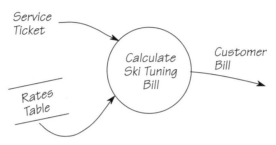

Figure 2.12.13: Process model for Terry's Ski Tuning Service system.

An incomplete but sufficient data dictionary is

Service Ticket = Technician + Day Of Week + Time Of Day
            + {Service Performed}

Rates Table = {Service Name + Service Rate}

Customer Bill = Service Ticket + Total Charge

Time Of Day = * Data element *
            ["Day" | "Night"]

Trail Guide

This Trail Guide is duplicated in Chapter 4.8 in case you don't return here after checking your answers.

● Easiest: Now it's time to return to the Piccadilly Project to specify some of the processes that you have identified. Your destination is Chapter 1.11 *Writing Mini Specifications.*

■ More Difficult and ◆ Most Difficult: This chapter was not part of your trail, but there is no penalty for reading it. Your Project work resumes in Chapter 1.11 *Writing Mini Specifications.*

❋ Promenade: You now know how to specify the requirements for a system. But there are other requirements that you'll meet in Chapter 2.13 *Modeling New Requirements.*

MODELING NEW REQUIREMENTS 2.13

Before You Reached Here ...

● Easiest and ■ More Difficult: You have modeled Piccadilly's essential requirements in Chapter 1.13 *More Events,* which used the existing system as a starting point. However, not all the requirements for the final system are part of the existing one. The users will certainly take advantage of your development effort to introduce requirements that have no current equivalent. It is this kind of requirement that we call, logically enough, a *new requirement.*

◆ Most Difficult: This is not on your trail. You should be building models for Piccadilly in Chapter 1.14 *Some New Requirements.*

❋ Promenade: New requirements are an extension of the modeling you've already seen. The difference is there is no current implementation of the requirement. New requirements can be discovered at any stage of a project. Indeed, new requirements are often among the first things the users want to discuss.

Defining New Requirements

New requirements are those not within the original context of the study. If at any stage of the project you decide to add a new requirement, it must be modeled along with all the others.

Let's look at some typical new requirements:

1. New essential requirements that are not part of the current system: Management sees a new business opportunity, and the current system cannot support it, nor can the system be modified to incorporate the new business requirement. Often this kind of need prompts the whole systems redevelopment effort and results in one or more new event responses.

    Similarly, new requirements "grow" during the analysis. Thanks to your analysis modeling efforts, users can see a better picture of their business. Now that they can communicate successfully with technical people, they will suggest all sorts of requirements for things they'd like to do, if only they had a system for doing it.

2.   New requirements that are outside the current context of study: Again, your analysis gets the users thinking, and they discover useful functionality in adjacent systems. By extending your context to include functions from other parts of the business, the final system will be far more useful to the business as a whole. These new requirements usually add several new event responses to your system.

3.   New requirements that are actually modifications to the way the current system responds to an event: These often appear to be completely new requirements, but your analysis reveals them for what they are. This type of new requirement is usually an addition or a change to an event response that is already within your context of study.

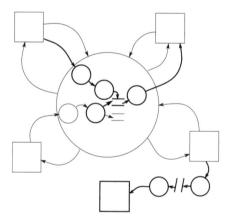

Figure 2.13.1: New requirements appear in a number of guises. Some are additions or modifications to the system you are studying, and some are (for the moment) outside your context of study.

We've noticed over the years the tendency of some project members to discourage their users from bringing up new requirements. The analysts feel that new requirements are an inconvenient interruption to the analysis. This attitude defeats the whole purpose of analysis: to discover *all* the requirements for a new system. New requirements are just as valid as any other, no matter how inconvenient. Eventually, you will have to incorporate new requirements into the system. To leave them until after implementation makes the project far more expensive than is necessary.

However, any requirement that alters the context of the study also alters the amount of time needed to complete the project. To ensure that you reflect the impact of all new requirements, you must change the schedule.

Modeling New Requirements

Often, new requirements are given to you along with an imagined implementation. It is temptation beyond endurance for user management to see sparkling new technology being promoted by a rival and not wish to have it themselves. "When we have the new system, we can issue credit cards printed with a photograph of the new offices, and we can offer itemized billing of the services, just like Compétiteur, Inc. Or maybe we should have a picture of the chairman's new car on the card . . . or maybe his dog." When you hear such dreams spoken aloud, you need to separate the essential requirement from the ideas for its implementation.

The best way to deal with a new requirement is to model it in the same way you model all the existing requirements. That is, build event-response process and data models for each new requirement.

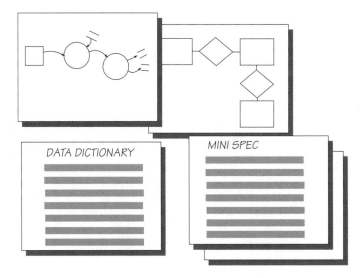

Figure 2.13.2: The event-response model for the new requirement is just like any other. There is no reason to model new policy differently from any other policy.

When you model a new requirement using event-response process and data models, some of the entities, attributes, and relationships may already exist in the system data model, just as some of the processes may be part of your current process models. Do not let this influence your approach. Treat each new requirement as a stand-alone mini system. This strategy helps you to focus on the details of the policy. Later on, you can worry about how and whether to integrate the new requirement with the existing context of study.

As an example of a new requirement, let's reconsider the user's statement: "When we have the new system, we can issue credit cards printed with a photograph of the new offices, and we can offer itemized billing of the services, just like Compétiteur, Inc. Or maybe we should have a picture of the chairman's new car on the card … or maybe his dog." If you ignore the obviously silly implementation ideas, there are several events that you need to model:

- *Customer applies for card*
- *Customer makes a purchase*
- *Time to bill the customer*

The essential event-response models look like those in Figures 2.13.3, 2.13.4, and 2.13.5.

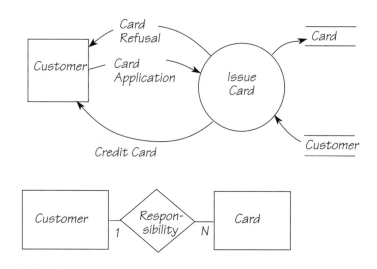

Credit Card = Card Number + Expiration Date + Customer Name

Card = Card Number + Expiration Date + Credit Limit

Figure 2.13.3: Essential event-response models for **Customer applies for card.**

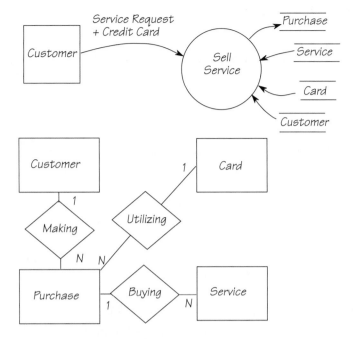

Figure 2.13.4: Essential event-response models for
Customer makes a purchase.

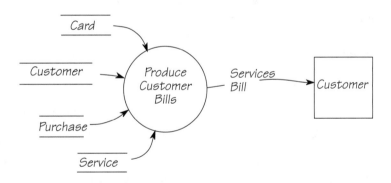

Services Bill = Customer Identification
  + { Card Number
  + { Date Of Purchase + Service + Amount }
  + { Total For Card }
  + Grand Total

The data model is the same as for the previous event

Figure 2.13.5: Essential event-response model for **Time to
bill the customer.**

Are the Requirements Really New?

New requirements affect the essential requirements of the system. For each new requirement, you must determine the following:

1. Do I need to change the context of the system because of a completely new boundary data flow? Sometimes, the flow is not really new, but is an existing flow disguised as a new one.

2. Do I need to store new data? Do the attributes exist in my current data model to support the new requirement, or must I create completely new ones?

3. Do I need to define any new relationships? Does the new requirement create or use relationships between new or existing data objects?

4. Do I need to define any new business rules or change any existing business rules? Sometimes, of course, the new requirement is really a modification of an existing piece of policy.

If the answer to any of these questions is yes, then the new requirement is a genuinely new piece of essential policy and you must incorporate it into the specification. If the answer is no, then you either have a duplication of something the system is already doing, or the user has been feeding you implementation constraints. Even though these constraints are not part of the essential requirements, you will want to make sure that you do not forget them. Trap them when they arise by adding them to the appropriate view of the system: the new physical model.

Let's look at the three example event responses modeled in Figures 2.13.3 through 2.13.5. Suppose there is no data flow anything like CREDIT CARD in the current context diagram. Nor are the attributes CARD NUMBER, EXPIRATION DATE, or CREDIT LIMIT stored by the current system. If this is the case, you can be sure that the first event response and the entity CARD are completely new.

In the case of the second event response, we discover that our context already covers the response to *Customer makes a purchase.* So the second event response is a modification of an existing event response for which you have already built a model. The credit cards are new, so anything to do with relating PURCHASE to CARD will modify an existing mini specification.

The third event response, *Time to bill the customer,* is new. The bill assembles the purchases by card, so consider this to be a new event.

Some of the stored data already exist in the system. For example, CUSTOMER must be part of any current system, as must PURCHASE and SERVICE. The new addition to the data model is CARD. So you must update your existing data model and data dictionary to show the entity CARD and the new relationships with the existing entities.

The part of the user's statement concerning the decoration of the charge card is an implementation detail and can be safely diverted to the new physical model.

Changes to the Context Mean ...

One of the first tasks of analysis is to define the context of the system. However, once defined, it rarely remains unchanged. Throughout the life of the project, change is inevitable: The business policy of the enterprise will change; external forces, such as alterations to the law or changes in the business environment, will also affect your systems.

Careful monitoring of new requirements helps to keep your project under control. Whenever there is a change to the context of the system, there must be a corresponding change to the amount of analysis effort. This means the project plan must be altered to reflect that the project is delivering more or less functionality than was thought when the delivery date was decided. The context of the system establishes the analysis effort. If the context changes, so does the effort.

Summary

New requirements are not part of the current system, but must be included in any future system. They may also be modifications to parts of the existing system. New requirements differ very little from any other requirement and are modeled in the same way. That is, event-response data and process models are built for each new requirement.

Similarly, just as you separated the essence from the implementation to build your essential event-response models, you also filter out any proposed implementation for the new requirements. While the implementation is not allowed to influence the essential policy of existing requirements, neither is it allowed to obscure the future ones.

Trail Guide

● Easiest and ■ More Difficult: To practice modeling the new requirements for Piccadilly, go to Chapter 1.14 *Some New Requirements.*

◆ Most Difficult: This was not required reading. We suggest you go back to work in Chapter 1.14 *Some New Requirements.*

✼ Promenade: You have now seen all the analysis of the requirements. We began with the current system and abstracted its essence, and then added the new requirements. Now it is time to change viewpoints again and to consider how the new system is implemented. Go to Chapter 2.14 *New Physical Viewpoint.*

NEW PHYSICAL VIEWPOINT 2.14

Before You Reached Here ...

● Easiest and ■ More Difficult: You have verified the integrity of your Piccadilly models in Chapter 1.15 *CRUD Check*. Now that you know you've captured all the events, it's time to think about the new implementation. In this chapter, we consider the system from the viewpoint of the devices used to implement it.

◆ Most Difficult: This is not part of your journey. Piccadilly's implementation waits for you in Chapter 1.17 *Piccadilly's New Environment*.

❋ Promenade: Now that the real requirements are known, it's time to think of the new implementation. The goal here is to find the most effective way of using the system's environment to implement the essential requirements.

Implementing the Essential Requirements

Your essential model is an abstraction that intentionally avoids anything to do with the implementation of the system. That's so you can understand and specify the real business problem. Now that you have defined the essential requirements, the time has come to bring them into the real world.

The new physical viewpoint acknowledges that people, computers, and many other devices are necessary to make the essential requirements function in the real world. The viewpoint makes the transition from pure essential requirements analysis to the beginning of design. It defines how the key interfaces will be implemented using the available technology.

The new physical model is built using a mixture of external and internal design skills. The external design is comprehensible to nontechnical people, because it defines the behavior of the interfaces from the users' point of view. The internal design, which requires technological expertise, is concerned with the details of how the technology works.

New physical models are built by teamwork. The users, analysts, and systems designers all must bring together their understanding of

- the current environment and how it will be changed by the new system
- the essential requirements, both process and stored data
- the available technology

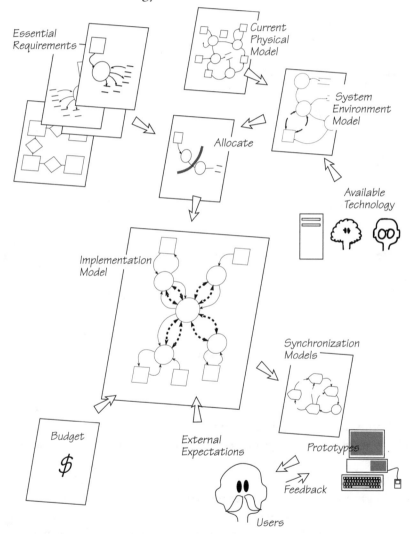

Figure 2.14.1: Building the new physical viewpoint involves several activities that take place more or less simultaneously. The essential requirements are mapped into the system environment by allocating each event response to the most appropriate device. At the same time, the behavior of the interfaces is agreed with the users by means of synchronization models and prototypes. The implementation model is the final picture of the proposed system; but before it can be finalized, the system must conform to the users' external expectations and budget.

Additionally, the users or clients of each development project usually have certain external expectations for the automated system:

- the goal for the automated part of the system
- the media and formats of the inputs and outputs
- the processors to drive the system
- the way the users want the system to behave

What's more, there are always implementation constraints, including

- budget limitations
- deadlines for the delivery of the new system
- expectations about the number of people needed to operate the system

The new physical viewpoint balances these expectations and constraints to make the optimum implementation of the essential requirements.

Defining the System Environment

The system environment is all of the technology—the hardware, software, and people—that is available to the designers of a new system. A system environment model defines the capabilities of each environmental component, specifically the component's speed, capacity, availability, cost, and interfacing capability. This model is the opposite of the essential requirements model in that it ignores application requirements and focuses on the technology. With this model, designers have a specification of the technological constraints. Since the model is independent of any particular implementation, it can be reused by any systems development project working within the same technological environment.

There are a variety of ways to build the system environment model. One method is to use a data flow diagram to summarize the environment, with each bubble in the diagram representing an environmental hardware or organizational component. Then, for each component, you draw a lower-level model to illustrate the software environments that run on the particular component and the technological interfaces between them.

Examples of software environments are programming languages, operating systems, database management systems, transaction processing monitors, word processing packages, and any other proprietary software that runs within your environment. The mini specification for a software environment bubble could be the manual provided by the supplier of that product. The interfaces between the environments can be defined in the form of data dictionary definitions.

Another way of building the system environment model is to simply list all the technology that is available to the system and provide pointers to manuals, documents, and people who have knowledge about the environment.

Consider your own organization and the variety of technology used. You probably already have some kind of informal system environment models made up of technological knowledge held by a variety of people and project groups. The disadvantage of this fragmented approach is that the knowledge is not centrally available, which usually results in duplication of effort. The best way to avoid this is to build centrally accessible system environment models using a standardized approach.

Processors

The portion of a system environment model in Figure 2.14.2 shows several hardware processors that could be used to implement a new system. Other types of possible processors are discussed below.

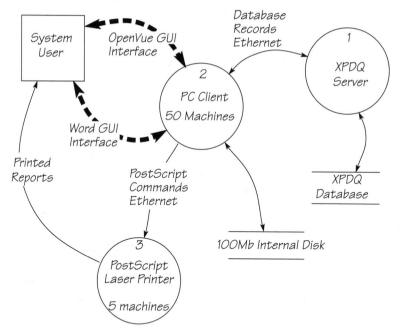

Figure 2.14.2: *This portion of a system environment model identifies the hardware processors and the interfaces between them. Interactive interfaces are shown by the heavy dotted arrows. The interfaces are defined in detail by means of data dictionary definitions or pointers to manuals and examples of their use. The processors can be further partitioned by software environment, as in Figure 2.14.3.*

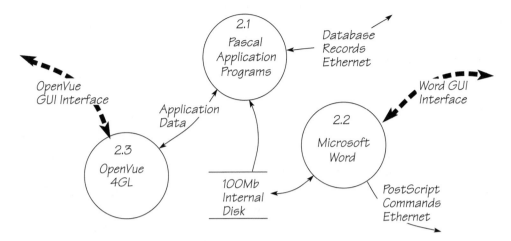

Figure 2.14.3: A more detailed look at the hardware processor called PC CLIENT. Again, every interface and software processor is defined by means of pointers to manuals or examples of their use.

Human Processor

Organizations normally employ people to carry out data processing tasks. They've been doing this for hundreds of years and, despite the arrival of computers, will be doing it for a few more centuries. The human has several advantages over a machine as an implementation processor.

- People are relatively fast to train. It is usually quicker to explain the task to a person than to develop the appropriate software.

- People are good at improvising. When faced with unforeseen circumstances, they can take the appropriate action. However, a computer's rigid program usually does not allow for ad hoc decisions unless the organization has invested a lot of time in building an expert system. For judgmental processes, it is cheaper to have a person do the job.

On-Line Computer

On-line computers are the workhorses of today's implementations. Most organizations want their people to have immediate access to the data and to have processing capabilities (spreadsheets, word processors, reporting, and query facilities) that support their tasks.

A problem with on-line computers is their capacity to adequately support the expected processing demands. It seems that each year brings a several-fold advance in the processing power of computers, but during the same period, the demand for processing increases even more.

Client-Server Architecture

Client-server architecture involves a number of machines. All (or most) of the data are centralized and are held by the central server machine, sometimes a mainframe, but often a minicomputer or one of the more powerful personal computers. Each of the system's users has his own machine, usually a personal computer, which is a client for the server's data. Clients can download central data, or access as much of the data as they need for their localized processing. Client-server architecture has the advantage of freeing the users from the restraints of queuing to use the central facility, while giving them the flexibility to implement their own systems locally without losing access to central data.

Batch Computer

A batch computer doesn't produce its response immediately, and users must wait for their output. Although this is old-fashioned computing, there are situations in which the use of a batch computer is desirable:

- The amount or type of processing is such that people don't want or need to wait for the output.

- The data storage method dictates a batch solution; for example, if large files are being updated sequentially, or if the input is delivered to the system in batches.

- The task should not have an immediate response; for example, waiting for the end of a pay period is preferable to printing a paycheck each time an employee records some hours worked.

Mechanical and Electromechanical Devices

Computers and humans are not the only performers of tasks for a system. Look at any office today, and you see copiers and scanners for processing images; telephones, answering machines, voice mail, faxes, and modems for handling voice and data; envelope openers and stuffers for regular mail; and so on.

Look at the wealth of new electronic devices that are available. Look at the variety of print and broadcast media available, including magazines that cover the wealth of electronic wonders available for data processing and storage. The rapid rise of portable computers has spawned an entire industry of electronic devices that can do an amazing amount of work, and still be carried around in a briefcase or a pocket. It seems that virtually the only limitation in electronics is your budget.

Data Containers

Data containers hold the system's stored data. If the volume of data is large or if the data must be accessed rapidly, there is little alternative to selecting computerized data containers such as databases and indexed files. Manual containers—such as books, papers, and filing cabinets—are used only when volumes of data and access times permit.

The choice of data containers is somewhat constrained, as when the chosen processor doesn't support all of the available data containers. For example, not all databases run on all machines. Also, there may be existing databases that must be used. In this case, your choice of processor may be dictated by the data container.

Data Carriers

Data carriers are the media for transmitting data from one processor to another. The carriers may be screens, paper, data lines, and so on. The significance of choosing the data carrier is that it determines the presentation of the data, as well as how the processor is used. For example, compare the interactions between a person and a computer when the output comes out on paper versus a screen. When the data carrier is paper, the user must submit all the input in batch mode before getting any response from the computer; when using a screen, the user enters data in small chunks and gets a response for each entry.

The terminators that surround the context may also have expectations about the carrier selected to convey their data, both to and from the system. You will almost certainly use a variety of data carriers for your implementation, so you must ensure that you select carriers for the boundary data flows that match the expectations of the terminators.

Allocating Processes and Data

Now that you have defined all the elements that make up the system environment, turn your attention back to the application. Your design task is to map the essential requirements into the available environment.

For each event response, allocate each process to the most appropriate processor, each data store to the most appropriate data container, and each data flow to the most appropriate data carrier.

Stating this rule is a lot easier than doing it. First, what does "most appropriate" mean? For our purposes, we mean the optimum mix of skill, cost, and capacity.

Skill is the ability to accomplish or perform a certain task. Accurate calculations and retrieval of large volumes of data are best done by a computer. Humans are better than machines at making judgments and interacting with human customers. (Of course, if your customers are not human, perhaps another device would be appropriate.) There are many other skills exhibited by other devices that can be assigned to perform certain system tasks.

The cost of a device naturally plays a major part in its selection. While the processing skills of large-scale computers may be attractive, their cost may be beyond the budget for the implementation. When the benefits gained by using a certain level of skill are not worth the cost, then it becomes necessary to settle for a lower skill level. So cost to some extent determines the amount of skill employed for the task.

The third factor is capacity. Before allocating a task to a processor or data container, the system designer must ensure that the processor or container has sufficient capacity for the processing throughput, or quantity of data, both now and in the foreseeable future. Once again, the greatest capacity is almost always associated with the greatest cost, so the designer has to determine the best capacity for an acceptable cost.

With these three factors in mind, the designer must allocate every process, store, and flow to the appropriate device. Allocation involves repartitioning and adding complexity to your model. To help stay in control, use your event partitioning and do the allocation event by event. The result of this work is an allocated event model that is made by repartitioning the essential event-response models to show the various processors that will be used to implement the event. Here is the step-by-step procedure:

For each event, build an allocated event model:

> For each process within the event response,

>> Allocate the process to the most appropriate processor in the system environment using the optimum mix of skill, cost, and capacity. Don't forget that some events will be allocated partly or completely to human processors.

If a process is fragmented between two processors,
    Then add a new interface between the two fragments.

For each essential data flow,
    Add implementation constraints to the data dictionary (volume, frequency, expected growth rate, triggering mechanism).
    Choose a data carrier from the system environment and annotate the data flow.

For each essential data store,

    Add implementation constraints to the data dictionary (volume, frequency, expected growth rate).
    Choose a data container from the system environment and annotate the data store.

For each data flow between two different processors,

    Do the external design and, when necessary, demonstrate it with a transaction synchronization model or a prototype.

Add the physical description of all flows and stores to the data dictionary.

Integrate the event with other events that use the same data. Be on the lookout for similar patterns. For instance, you may find that part of the design for one event response is identical to that for another event response. In that case, you can avoid duplicating your design effort by reusing the common characteristics.

Add the allocated event to the implementation model.

Implementing External Expectations

External design is when you implement the external expectations, those requirements concerning the physical appearance of the implemented system. Using Piccadilly as an example, let's say that management wants the sales staff to be able to see the up-to-the-minute status of the breakchart. To achieve this expectation, you must allocate the breakchart maintenance and the sale of spots to a processor that can display the results on-line to all executives. The display may be a large, centralized one, or it may be done via personal computers on the sales executives' desks.

Note the close link between the external expectations and the data carriers, and between the data carriers and the processors. Whenever the medium of data is specified, the data carrier has to be made available along with the necessary processing capabilities. Similarly, when the external expectations require the system to process two thousand transactions a minute, then it automatically means using one or several processors with the necessary speed to fulfill the expectation.

Introducing the Environmental Processes

While you were building the essential models for Piccadilly, you were able, indeed encouraged, to put aside the harsh realities of the physical world—of corrupted data, incomplete or incorrect data, machine failures, network breakdowns, software glitches, and the several hundred other facts of life that plague automated systems. Now you must deal with these realities.

While you are introducing your perfect essential requirements into this harsh world, you also must introduce the processes that will protect your system from the dangers of the real world. But note the order in which you are doing this. First, determine the business requirement, and then determine how it is to be protected from physical failure. There is little point in being concerned with database recovery procedures before you know what data must be stored.

Many of the environmental processes are introduced after the processors have been selected. For example, if you select two different processors to implement one event response, you will probably want to install a border guard to edit the data that enter the second processor. Just as a border guard may check your passport as you leave a country, you may need to install a border guard process to check for errors as the data leave a process (see Figure 2.14.4).

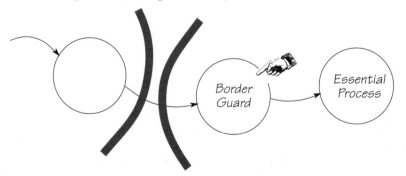

Figure 2.14.4: This portion of an allocated event response shows two processors. The heavy lines are the processor boundaries. The process called BORDER GUARD receives the data flow and edits the flow's contents to ensure that it contains expected and legal data.

In addition to border guards, you'll need postmen to put the data into a form that is understandable to the next processor, and to send the data to that processor. For example, keying data into a computer is a postman process, which exists to convert the data into machine-understandable form, and to transport the data to the next processor.

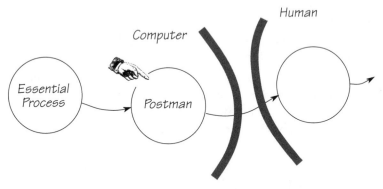

Figure 2.14.5: One processor here is a computer, which works in binary. The other processor is a person who works in English. This necessitates that a postman be installed to translate the computer's output into a human-readable form.

External Design

External design determines the behavior and appearance of the system from the point of view of someone who is not interested in how that behavior is achieved. Typically, external design is concerned with building models of the intended inter-action between two processors, often the end user and a computer.

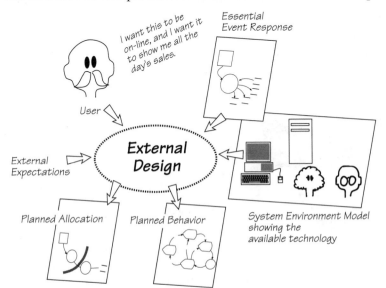

Figure 2.14.6: External design brings together all of the factors that determine the physical behavior of the implemented event response. External design must use the best of the available technology to make the essential requirements work as closely as possible to the users' expectations.

317

External design is closely linked with the selection of data carriers. For example, when screens are selected to carry data from a computer to a human processor, the external design must make the system work optimally using screens. Similarly, the choice of processor has a large effect on the external design.

User Orientation

Perhaps the largest single influence on the external design is the way the users regard each task in the system. Sitting with the users to observe their work is rare for software systems designers, but is normal for other types of businesses. For example, several large U.S. fast-food companies send their head-office executives to the field once a year to work for a few days in one of the outlets. These executives cook hamburgers and French fries, serve customers, and generally get to know that end of the business.

These fast-food companies depend on having good relationships with their customers, and they realize that the head-office executives must understand the retail part of the business if their decisions are to make for better organizational systems as well as better relationships with the customers. The idea of understanding the system from the end user's viewpoint applies to designers of any kind of system. Time must be spent observing the business that the system is intended to support.

For instance, Figure 2.14.7 illustrates how a user thinks a system responds to his wanting to rent a car. The user is concerned with the car model, the date he has to return the car, the agreement for driving the car (he must be a licensed driver), and the form of payment for the car rental. For a systems design to be successful, it should provide for an automated system that imitates this view. In other words, if you are the designer, you must provide an implementation that allows the user to manipulate cars, drivers, payments, and driver's licenses—objects that reflect the user's way of working. Failure to do this usually results in post-implementation modifications to make the system fit the way the user does his work.

Figure 2.14.7: The user's view of a car rental event response.

In another example, our house in London was being renovated several years ago, and we had the opportunity to design a completely new kitchen. The principles of user orientation told us that we could design a suitable kitchen only if we carefully observed how we prepared our food. We knew that a standard kitchen would not be satisfactory, nor would it be acceptable to use other people's observations of food preparation, since they eat different food and use different methods of preparation than we do. A kitchen designed for other people would be unsuitable for the kind of  cooking that we do. Our diet is largely Asian and vegetarian. Meat-eaters would require more frequent access to the oven and stove top, and less preparation area.

Several weeks of watching each other resulted in a list showing the usage frequency of most items in the kitchen. The cook at our house touched the following items in order of frequency of use: chopping board, knife, water, refrigerator, garbage disposal, other cooking implements, stove top, and pan storage. The kitchen was then designed to put the counters, sink, refrigerator, knife rack, garbage disposal, and other tools in positions such that the most frequently touched items would be close at hand. Items used in an oft-repeated sequence were related to one another. For example, the sequence of taking something from the refrigerator, washing it, and then cutting it was repeated often. This indicated that the refrigerator, sink, and chopping board should form a convenient triangle, with the garbage disposal near the cutting surface. We noted all actions, their frequency and sequence, and the correlation of these gave us the design for the new kitchen.

The design produced a kitchen unlike any we had seen before, and very much unlike the ideas we had before the observation exercise. Cooking in this kitchen is a pleasure. To prepare a meal, we need not take unnecessary steps; the most commonly used storage, implements, and facilities are within reach or one or two steps away. Nothing is hard to find, nor difficult to get. But this has not been achieved by making the kitchen small. On the contrary, it is rather a large room. Washing the dirty dishes takes place in a different part of the kitchen from the food preparation. It has its own traffic flow. To get dirty dishes from the table to the dishwasher, to the storage shelf, and back to the table can all be done without interrupting the cook. The in-sink garbage disposal is accessible to both food preparation and to dish cleaning. The kitchen is usable because it was built only after closely observing the intended task.

Eccentricities and Idiosyncrasies

Personal idiosyncrasies relate to the way someone does their task. The task itself is neutral. We strongly advocate that you observe the users' work, so that you thoroughly understand the work and the users' metaphors. But your design should not slavishly imitate the users' every idiosyncrasy.

> Personal eccentricities and idiosyncrasies of the designer or the user must not be allowed to influence the design, nor must anyone's preconceptions get in the way of orienting the software to the user.

Preconceptions can distort implementations. Sometimes, the users don't understand or are unaware of the available technology. For example, users who have only worked with a menu-driven system will tend to see any new system as a menu-driven system. The users who are comfortable with manual systems will expect a computer system to mimic a manual one. The external designer has to eliminate the users' preconceptions by minimizing any existing technological bias.

Road Signs

When we drive along a highway, we expect to see road signs. These signs, typically at intersections, point to various destinations. Now consider the alternative.

Suppose all road signs were removed. Navigation on any but the most familiar roads would be difficult, and most drivers without a map would become hopelessly lost. Why is software any different? We cannot expect users to navigate through computer systems without road signs to guide them. Using an interactive computer system is somewhat similar to driving cross country.

Most transactions in business systems require several smaller tasks to be done in order to complete the transaction. Sometimes, users must choose between alternate tasks, and subsequent tasks usually depend on the results of previous ones. At any stage of the processing, they have a right to expect road signs to inform them of where they are in the process, where they are going, and the available options.

Figure 2.14.8: "From here, turn right to go to Meldon and Whalton. Turn left to go to Needles Hall Moor and Netherwhitton. Straight ahead is Mitford, and behind you is Hartburn." This is a lot of information, but no more than is reasonable to expect from a road sign.

Computer road signs can be implemented by means of intelligent menus that appear on screen. As an example, let's say a user adds a new customer to the database. Before returning to the main menu, the system shows the user a menu of possible actions: Add the new customer to the mailing list, or send the optional

> The system should display what it has to offer, as well as keep the users informed about progress. It should provide road signs.

"Welcome aboard" letter, or quit. These options are much more accessible on the screen, of course, than buried in some user manual. Similarly, screens can display status lines to inform users of the exact state of processing, the results they have gained, and any processing exceptions.

Consistency

Each application written for the Macintosh® computer must follow the interface standard as defined by Apple. A similar situation exists for Microsoft Windows™.

> Be consistent across all systems. Use design templates so that every application reflects the same interface design philosophy.

This means that every piece of software, regardless of who writes it, has the same appearance to the users. Consider the advantages. Mac users need not learn each new piece of software, nor spend time with the manual; they need only open the application and start using it.

This is really quite an extraordinary feat. The thousands of companies that build software for the Macintosh have proved they can make their programs look and act the same. Why is it, then, that within one company user interfaces can be so radically different? Why does the interface get reinvented for each new system, and sometimes for each new transaction?

One way for organizations to address this problem is that they build reusable templates for the use of all their designers. These templates provide application-independent patterns that can be used to design interfaces. The use of common templates speeds up the design as well as the implementation process. The consistency of behavior patterns between systems thus enables users to learn and use new systems more easily.

Behavioral Models

When you are negotiating with your users about the proposed implementation of the system, you will build several *behavioral models* to demonstrate how the users will manipulate the system. A behavioral model is a simulation of the yet-to-be-

321

implemented computer system, and it can be either an automated prototype or a paper model called a transaction synchronization model. An automated prototype mimics the behavior of the proposed system and is built with special software tools that allow for the rapid construction of a system. The user tries out the simulation, then agrees to it, or asks for modifications. In this way, behavioral models allow the user to try out several alternatives before selecting a final version. The behavioral model then becomes part of the specification for the production version.

Automated prototyping software has become very popular in recent years, and does offer some distinct advantages:

- Prototypes help in communicating with users who do not relate well to the more abstract data flow and data models. Those people are more physically oriented and less able to think in abstract terms and thus cannot help much to verify pure policy models.

- Prototypes are extremely useful in demonstrating possible implementations of the system. Once the essential requirements are known, the analyst can build a physical simulation of how these requirements might be automated. In this way the prototype is used to demonstrate the automation interface, and so contribute to the negotiation of the final system.

- Prototypes can serve as an adjunct to gathering requirements. When the analyst builds a prototype of the system and shows it to the users, he is inviting criticism and suggestions. The prototype is a representation of the users' work, and so the users are highly motivated to contribute requirements and suggestions.

Another way to simulate the behavior of the system is to use a paper model called a transaction synchronization model.

Developing the Transaction Synchronization Model

A *transaction synchronization model* is an elaboration of the data flow diagram. It is built by analysts or designers to demonstrate the intended interaction between a user and a machine. Later, it is used as a basis for designing the interactive software.

We've found that nontechnical users can comprehend the transaction synchronization model provided we talk them through it carefully to illustrate the dynamics. The intention is to acquaint users with how they will interact with their future system in order to avoid any misunderstandings that would necessitate changes to the installed system.

The transaction synchronization model is also a transitional model. When dealing with a complex interactive business system, even the most intuitive designer can have trouble making the leap from an essential data flow diagram to a design model. However, it is a relatively straightforward step to convert the synchronization model to a detailed software design.

For example, the transaction synchronization model shown in Figure 2.14.9 portrays a simplified flight reservation system. This model shows that the system starts with the FLIGHT REQUEST SCREEN, which the operator sees when the reservation transaction is selected. He uses the screen to enter the PASSENGER DETAILS.

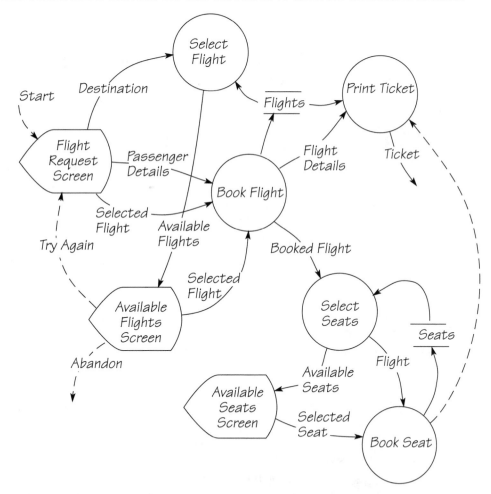

Figure 2.14.9: A sample transaction synchronization model of a flight reservation system. This scaled-down model illustrates the behavior of the system during the time that one event is taking place.

Next, the operator books a flight for the passenger. If the passenger knows which flight he wants to travel on, the operator keys in the SELECTED FLIGHT with the PASSENGER DETAILS. Otherwise, only the DESTINATION is entered.

When the system is given only the DESTINATION, the process SELECT FLIGHT retrieves all flights to that destination showing seat availability. The operator uses the AVAILABLE FLIGHTS SCREEN to search for one that is suitable for the passenger, reading the options to the passenger over the telephone until the passenger selects one. The SELECTED FLIGHT is recorded by the BOOK FLIGHT process. If no flights are acceptable, the operator can return to the FLIGHT REQUEST SCREEN and enter a new destination.

After booking the flight, a seat for the passenger is selected and booked in a similar manner to the flight. When this has been done, the process PRINT TICKET produces a TICKET. The boarding pass showing the seat will be produced later when the passenger checks in for the flight.

Notation for the Transaction Synchronization Model

The *interface process* represents an action carried out by a collaboration between the computer and the operator. The screen symbol is used because the computer has a screen to display its information. This symbol represents the display of information, manual decisions that are part of the process, and the act of entering data into the computer.

The interface process may include elementary data validation. Since most modern programming languages have a capability for on-screen field verification, think of the interface process symbol as representing such simple syntactical editing tasks in line with the screen-handling capabilities of the target language. However, the interface process should not be attributed with other automated processing capacities. For example, file lookups and data transformations are considered outside the scope of the interface process. These should be shown as normal processes, as discussed below.

The computer's screen can remember what has been displayed. Subsequent screens can add to the display and selectively delete information. Each interface process notation need not be implemented as a completely new screen; rather, it may be a variation on a previous display.

A *data flow* in the transaction synchronization model has the same meaning as in a data flow diagram. It carries data from one process, interface process, or data store to another. Note that in the transaction synchronization model, the data flow coming from an interface process means the operator has entered that data. The system selects a process depending on the content of the data flow. In the flight reservation example (Figure 2.14.9), the system activates either SELECT FLIGHT or BOOK FLIGHT, depending on the data entered by the operator. Its ability to activate processes means that the data flow both carries data and has a control component.

Now look at AVAILABLE FLIGHTS SCREEN. The data flow SELECTED FLIGHT emerges from the interface process if the operator finds a suitable flight. If he cannot find one, he enters either TRY AGAIN to activate the starting screen, or ABANDON to terminate the transaction.

A *control flow* indicates that some processing condition has occurred and that the system is to activate the process pointed to by the arrow. The label attached to the control flow indicates the condition. In the above example, TRY AGAIN would result if the operator uses a predetermined signal (menu choice, function key, or something else) through the keyboard.

Control flows give sequence to the model. They are used when it is necessary to invoke a process when data flow is needed. For example, the unlabeled control flow between the processes BOOK SEAT and PRINT TICKET indicates that the ticket is printed only after a seat has been booked. The amount of control shown in the model is varied to suit the reader. You can, of course, make a diagram unreadable by including every possible control flow. Instead, the convention is that if there is a data flow, it is unnecessary to also show a control flow. For instance, the user enters the data DESTINATION on the FLIGHT REQUEST SCREEN. It is true to say that the implemented design will also contain a flow of control from the screen to the SELECT FLIGHT process, but for the purpose of behavioral modeling, the data flow is sufficient. The control notation is used only when it adds to the reader's understanding of the external behavior of the system.

A *process* is the same as in a data flow diagram. When used in a transaction synchronization model, a process is either a fragment of the essential model, or a new process added because of the system implementation environment. Processes can be allocated to either a human, computer, or any other device. If you want to draw the reader's attention to the type of processor that will be used for the implementation, then annotate the process with the name of the device.

A *data store* (enclosed in parallel lines) also has the same meaning as in a data flow diagram. Data stores contain the data used by each process. Sometimes, showing all the data stores makes the model unnecessarily cluttered. In this case, group the data stores into higher-level stores such as are used in the flight reservation system model.

Physical Descriptions of the Data

Now that you are in the preliminary design stage, you can properly think about the implementation of the data. For example, considering the volume, format, size, frequency, and medium of data items is necessary in order to make decisions about their allocation. The physical appearance of the data is now an issue.

Add the physical details to the data definitions in the data dictionary. For example, the data from the Nelson Buzzcott Employment Agency system would be revised as follows:

Applicant = Applicant Name + Applicant Date Of Birth + Applicant Address
 + Date Registered + Salary Required
 * Estimated volume: 500,000. Medium: DB2 databases. *

Applicant Details = Applicant Name + Applicant Address
 + Applicant Date Of Birth
 * Estimated frequency: 100 per day. Medium: on-line
 transaction. *

Applicant Address = * Size: 4 lines, 20 bytes each. *

Applicant Date Of Birth = * Format: yyyymmdd. Earliest year: 1900. *

Applicant Name = * Size: 20 bytes. See Boyd's rules for names. *

Figure 2.14.10: *Data dictionary entries updated with the appropriate implementation information. Be careful to add only those details that are necessary for you to make allocation decisions involving that item, but also include the results of those allocation decisions.*

The data dictionary reveals what data are to be displayed and their format, and the transaction synchronization models specify the operator's manipulation of the data. At this stage, you also need to think about how data are to be presented on the screens. We suggest that you use some kind of screen painter or prototyping tool to build mock-up screens for the users to see how their data will appear. In the case of critical or complicated transactions or transactions whose behavior pattern is difficult to decide, it's a good idea to give the users a chance to give the behavior pattern a detailed test. Build a working model prototype by using a prototyping tool to simulate the behavior you have specified in the transaction synchronization model.

When a data flow from the automated system is allocated to be a report intended to be read by a person, the users need to know how that report will look. You can accomplish this with a standard report layout.

Developing the Implementation Model

An *implementation model* is a high-level view of how the new system will be implemented. It shows the organization of the major processing components of the system—the people, computers, and other devices—and the interfaces between them. This model is also a high-level summary of the allocated event-response models (Figure 2.14.11).

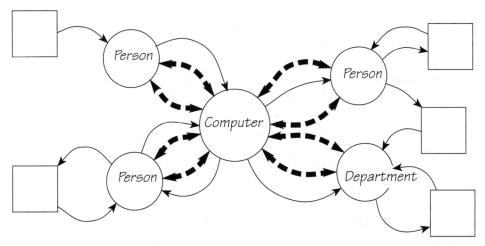

Figure 2.14.11: A typical implementation model, which shows a central computer accessed by the surrounding people. Each processor shown in the model contains a number of processes and sometimes stored data. The two-way heavy dotted arrow indicates an interactive transaction between a human and an on-line computer.

The details of interactions shown in implementation models can be specified in one of three ways: a transaction synchronization model, a prototype, or a data dictionary definition. In every case, the specification component bears the same name as the two-way flow in the implementation model.

Unbroken data flows in the implementation model indicate a batched activity. For example, a data flow from the computer to a person is a report that can be produced in batch mode. Data flows into the computer are also batches of data being entered without the need for any interaction with the operator.

The data flows to and from the terminators are critical. First, they are the boundary data flows from the context diagram; their content cannot be changed without permission from the terminator. Second, their physical medium cannot be changed unless agreed to by the terminator. Third, if you do not specify every detail of a boundary flow and get approval from the users, then no matter how good your system is, it will exist in a vacuum because it will not fit into the real world. The terminators have expectations about how they deliver their data to the system, and how it is delivered to them. For example, a flow of data from a payroll system to a terminator entitled BANK CLEARING SYSTEM is expected to contain specific data formatted in a specified way and carried on a specified medium. So there is no point in designing an on-line data transfer system if the BANK CLEARING SYSTEM terminator is expecting a magnetic tape.

The implementation model is a high-level guide that shows all the different design tasks and the connections between them. For instance, if you see an interaction between an automated processor and a user department, you know there are two detailed design fragments that must eventually connect to each other. The software designer must design some software using the software environments available within his processor. The users will design some procedures using the organizational knowledge within their processor. Eventually, the products of these two very different design tasks must fit together. The implementation model is a way of managing all the pieces of the project.

Checking the Result Against Expectations

Now that you have a model of the system to be implemented, it's time to reflect on whether this is the system that was originally envisaged at the beginning of the project, or whether the requirements analysis revealed an altogether different system. This is also the time to check whether the intended costs and benefits for the development effort will be achieved.

Some of the processes from the event-response models have been allocated to a computer. In *Controlling Software Projects* (see the Bibliography), DeMarco gives a comprehensive method of estimating software development costs based on the data to be processed by the bubbles. Your event-response models already show all the information needed for this estimation technique, so it can be a straightforward task to determine the cost of implementing the software component of your system (so straightforward in fact that we believe this kind of estimation should be provided as an integral part of every CASE tool).

With these estimates, you have a cost of the complete software building effort that is based on small, tangible pieces. If the cost is greater than originally anticipated (it almost always comes as a nasty surprise), you cannot make it cheaper by wishful thinking, nor can you say that you will have to "work smarter" and get it done more quickly. However, one thing you can adjust is the amount of software to be written.

You can do this by altering the automation boundary. If there are fewer processes inside the computer system, it will cost less. You can also negotiate with the users and management for a system containing less functionality.

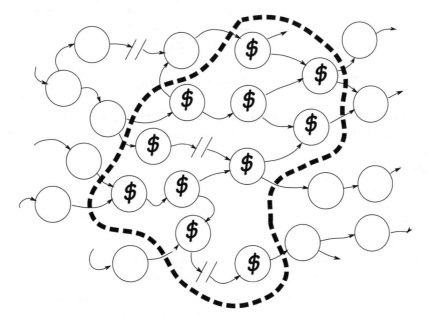

Figure 2.14.12: A straightforward way to determine the cost of the software is by estimating the cost of each process inside the computer.

Another way to reduce the cost of building software is to make use of the fact that in most commercial computer systems, there is a large amount of similarity between transactions. Look closely at all the event responses. How many of them act in a similar way? Ignore for the moment that they use different data, and consider whether it is possible to construct a series of program templates that can be reused, with small modifications, for different transactions. By carefully designing the templates, you will find that reuse can significantly cut your software construction cost.

Now is the time to examine the capacity of the target computers to ensure that they can adequately handle the processing and data storage demands. The total traffic the system has to bear is determined by the number of end users and the frequency of their transactions, taking into account the transactions' demands for stored data.

The capacity of the data containers should also be examined. At this stage, the database designers have enough information to review a database design, and to determine that it will be able to handle the anticipated volumes.

You should audit the external expectations for the system, too, at this stage. Are there expectations that have not been met? Is the cost of implementing them (now that you know what that cost is) beyond the anticipated benefit? Answering these questions will probably uncover discrepancies that must be resolved with the users and management before implementing the system.

Finally, you may want to investigate the emerging discipline of *risk management,* which advocates determining the greatest risks associated with a project. For example, suppose that you have never before implemented a particular database, and yet the whole project relies on its working correctly. Risk management says that the first thing to do is implement the database; if that doesn't work, there is no point in doing anything else. The current literature gives techniques for assessing risks, and we particularly recommend the tutorial edited by Boehm (cited in the Bibliography).

Summary

The new physical viewpoint shows the system that you intend to build. As with the other viewpoints, you select whichever models you need to accurately portray this system, and build as much as you need of them and no more.

A system environment model is constructed to illustrate the technical and organizational details of the environment in which the system will operate. While this model can really be seen as a view of the current system, its use is mainly directed toward determining how the new system will be implemented.

New physical modeling starts by allocating each of the essential event responses to the most appropriate processor and data containers in your system environment. This allocation process determines the interfaces between processors. For each of these interfaces, you need a behavioral model to demonstrate to the users how they will manipulate their automated system. Behavioral models can be transaction synchronization models or automated prototypes. You also need to design its external appearance using a screen painter or report layouts.

The data dictionary definitions for all the flows and stores are enhanced by the addition of physical characteristics and implementation constraints (volume, frequency, and expected growth rate).

The highest-level model of the new physical viewpoint is the implementation model. This model summarizes the implementation by showing all the processors and the interactions between them. The details of the new physical viewpoint are contained in the allocated event-response models supported by the data dictionary, mini specifications, and behavioral models.

Once you know the amount of processing that is to make up the new computer system, you can estimate the cost of implementing the software. Be prepared to negotiate when the costs are greater and the risks higher than anticipated. At this stage, you should also evaluate the high-risk items and implement them first (if you have not already done so).

Although we have presented the new physical viewpoint in procedural terms, we do recognize there are many factors influencing this view. One factor is the time when you build this model. Although we have suggested waiting until all the essential requirements are known, time and resources dictate that most projects need to have activities going on in parallel. While you cannot finish this view until you know all the requirements, you can choose to start it much earlier in the project than we've indicated. It is worthwhile to build parts of a preliminary new physical model as soon as you have a basic understanding of the system. For example, as soon as you understand the intended environment, record it in a system environment model, which can help you in planning the implementation activities and in making early estimates about the implementation. Once you have done some essential modeling, you can allocate selected events to give users an early indication of what is possible. This early model can also be used for gathering implementation ideas as the project progresses.

What to Do
The environment for your next system will be different from any that we can describe, so we feel it unfair to make you go through an exercise to build a new physical model. Instead, we propose to walk you through the construction of the new physical model for the Piccadilly Project.

Trail Guide
● ■ ◆ ✵ All trails: Go to Chapter 1.16 *Strategy: Toward Implementation* to see how this viewpoint works for Piccadilly.

2.15 OBJECT-ORIENTED VIEWPOINT

Before You Reached Here ...

Regardless of your trail, ● ■ ◆ or ✳, you have read about the implementation of the new system in Chapter 1.17 *Piccadilly's New Environment,* and you have arrived here because you decided to follow the optional fork in the trail. First let's review why you should read this chapter. Object-oriented programming languages represent a fundamental change from third- and fourth-generation languages, and their proponents speak of revolutionary ways to build and maintain systems. As the popularity of object orientation is growing and we believe will continue to grow, it is important for you, a systems developer, to understand how it affects your job.

To take advantage of the enhanced capabilities of object-oriented languages, you must use a different design approach. Regardless of the design method, the analysis that precedes it must be done in such a way that you capture all the essential requirements. How? This book describes how you do that.

In this chapter, you'll learn about the meaning of the object-oriented paradigm and how it affects the work of systems analysts and systems designers. In particular, we explain how the object-oriented approach relates to the analysis approach that you have been using in this book.

What Is Object Orientation?

Traditionally, systems have kept their data and their processes separate. This separation began with the first programming languages in the 1940s. The metaphor that language designers had in mind was the clerk or mathematician, whom the computer was intended to replace. They viewed a computer program as the equivalent of a set of instructions that could be given to a person. People kept their data in filing cabinets or books, retrieving the information when needed. So, too, the early programming language designers had the computer keep their data in files separate from the processes inside the computer.

The object paradigm changes this tradition by *encapsulating* items of data along with the processes that operate on that data. This encapsulation is called an *object*. To illustrate, let's think of an object from the Piccadilly Project. Suppose you have an object called SALES EXECUTIVE, which is an abstraction of everything you know about a sales executive: the data that describes the sales executive (the attributes), together with the processes that manipulate the sales executive's data, are packaged together in this object (see Figure 2.15.1).

```
SALES EXECUTIVE
     OBJECT
MY PRIVATE DATA
   Executive Name
   Executive Address
   Executive Review
MY PRIVATE PROCESSES
   Create Executive
   Delete Executive
   Reference Executive Name
   Update Executive Name
   Reference Executive Address
   Update Executive Address
   Reference Executive Review
   Update Executive Review
```

Figure 2.15.1: The SALES EXECUTIVE object is an encapsulation of everything that we know about the object.

Encapsulating the data and processes means that the object is able to function as a self-contained unit, very much like the real-world sales executive. The details of *how* it performs its services are hidden from any other object, although *what* services it offers to other objects are known. Again, this approximates the behavior of the sales executives as you would see them in their natural state. For every sales executive employed by Piccadilly, an object-oriented system would have an instance of the SALES EXECUTIVE object. Each one is an encapsulation of its own data (sometimes called "state"), and its own process (also called "behavior").

Other objects can send *messages* to the SALES EXECUTIVE object requesting it to perform one of its services. Suppose that an ADVERTISING AGENCY object wants a sales executive's name. Such data are private, being hidden from every object except the SALES EXECUTIVE object. How does the ADVERTISING AGENCY object get what it needs? It sends a message to the SALES EXECUTIVE object, requesting the SALES EXECUTIVE object to carry out its REFERENCE EXECUTIVE NAME process and to return the EXECUTIVE NAME to the ADVERTISING AGENCY object (Figure 2.15.2).

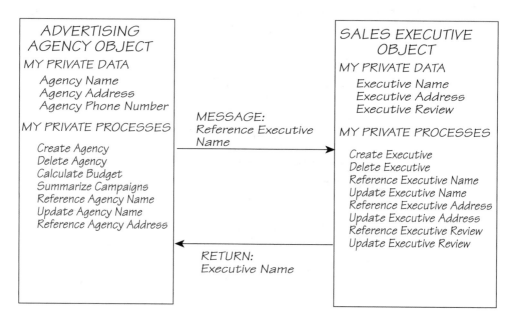

Figure 2.15.2: The ADVERTISING AGENCY *object sends a message,* REFERENCE EXECUTIVE NAME, *to the* SALES EXECUTIVE *object, which responds by carrying out its private process* REFERENCE EXECUTIVE NAME.

In order to perform one of its processes, it might have to send messages to other objects, which might in turn send messages to other objects, and so on, until all of the messages have been resolved. Suppose that another object sends a SUMMARIZE CAMPAIGN message to an ADVERTISING AGENCY object, which has a process called SUMMARIZE CAMPAIGN. It responds to the message by activating that process. Now further suppose that the SUMMARIZE CAMPAIGN process needs to know the average predicted television ratings for the period of the campaign. To get these ratings, the ADVERTISING AGENCY object sends another message, CALCULATE AVERAGE RATINGS, to a PREDICTED RATING object. Figure 2.15.3 illustrates the messages flowing between objects.

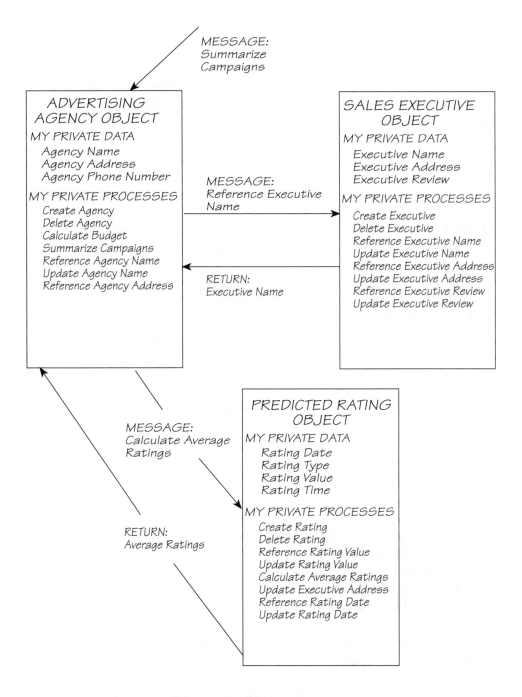

Figure 2.15.3: An example of messages flowing between objects.

If you continued with this example, you would discover that the PREDICTED RAT-ING object and the SALES EXECUTIVE object in turn send messages to other objects. Even with this small example, you can get an inkling of the possible complexity of message flows. Now you can see that it is simplest to think of an object-oriented system as a network, made up of encapsulations of process and data, sending messages to each other.

Classes

Objects mimic their real-world counterparts by being uniquely identifiable. For instance, each SALES EXECUTIVE working for Piccadilly has a unique value for SALES EXECUTIVE NAME. However, even though each SALES EXECUTIVE object has a different value for SALES EXECUTIVE NAME, every instance of SALES EXECUTIVE object has a SALES EXECUTIVE NAME. So we can say that sales executives share the characteristic of having a sales executive name, and this leads to the concept that individual objects with the same characteristics can be categorized into *classes*.

The SALES EXECUTIVE class is a generalization of the data and processes that are common to all sales executives. For example, the SALES EXECUTIVE class knows that every SALES EXECUTIVE object must have a piece of data called SALES EXECUTIVE NAME. For every sales executive employed by Piccadilly, the object-oriented system holds an instance of the SALES EXECUTIVE object. Every new SALES EXECUTIVE object that is added to the system is cloned from the SALES EXECUTIVE class. The class acts as a template to ensure that the new object contains all the data and processes appropriate to an instance of that class.

Let's extend the concept of classes a little further. Another class, MANAGER, has some characteristics in common with the SALES EXECUTIVE class. For instance, both classes need a name, address, employee number, and so on, together with the processes necessary to manipulate those data. Because there are data and processes common to both classes, we can create a more abstract class for the common characteristics. The manager and the executive are both employees of the company, so we can have a more abstract class called EMPLOYEE. We can factor out the common data and processes and assign them to this new super class. We can arrange the manager class and the employee class into a *class hierarchy*. This example is shown in Figure 2.15.4.

The EMPLOYEE class holds the characteristics common to all employees. The MANAGER and the SALES EXECUTIVE classes have characteristics that are unique to their specialized roles in the company.

The advantage of arranging the classes into a hierarchy is that any class can *inherit* the data and process of the classes that are above them in the hierarchy. For example, Figure 2.15.4 shows two subclasses of SALES EXECUTIVE. The junior sales executive has a specialized role, and requires its own unique data and process. However, it also has to have all the properties of a SALES EXECUTIVE, so it inherits them from above. For example, suppose that a JUNIOR SALES EXECUTIVE object receives a message REFERENCE EMPLOYEE NUMBER. The JUNIOR SALES EXECUTIVE doesn't have such a process. So JUNIOR SALES EXECUTIVE simply asks upstairs. Since the SALES EXECUTIVE class doesn't have this process either, it in turn asks above it. The EMPLOYEE class has the required process, and so JUNIOR SALES EXECUTIVE can make use of it.

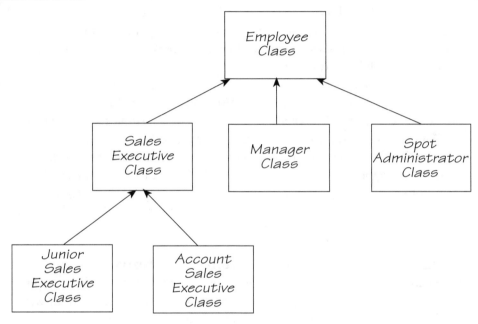

Figure 2.15.4: A class hierarchy. The classes toward the top are more abstract, while the lower ones are more specific.

The benefit of inheritance is that once you have assigned a process or data to the correct class, either can be reused by any class below it in the inheritance. Thus, you avoid repeated implementation of the same data and process. Do it once, and reuse it.

Now suppose your system needs another type of EMPLOYEE, a spot administrator. You can extend the class hierarchy to include the new class. It has its own unique characteristics, such as an authority rating for preempting commercial spots. The class inherits everything else it needs from the EMPLOYEE class above it. Therefore, new classes, provided they are correctly located in the hierarchy, can be added with the minimum of new characteristics.

It is this notion of *extension* that gives the object approach a significant advantage. The existing classes are stored in a class hierarchy, which acts somewhat like a library for future programmers to make use of whatever they find. Object-oriented designers start their work by browsing the class hierarchy, looking for classes that can be used as they are, or ones that can be extended to satisfy the needs of a new application.

As the hierarchy is based on the data used by the system, it tends to be more stable and provide more reusability than libraries of reusable code modules that are based on functionality alone. Should Piccadilly decide to build a system for the Personnel Department, for instance, the developers can reuse the EMPLOYEE, SALES EXECUTIVE, MANAGER, and SPOT ADMINISTRATOR classes that you discovered when analyzing the Piccadilly Airtime Sales system.

The classes we have discussed here are *application classes:* They all are related to the essential subject matter of the business. An object-oriented design environment also provides you with many *environmental classes:* classes relating to the physical side of the business, like window, button, pointer, menu, table, line, page, counter, and so on. The job of an object-oriented designer is to assemble a system making optimum use of the application and environmental classes that already exist. When a class does not exist and cannot be bought, the designer writes a new class and adds it to the hierarchy. This new class becomes available for future systems development.

Relating Analysis to Object-Oriented Systems

As we've said, the object-oriented design strategy is quite different from traditional methods. But what about the analysis of object-oriented systems? To answer this question, we can look to the history of systems analysis.

To repeat, systems analysts have traditionally separated the modeling of process and the modeling of data. These two activities were seen, if not exactly as natural enemies, at best as distant relatives who didn't speak to each other but only met at funerals. Early structured analysts considered (wrongly) that process analysis could be independent of data analysis, and the data analysts believed (again wrongly) that process was something so inconsequential it could be left for the programmers to invent as they chose.

The chasm between the two disciplines was narrowed by McMenamin and Palmer's essential analysis approach, and by the growing awareness of the adherents of Chen's and Flavin's entity-relationship approach that process was important, too. (All three of these references can be found in the Bibliography.) Since our course "Modern Structured Analysis" was first offered in 1984, we have emphasized the importance of both data and process by teaching both types of model. Indeed, the exercises in this book give you first-hand experience of modeling each event response using both process and data models, so that you can see the strong link between the system's process and data and thereby gain a complete and rigorous understanding of the system.

Consider the context diagram that you built at the beginning of the Piccadilly Project. The incoming boundary data flows deliver all of the data to be used or stored by the system. These flows are the source of the objects; they provide the raw data that you formed into entities in the data model (Figure 2.15.5).

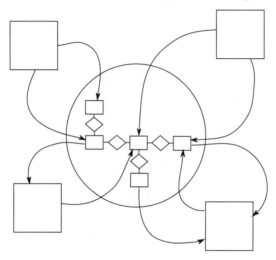

Figure 2.15.5: The system's context. The incoming data flows deliver all the data to the system. You can determine the entities and relationships from the boundary data flows.

What is the difference between the entities in the data model and the classes in an object-oriented system? Very little, once you have collected the processes necessary to complete the encapsulation. Identifying the entities is the first step in discovering the classes for an object-oriented implementation.

Next, let's see how the analysis of processes differs in an object-oriented system. In the Piccadilly Project, you used a combination of current physical modeling and event-response modeling to collect the system's processes (see Figure 2.15.6).

Once the event responses have been modeled, they can be assembled around the entities that they access to make the classes. In other words, the entity becomes the anchor for various event responses. For instance, using the event model in Figure 2.15.6 as an example, EXECUTIVE NAME is an attribute of the SALES EXECUTIVE entity. It becomes one of the data items belonging to the SALES EXECUTIVE class. Along with that, all the process fragments that use the EXECUTIVE NAME (CREATE A NEW SALES

EXECUTIVE, CHANGE THE NAME IF SHE MARRIES, DISPLAY THE NAME, and so on) would become part of the SALES EXECUTIVE class. Meanwhile, other data and process fragments, such as AGENCY NAME and AGENCY CREATE, have nothing to do with our SALES EXECUTIVE, so they would be attributed to another class, probably called ADVERTISING AGENCY. In order to trigger the event response in an object-oriented environment, you would design the message flow between the ADVERTISING AGENCY and SALES EXECUTIVE objects that are concerned with the event.

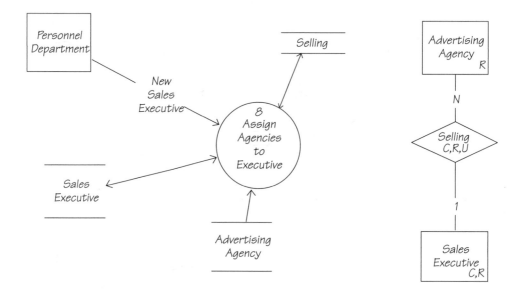

Figure 2.15.6: The event-response process and data models for **Personnel hires a sales executive.**

Analysis in an Object-Oriented Environment

You can start your analysis by identifying classes. This is similar to the strategy you used when you built a high-level or first-cut system data model very early in the Piccadilly Project. The question to ask is whether the classes that you discover are *relevant* to the problem that you are analyzing. This leads to the need for setting your context so that you can verify that a given class or entity type is relevant within the context of your project. This does not mean just verifying that the name of the class is relevant. You need to prove the relevancy of each fragment of data and process. The way to keep control of your analysis is to use a strategy that you are already familiar with: event partitioning.

As you have seen in the Piccadilly Project, your system is made up of a number of small systems, each of which is represented by an event. Event partitioning helps you identify the classes that you already have, identify the classes that you need to build, and design the message flows necessary to support the requirements within your context. Figure 2.15.7 summarizes the strategy to use if you are doing analysis in an object–oriented environment.

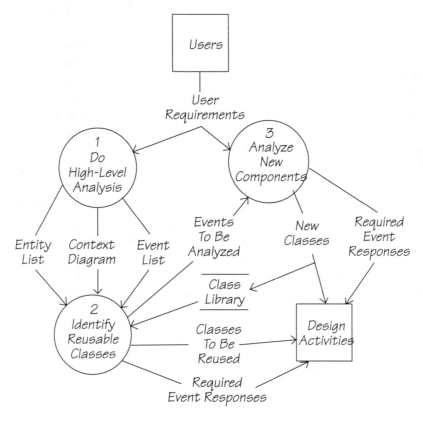

Figure 2.15.7: Use event partitioning to control the identification and production of new classes.

When you do the high-level analysis (process 1), you use the strategy you have been practicing in this book to set the context, make a list of potential entity types, and build an event list. Then you map the results of your high-level analysis to identify existing classes that can be reused to satisfy the requirements of the system within your context (process 2). The events or partial events that are not provided by the existing class library are analyzed in detail (process 3), new classes are added to the class library, and the required event responses are the input for designing the message flow for each event. By using event partitioning, you have the freedom to iteratively analyze, design, and implement parts of the system while retaining control of your work.

Use of the approach that we recommend here will protect you from a number of dangers. First, if you risk doing analysis without defining your context using boundary data flows, you'll certainly have a problem building a relevant and complete system. In our experience, the lack of a well-defined context is the most common reason for project failure.

The second problem area is the essential view. Your analysis technique must allow you to clearly differentiate between the essential requirements (represented by application classes) and the implementation constraints (environmental classes). Without this separation of views, you will perpetuate all of the long-term problems associated with systems that are difficult to understand and maintain.

The third and perhaps the worst problem is the fragmentation of information. Remember that a class is the encapsulation of data and the attendant processes. Now recall the example of the SALES EXECUTIVE class. This class is referred to by events that are the concern of both the Sales Department and the Personnel Department, each department being interested in sales executives for different reasons. They both process data about the executive, but they are probably unaware of the other's processes. In other words, the processing for a single piece of data is distributed. There is no natural encapsulation of data and process, so it's difficult for an analyst to find complete and relevant classes without having some way of controlling the analysis. Our approach to analysis provides you with the control and flexibility that you need.

Object-Oriented Systems Development

To map your essential requirements into an object-oriented environment, you group each entity and some of the relationships in the data model into a class. The fragments of processing from the event-response process models are allotted to a class according to the data that the process uses. The derivation of classes and objects can be organized by applying templates that are tailored to your own environment (see Figure 2.15.8). Templates are a way to make use of the repetitive nature of many design situations. Each template typically is a set of rules that help you to recognize characteristics of entities and to translate them into a design for the class. (Our partner John Palmer and we have developed a collection of analysis to object-oriented design templates as the subject of our workshop "Object-Oriented Design: The Essential Strategy."

While we are on the subject of design, let's step back to look at how design for an object-oriented environment relates to design for other environments. A design path is the name for the kind of design that must be done before the system is built. A developer's design paths are determined by the types of technology within the system environment. For example, if the environment contains a number of people with different skills and an object-oriented programming language, the designer must take both the organizational design path and the object-oriented design path. In other words, the design technique varies according to whatever is being designed. However, no matter which or how many design paths are taken, the designs must be coordinated and connected before the new system can work.

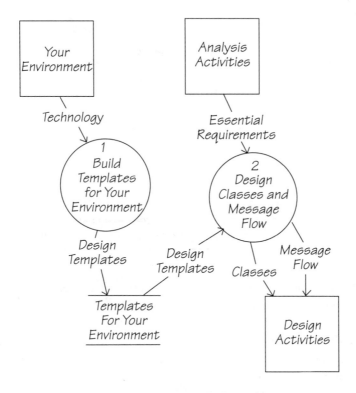

Figure 2.15.8: Templates specific to the system environment are used to derive classes and message flow from the essential requirements specification.

In this chapter, we have discussed how essential requirements relate to object-oriented systems. As shown in Figure 2.15.9, the essential requirements, developed according to the strategy you have been using in this book, provide a complete and precise requirements specification to design using any combination of technologies.

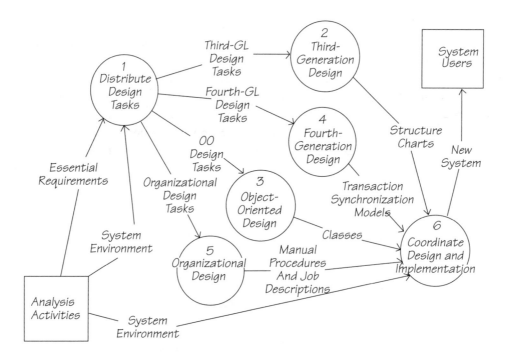

Figure 2.15.9: *Note that the analysis is the same regardless of the implementation, whereas the design path differs according to the types of technology within the system environment.*

Summary

The overwhelming advantage of object orientation is reuse, hence increasing the speed of building and changing systems. The object, a logical encapsulation of data and process, provides a well-specified component that could be reused in other systems interested in the same subject matter. This chapter is mainly concerned with discussing the object-oriented paradigm and illustrating how our approach to analysis maps with that paradigm.

We've not attempted to treat the object view in great detail. There are several good books on the subject, and we can't do justice to this important subject in a single chapter. As the analysis thinking for object-oriented systems is the same as for any other kind of implementation, we chose to concentrate in this book on analysis and refer you to the Bibliography for more object-oriented reading (see Booch, Jacobson, Rumbaugh et al., and Shlaer and Mellor).

Trail Guide

● ■ ◆ ✳ All trails: Go to Chapter 1.18 *Analysis Strategy.*

SECTION 3

Project Reviews

REVIEW: START WITH THE CONTEXT 3.1

Before You Reached Here ...
You have done the problem set in Chapter 1.2 *Start with the Context*.

The Context Diagram for Piccadilly Television

The context diagram that we derived from the description of Piccadilly Television is shown in Figure 3.1.1. Compare your diagram with it.

The Case of the Missing Users

In some cases, we've interpreted something that was said in a user interview or statement, and consequently your names may vary from ours. As long as the intention is comparable, though, your answer is acceptable. Of course, when you're doing real-world systems analysis, you'll need more than just good intentions. You'll need the users to help you out.

Only the users can confirm the factual substance of your models. The goal of analysis modeling is to have a close collaboration between the users and the analyst. Because you are working from a book and there are no users sitting with you at your desk, you have to compensate. For this Project, concentrate on the form of your answer and its intent. There are no users present, and so your interpretation of the substance is, within reason, as good as ours.

The Boundary of Your Project

Probably the hardest part of the exercise is determining precisely what is inside and what is outside the system. When you draw the context diagram of the system, you are determining exactly what you are going to study. For our solution, we looked for those things that Piccadilly controlled and that could be studied reasonably by an analyst. We also chose to study parts of the business that we thought we could affect if we devised some way to improve the process. We declined to study anything that was owned by other organizations or any part that we couldn't influence or improve.

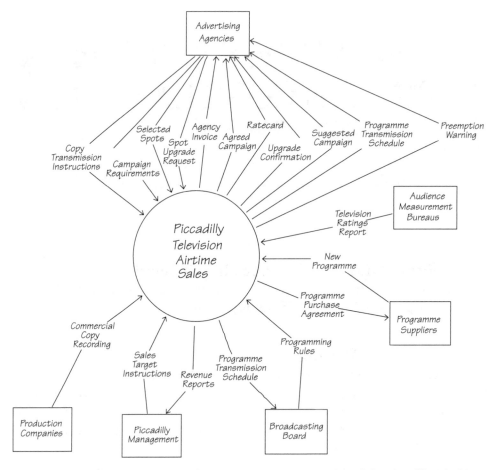

Figure 3.1.1: Our interpretation of the Piccadilly Television context diagram, which summarizes the scope of the Project. The names used for the terminators and data flows are, as much as possible, those used in the text.

Interpreting the Business

You had a written description of the operations of Piccadilly as input to the problem. Like all pieces of natural language prose, it was ambiguous. When you drew the context diagram, you imposed an interpretation on it. Whether or not you made the right interpretation is now to be decided by you and your users. By showing the users your context diagram, you are saying, "This is where I think the project boundaries are. These are the data flows that are at the limits of the system's responsibilities. Now let's go through each of them to see if we all share the same understanding of the scope of this system."

For example, the information communicated between the advertising agencies and Piccadilly sales executives may have been ambiguous as it was presented in the text. On the other hand, the context diagram states very clearly that the ADVERTISING AGENCIES send CAMPAIGN REQUIREMENTS, SELECTED SPOTS, SPOT UPGRADE REQUEST, and COPY TRANSMISSION INSTRUCTIONS to the Piccadilly system. This means that the system you are studying begins its responsibility when it receives these data flows. So you are declaring that you intend to study how the sales executives process these data. If you and the users do not intend to study this part of the system, you must remove these data flows from the diagram.

You proceed the same way for the data flows leaving the context. The data flow AGREED CAMPAIGN leaves the system and goes to the ADVERTISING AGENCIES. The diagram states that your analysis will study how that flow of information was created. However, the responsibility of the system ends once the data making up AGREED CAMPAIGN have been produced. What the agencies do with the data is outside the scope of your study.

The text mentions that the ADVERTISING AGENCIES communicate with film PRODUCTION COMPANIES. The agencies must be telling the production companies which commercials to make and where to send the commercial copy. Why doesn't this communication show up in the context diagram? Consider the other data flows between these two terminators and your context. They all point to the fact that the commercial has already been made by the time Piccadilly hears about it. If your system were interested in the production of commercials, these boundary flows would be very different. So you must rule out any interest in the communication regarding the production of commercials. Your interest in the PRODUCTION COMPANIES is limited to the fact that they are the ones that send COMMERCIAL COPY RECORDING to the system. Your diagram states that you intend to study the processing that takes place after Piccadilly has received this data flow.

It's the Message, Not the Medium

Did you include the data flow PREEMPTION WARNING in your diagram? You should have. Recall that the sales executive telephones the agency to warn his contact about a possible preemption. Even though the data are traveling by voice and telephone line, the information is still a data flow. You will have to find out what processing and data are used to create this warning.

Your context must include all of the interfaces between your project and the terminators, not just those interfaces in the form of documents or other tangible media.

Internal or External?

The terminator PICCADILLY MANAGEMENT receives REVENUE REPORTS and produces SALES TARGET INSTRUCTIONS. Our diagram says that you will study how the revenue reports are produced and how the sales target is used. By treating PICCADILLY MANAGEMENT as a terminator, you are stating that you will not study either what managers do with the revenue reports or how they set the sales target.

> The context of your study is not defined by the terminators; it is defined by the data flowing to and from the terminators.

Although PICCADILLY MANAGEMENT appears as a terminator, that doesn't exclude some of the processes within your study from being carried out by the same managers. For instance, you heard that managers are concerned with setting new rates so that a new RATECARD can be sent to the agencies. When you study the Project in detail, you'll find that management is involved in one of the processes inside your context of study.

Your interpretation may well differ. Suppose, for example, you decided that your project will study the rules for setting the sales target and using the revenue reports. Then, both data flows, SALES TARGET INSTRUCTIONS and REVENUE REPORTS, would disappear from the context diagram. They would become data flows inside the system, and they would reappear when you studied the details of the project.

Which of the above interpretations is right? The answer is they both are. Almost all of the information you receive when analyzing a system can be understood in more than one way. This problem of multiple interpretations is the reason to build a context diagram. The context diagram totally eliminates the ambiguity because it can be interpreted in only one way. While this doesn't make the diagram correct every time, it does mean that because the users can see your precise understanding of the system, they can either agree that you show it as they mean it, or disagree and ask you to change it. The point is that you make it possible to raise questions and to work toward a consensus.

No doubt there will be changes to the context as work on the system progresses. New requirements will be added to the system, while other requirements, previously thought indispensable, will be deleted. Many adjustments will result when you start to do the detailed analysis and everyone realizes just how much work is involved. The only real constant is that the context diagram will continue to display the precise boundaries of the system. How and where the boundaries are set is negotiated by you and the users.

Naming the Flows

Always try to name the data flows with the names the users know them by, since it makes your models much more recognizable to the users. However, sometimes you'll find names with buried meanings. For instance, the first time we talked to the Piccadilly users, they referred to the "green book." Since that is not too informative, we needed to clarify the meaning: "What do you mean by 'green book'?" we asked. "What do you use it for? What data does it hold? Show us a green book." The buried meaning turned out to be the data flow TELEVISION RATINGS REPORT. (The book, as we found out, wasn't even green at all. At some time, the procedure changed and the ratings came in a yellow binder. This did not, however, cause people to change their terminology in the least.)

> Try to use the same names as the users do, but make sure the names reflect the real business purpose of the data flow.

Were you tempted to abbreviate the names? Maybe you used C REQ instead of CAMPAIGN REQUIREMENTS. Don't do it. Otherwise, you create your own buried meanings just as obscure as some of those the users have invented. Remember, your objective is to understand the business requirements so that you can raise relevant questions. You want to uncover misunderstandings before a lot of detailed work has been done. That means adapting to the language of the business you're studying.

With a general description of the users' business, you've turned it into an unambiguous statement of the Project's context of study. You now have a well-defined starting point. Later, your detailed analysis may reveal things that change the context. However, the context diagram helps to keep the Project under control. Any change to the context means a change to the size of the Project. Each change can be measured for its impact on the Project, as we'll discuss.

✚ Ski Patrol

The ✚ Ski Patrol can help if you are having trouble with the Piccadilly Project, especially with the technical aspects of the model. If you are satisfied that your diagram conveys your intended meaning, that a user could understand it enough to question it, and that you can produce a reasonable context diagram for your next project, you have no need of the patrol. Proceed directly to the Trail Guide below.

If you have a technical problem with your model, read on. First, your model should have the same form as ours. There should be only one bubble, and about the same number of terminators as in ours. If not, we suggest you review the discussion in Chapter 2.2 *Data Flow Diagrams* about context diagrams. If your model shows several bubbles, you have already started to partition the system. While this is not necessarily wrong, our experience has shown that it always pays to have the context reasonably decided before breaking the system into its components. At this stage, you should be concentrating on the boundary data flows, instead of the functional areas.

We trust that you've not committed the crime of leaving any of your data flows unnamed. If any of the other data flow conventions gave you problems, again, return to Chapter 2.2 *Data Flow Diagrams,* work through the exercises, and study the sample answers. There will be more on data flow diagrams later in the book (Chapters 2.6 *More on Data Flow Diagrams* and 2.7 *Leveled Data Flow Diagrams*).

Trail Guide

● Easiest: Go to Chapter 2.3 *A Variety of Viewpoints* for a discussion of the viewpoints used to build analysis models.

■ More Difficult: Go to Chapter 2.3 *A Variety of Viewpoints.* This trail assumes that you already know how to build models, so it focuses on the viewpoints that you use to build them.

◆ Most Difficult: Go to Chapter 1.3 *What About the Business Data?* for more work on the Piccadilly Project.

✳ Promenade: The Project chapters are not on your trail, but if you have found this interesting, you might consider switching trails. Otherwise, you can pick up your own trail in Chapter 2.1 *Analysis Models.*

REVIEW: WHAT ABOUT THE BUSINESS DATA? 3.2

Before You Reached Here ...

You have built the first-cut data model for Piccadilly Television. This chapter gives you a sample answer and a discussion of how we derived our model. While we asked you for a first-cut model, we have cleaned up the model to an extent by eliminating entities and relationships that you may have included on your first pass.

The important task in this chapter is not for you to see if your model is a precise match to ours, but for you to be able to rationalize all the decisions that you made to build your model and to understand our reasoning. In some cases, we admit we have cheated. Since we have already analyzed this system, we possess background knowledge that you don't have. Even if we had tried to give you all this background knowledge in the problem statement, there still would be variations between your model and ours. It's more important that you make a model of how you interpret the policy, and that you raise questions for the users, than to simply grind through a mechanical translation into a data model. In doing systems analysis, you can never get a thorough and complete statement the first time. However, once you have a first-cut model, you do have some control over the analytical tasks of questioning and improvement.

Looking for Potential Entities

We gave you a rule of thumb that says when you have a description of the business, look for nouns to indicate potential entities. Let's apply it to part of "The Story of Piccadilly Television" from Chapter 1.2 *Start with the Context.*

"The advertising agencies buy commercial spots that make up campaigns to advertise the products they represent. Each agency sends its campaign requirements to the Piccadilly sales executive who deals with that agency. The executive then models the campaign by selecting commercial breaks for the spots to occupy that will be profitable to Piccadilly, and that will deliver the required ratings to the advertiser. When the executive is satisfied with his selections, the suggested campaign is communicated to the agency. The agency responds by selecting spots from the executive's suggestions and informing him of the choices. The executive finalizes the deal by sending the agency written confirmation of the agreed spots that make up the campaign."

The nouns in the description are *advertising agency, commercial spot, advertising campaign, product, campaign requirement, sales executive, commercial break, rating, advertiser, selection, suggestion, choice,* and *confirmation.* Note that in this listing, we have expressed the nouns as singular and used more descriptive names for *advertising agency* and *advertising campaign.* So let's use these nouns as names for potential entities.

Are These Entities Relevant?

Testing for relevancy means asking if the system has a genuine business need to remember the facts that describe that entity. Ask, for instance, if there is a legal reason for storing those facts. Are they used by people or computers to do their jobs? Are they retrieved for reports? To illustrate a test for relevancy, you could ask if Piccadilly needs to remember anything about the potential entity ADVERTISING AGENCY. You already know that Piccadilly sells commercial airtime to the agencies, so there is a need to remember who the customers are. At the very least, Piccadilly is interested in the name and address of the advertising agency in order to mail the invoices. Given this need to remember, add the entity ADVERTISING AGENCY to your data model. You can test this entity by looking at the boundary data flows for data elements that must be remembered about agencies. If there are any, these data elements become attributes of ADVERTISING AGENCY, and confirm that this is a relevant entity.

The rule of thumb gave us a potential entity called CAMPAIGN REQUIREMENT. These are sent by the agency to the executive who uses them to plan and negotiate the campaign. Once the campaign is settled and the agency has agreed to the spots, there is no requirement for the executive to remember the agency's original requirements. (You would have to get this decision from the users.) Any facts that must be remembered are attributes of ADVERTISING CAMPAIGN (there is a need to remember campaigns), so CAMPAIGN REQUIREMENT fails the relevancy test and can be omitted from the model. Similarly, ADVERTISER, SELECTION, SUGGESTION, CHOICE, and CONFIRMATION are potential entities that the system has no reason to remember. They fail the relevancy test and they, too, are omitted from the data model.

The entities we determined to be relevant from this test are shown in Figure 3.2.1. The next step is to find any relationships between these entities.

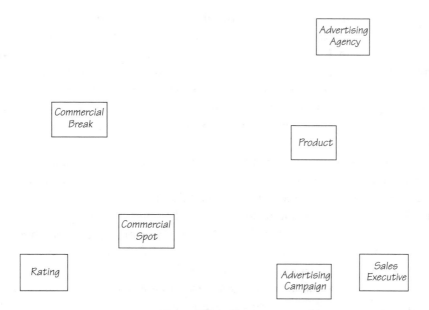

Figure 3.2.1: These seven entities are a starting point for the data model. Detailed analysis will specify the precise use of each entity and will add to and refine the data model.

Finding Relationships

Our rule of thumb also says that verbs in business descriptions indicate relationships. If we look again at the same paragraph:

"The advertising agencies buy commercial spots that make up campaigns to advertise the products they represent. Each agency sends its campaign requirements to the Piccadilly sales executive who deals with that agency. The executive then models the campaign by selecting commercial breaks for the spots to occupy that will be profitable to Piccadilly, and that will deliver the required ratings to the advertiser. When the executive is satisfied with his selections, the suggested campaign is communicated to the agency. The agency responds by selecting spots from the executive's suggestions and informing him of the choices. The executive finalizes the deal by sending the agency written confirmation of the agreed spots that make up the campaign."

The verbs and gerunds (words ending in "ing" that are the noun form of a verb) in the above are *buy, make up, advertise, represent, sends, deals, models, selecting, occupy, deliver, satisfied, communicated, responds, selecting* (again), *informing, finalizes, sending, make up* (again). Let's treat each of these as potential relationships, and test them for relevancy. For each relationship, test its relevancy by asking, "What is the reason for this relationship? Does the system need to remember this relationship?"

Examine the first potential relationship: "The advertising agencies *buy* commercial spots ..." First, let's discuss the reason for the relationship. You can see in the context diagram (Figure 3.1.1) a number of data flows between advertising agencies and the system. CAMPAIGN REQUIREMENTS, SELECTED SPOTS, and SUGGESTED CAMPAIGN exist because the agency is deciding to buy some spots. Selling spots to the agencies is a necessary part of Piccadilly's business, so you can say that the reason for relating an agency to its commercial spots is that the agency is buying them.

The context diagram also yields the answer to the second question about the need to remember the relationship. After transmitting spots, Piccadilly sends an AGENCY INVOICE to the agency. The invoice lists the spots the agency is buying. Because Piccadilly must be able to identify which agency is buying which transmitted spot, this establishes the need to remember a relationship between agency and spot. The result of the relevancy test is that now you can say an agency and a commercial spot participate in a BUYING relationship.

These verbs—*sends, deals, models, selecting, satisfied, communicated, responds, selecting* (again), *informing, finalizes, sending*—did not pass the relevancy tests. "Each agency *sends* its campaign requirements to the ..." You've already omitted campaign requirements from your model, so there can be no relationship with it. "The executive then *models* the campaign by *selecting* commercial breaks ..." Does Piccadilly have a reason to remember which executive models a campaign, and which modeling effort selects which commercial breaks? It is highly unlikely. However, there is a strong reason for remembering which spots *make up* the final composition of a campaign, but modeling the campaign is a temporary state, and not worth remembering. Similarly, Piccadilly has no need to remember which breaks are part of the executive's model: They aren't invoiced, and sales executives cannot reserve spots for their clients. Similarly, *communicated, satisfied, selecting, responds, informing, finalizes,* and *sending* are all transitional in nature, and there appears no good reason to remember them. If you have included any of these in your model, don't be too worried at this stage; remember that these are still *potential* relationships. Later analysis will confirm or deny the need for them and discover any that are missing, as you'll see later in the Project.

After examining all of the potential relationships, we have added the relevant ones to the model (see Figure 3.2.2).

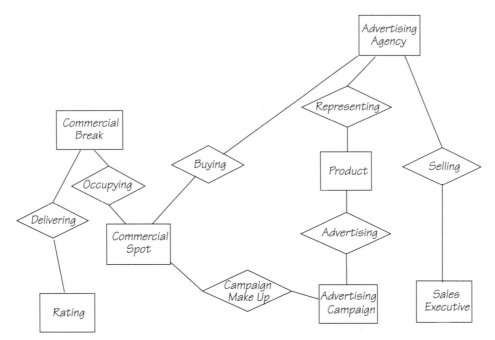

Figure 3.2.2: The data model with relationships added. We derived the relationships initially by identifying the verbs and gerunds in the Piccadilly description, and then testing each one for relevancy.

Adding Cardinality

Adding cardinality to each relationship in your data model is a way of learning more about the data. For each relationship in the data model, ask this question of the entities at either end of the relationship: "For one instance of this entity, how many of the other entity can participate in this relationship?"

For example, you know that the "advertising agencies buy commercial spots that make up campaigns to advertise the products they represent." This tells you that for one agency, there are potentially many spots. From your knowledge of Piccadilly, you know that one campaign is made up of many spots, and you suspect that one agency has many products. If you question the entity at the other end of the relationship, you may find that one spot can be bought by only one agency. (If another agency wants it, the second agency must pay a higher price and preempt the first agency.) Also, a spot may belong to only one campaign, and a product is represented by only one agency at a time. Piccadilly keeps a credit history on previous purchases by agencies that owe money because Piccadilly needs a way to follow up on bad debts.

Only one sales executive is assigned to an agency, and although an executive may have more than one agency, you are not given any evidence of it. If you ask the executives, they each will confirm that they do indeed deal with more than one agency. You can reasonably assume that a product will have more than one campaign, and that one break will carry several spots. Breaks are two or three minutes long, and Piccadilly has to ensure that the executives fill the breaks with spots.

The model in Figure 3.2.3 shows the cardinality that results from testing the existing relationships.

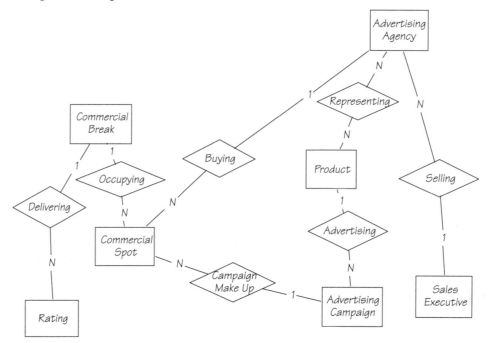

Figure 3.2.3: To determine the cardinality of relationships, ask how many of each entity participate in a relationship. Knowing the cardinality gives you a better understanding of the data.

Building a data model always raises interesting questions, and you can learn a lot about the system from answering them. For example, we said that a product probably has more than one campaign, but can an advertising campaign be for more than one product? The answer is no. Piccadilly considers it too difficult to get paid for multi-product campaigns. While an advertiser may use copy that promotes several products, Piccadilly writes its contracts as if there is only one product.

Can a product be represented by more than one agency? The answer is that only one agency can represent a product. However, sometimes a product changes its agency. In that case, Piccadilly keeps track of the agency that previously represented the product because of the need to collect all its debts.

Can an agency be in a selling relationship with more than one sales executive? The users' answer to this one is no. Piccadilly management is only interested in which sales executive has the current responsibility for selling to an agency. The answers to these questions show up as cardinal operators in the data model.

So far, the data model has defined only a subset of your knowledge about Piccadilly. Let's look at some of the other parts of the policy statement from Chapter 1.2 *Start with the Context.*

"Piccadilly produces some of its own programmes, and buys others from a variety of programme suppliers both in England and overseas. These programme suppliers inform Piccadilly of their offerings, which include first-run films, sporting events, documentaries, talk shows, and old movies. Some of the programmes, such as the talk shows and documentaries, may be a series with a number of episodes."

Now you have some more entities: PROGRAMME SUPPLIER, PROGRAMME, and EPISODE. Not all programmes have episodes. For example, a movie is a stand-alone programme. However, television broadcasters like to show movies under an umbrella programme name like "Prime Time Movie," "Movie of the Week," and so on. Therefore, it's easier to think of *all* programmes as having a number of episodes. The entities and relationships for this piece of the policy are shown in Figure 3.2.4.

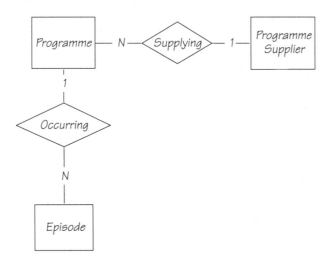

Figure 3.2.4: These entities and relationships show what Piccadilly remembers about the programming part of the system.

Let's go on: "Piccadilly's programme schedulers have the complicated job of decid-ing the date that each programme should be transmitted, and where in the pro-gramme the commercial breaks should be placed. To make these decisions, the schedulers use the weekly ratings that are supplied by the audience measurement bureaus and that tell them how many people are watching which programmes. The schedulers must also follow the Broadcasting Board's rules for placement of pro-grammes and for the number and placement of commercial breaks within those programmes. Four times a year, the schedulers set a new programme transmission schedule for the coming quarter."

There is nothing in this description to tell you why this system needs to remember anything about the programme schedulers. Perhaps another system such as payroll may need to, but unless this system is to report on schedulers' perfor-mance or suchlike, then you can eliminate the scheduler as an entity. The result of the schedulers' efforts is the quarterly schedule. While there is a need to remember it, the published schedule is made up of the programmes and commercial breaks that the schedulers have decided. As there is only one schedule for this system, and as we already have entities and relationships that provide all the necessary informa-tion (PROGRAMME, EPISODE, COMMERCIAL BREAK), there is no need to have an entity called "schedule."

The term "programme" is being used loosely here. If we are thinking there are many episodes of each programme, it's the episode that is broadcast, not the programme. This means that COMMERCIAL BREAK relates to EPISODE, as the break's proximity to the programme content of the episode is what attracts advertisers. The EPISODE will be one of many broadcasts on a given date.

The schedulers use RATING to anticipate the audience that each episode will attract. If an existing programme is continued in the new quarter's schedule in the same time slot, its ratings will be similar, with allowances made for seasonal factors (ratings are higher in the fourth quarter of the year), and for competing channels' programmes. For new programmes, the schedulers look at the actual ratings of simi-lar programmes to work out the predicted ratings. The sales executives use the pre-dicted ratings when they are selling commercial spots. So there are two types of rat-ings: predicted and actual. The upshot of this is that RATING relates to EPISODE, and that RATING has two subtypes: PREDICTED RATING and ACTUAL RATING. From this part of the business, we deduce the entities and relationships shown in Figure 3.2.5.

The entity DATE isn't shown in Figure 3.2.5, and you may not have it in your model. After all, the date of transmission could be an attribute of EPISODE. However, the users will tell you that there is a "day of transmission" or, as Piccadilly calls it, "breakchart day." Usually, the station starts its broadcasting day with the morning talk shows, and finishes some time in the early hours of the following morning with the "Late, Late Show." (We are just as perplexed as you why something in the early hours is called "late.") While this constitutes a *day's* transmission, it actually covers two *dates*. Look again at the ratecard in Chapter 1.2 (Figure 1.2.1) and answer this: If an advertiser buys a spot in the late Friday segment, and the spot is not transmitted until after midnight, what rate does he pay? The answer is the weekday 23.40-to-close rate. So the breakchart day is different from the actual date. We thus change the name of this entity to make it more appropriate to the system.

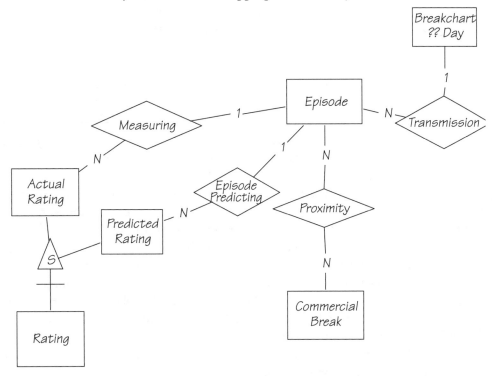

Figure 3.2.5: A partial data model for the scheduling part of the system. Note that some of the entities in this model have already appeared in other partial models.

How do you show the rules provided by the Broadcasting Board in a data model? One approach is not to show them: They could be considered as part of the scheduling process policy, and not as stored data at all. However, in this case, the PROGRAMMING RULES are shown in the context diagram as a data flow entering the system. So there must be a process like the one in Figure 3.2.6.

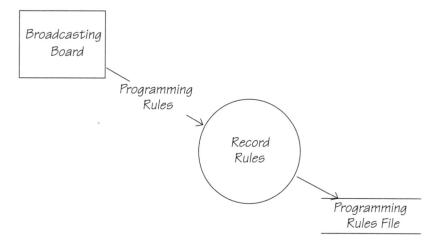

Figure 3.2.6: *The programming rules from the Broadcasting Board are stored for use by the schedulers.*

You can safely assume that the rules will change from time to time and that there may be special rules that apply to special seasons. So it makes good business sense to have the rules remembered. This gives an entity that we call BROADCASTING RULE because it is concerned not just with programmes but also with commercial breaks. But to what does it relate? The statement "The schedulers must also follow the Broadcasting Board's rules for placement of programmes and for the number and placement of commercial breaks ..." indicates that the rules relate to the scheduling of breaks within and around programmes. In other words, an entity relates to a relationship. This seems a little mind-boggling, but we can neatly solve the problem by having a three-way relationship between the participating entities (see Figure 3.2.7). Similarly, the rules apply to scheduling episodes of programmes, and so they can participate in the SCHEDULING relationship. The way to make sense of this relationship is to view it in three different ways:

- For each instance of one EPISODE and one COMMERCIAL BREAK, there are many instances of BROADCASTING RULE.

- For each instance of one COMMERCIAL BREAK and one BROADCASTING RULE, there are many instances of EPISODE.

- For each instance of one EPISODE and one BROADCASTING RULE, there are many instances of COMMERCIAL BREAK.

As you can see, *n*-ary relationships are more complex to understand than binary relationships. Often, they indicate that an analyst doesn't really understand a piece of policy, so he has related a clump of entities to each other. If this is the case, you'll find it particularly difficult to give the relationship a meaningful name. The best advice we can give you is always look for binary relationships first. If you cannot define your meaning using binary relationships, that is an indication you have a real *n*-ary relationship.

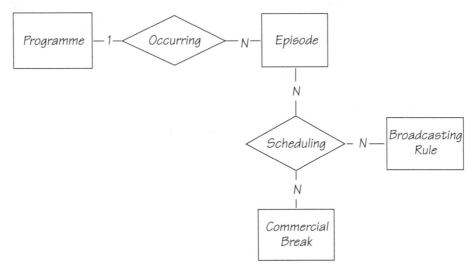

Figure 3.2.7: The programming rules participate in the relationship between the breaks in and around an episode, and the occurrence of episodes of a programme.

Bear in mind as you read through our explanation that your model does not have to agree entirely with ours. After all, you may have interpreted the statement differently and anticipated different answers from the users. The point of this discussion is to show you the data model that results from one interpretation of the business policy.

Let's go back to the context diagram in Figure 3.1.1 to see the details of what data enter the system and need to be remembered. Take the data flow COMMERCIAL COPY RECORDING, which comes from PRODUCTION COMPANIES. The system needs to remember the copy and where it came from. This gives two entities: COMMERCIAL COPY and PRODUCTION COMPANY, with a FILMING relationship between them. The COMMERCIAL COPY entity must also relate to ADVERTISING CAMPAIGN. Now consider this from "The Story of Piccadilly Television": "Some advertisers use several different commercials in a campaign, and the agency must send instructions on which copy is to be transmitted in each spot." The boundary data flow COPY TRANSMISSION INSTRUCTIONS delivers the data, which must be remembered. Digging a little deeper, we will discover that the reason for the copy instructions is to make sure that Piccadilly transmits each copy in a defined sequence. So Piccadilly saves the dates in the COPY TRANSMISSION INSTRUCTIONS as attributes of the COMMERCIAL COPY entity. The spot manipulators use the copy transmission dates when doing the ALLOCATING of COMMERCIAL COPY to COMMERCIAL SPOT.

There are two relationships, OCCUPYING and TRANSMITTING, between COMMERCIAL BREAK and COMMERCIAL SPOT. The reason is that OCCUPYING represents a temporary placing of the spot within the break. Remember that for some spot rates, there is no obligation to keep it in the break, nor indeed to transmit the spot. The TRANSMITTING relationship is established when the spot is broadcast. So this relationship represents an actual happening and is used for invoicing.

When you assemble all the fragments of the data model, you have a complete (to date) description of the business; the assembled model is in Figure 3.2.8. Compare it to your own, and make sure before going on that you can reconcile all of your (and our) decisions on how to portray the business policy.

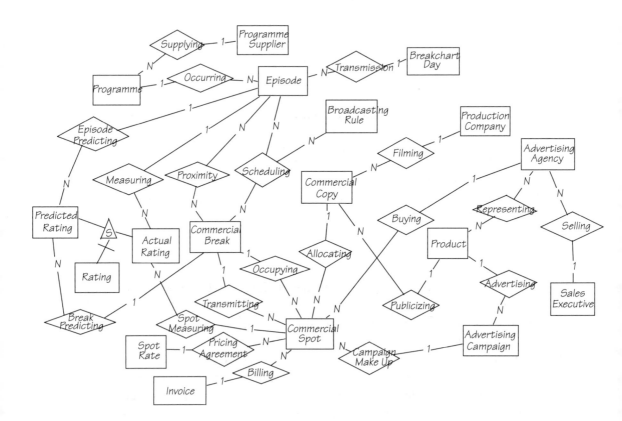

Figure 3.2.8: This first-cut data model serves as a guide for your detailed analysis. As you progress with the Piccadilly Project, you will define the data in each entity and relationship, as well as the processes that store and retrieve those data.

Defining Your Entities

We asked you to write down the attributes for the entity ADVERTISING CAMPAIGN. Some of the attributes are established by the executive when he uses (the data flow) CAMPAIGN REQUIREMENTS to plan the campaign. First consider what data make up the flow. When the agency gives Piccadilly the campaign requirements, the agency also must tell Piccadilly its name. So the name of the agency will be part of the flow, along with the name of the product to be advertised. The executive needs to know at least the budget for the campaign, the target audience, the ratings the agency wants to achieve, the length of the campaign, and, for scheduling purposes, the duration of each spot used in the campaign.

In the data dictionary, you can write the above description as

Campaign Requirements = ✳ Data flow. An agency's description of require-
        ments for an advertising campaign. ✳
        Agency Name + Product Name + Campaign Budget
        + Target Audience + Target Rating Percentage
        + Campaign Duration
        + {Required Spot Duration}

The braces around REQUIRED SPOT DURATION indicate there are many of them, whereas there is only one occurrence of each of the other data elements. (This notation is explained fully in Chapter 2.9 *Data Dictionary*. If you aren't comfortable with it, look ahead, but for now just accept it as a convenient shorthand.)

Which of the items in this data flow need to be part of the ADVERTISING CAMPAIGN entity? For the moment, we think this is a fair definition:

Advertising Campaign = Campaign Budget + Target Audience
        + Target Rating Percentage + Campaign Duration
        + {Required Spot Duration}

Note that each of the attributes describes an advertising campaign and only an advertising campaign. Other items from the data flow, such as AGENCY NAME and PRODUCT NAME, are not included as attributes of the entity ADVERTISING CAMPAIGN as they describe other entities. AGENCY NAME is an attribute of the entity ADVERTISING AGENCY, and PRODUCT NAME is, of course, an attribute of the entity PRODUCT. If you look at the data model, you'll see that ADVERTISING CAMPAIGN has a relationship called REPRESENTING with PRODUCT. The relationship links the product to its campaign, so that the system knows which campaigns belong to which products.

As the analysis progresses and you learn more about the system, add what you know to your models. You should write definitions for the entities and relationships as soon as you have something to say about them. At this stage, an examination of the boundary data flows reveals many data elements that you should attribute to your entities and relationships. It is also helpful to write a short statement explaining the purpose of each entity and relationship; this explanation is enclosed by asterisks.

Advertising Agency = * Entity. Buyer of commercial spots for advertising campaigns. *
Agency Name + Agency Address + Agency Phone Number

Representing = * Relationship. Keeps track of which agency is responsible for a product. Keeps track of historical relationships between agencies and products for collecting bad debts. Cardinality: for each Advertising Agency, there are many Products; for each Product, there are many Advertising Agency(s). Participation: Advertising Agency mandatory, Product mandatory. *
Representation Start Date + Representation End Date

Notice that the relationship definition specifies the purpose of the relationship, as well as the cardinality and participation rules for each entity involved in the relationship. The representing relationship also has two data elements attributed to it because both representation start date and representation end date truly describe the relationship. However, there will be many cases in which a relationship does not have any attributes.

The reason for writing these definitions is that they almost always raise questions. By asking these questions and getting answers, you'll learn still more about the system. Adding that knowledge to your models raises more questions, and … Believe us, it does end eventually.

Look at the other entities in the data model and consider the attributes they each contain. Later, after reading the data dictionary chapters (2.9 *Data Dictionary* and 1.5 *Building the Data Dictionary*), you will define them completely.

Another Way to Build the Data Model

Having just put you through the exercise of building a data model from a statement of the users' business, we now confess that this is not the only way to build such a model. Our reason for doing it this way is to give you some practice with data modeling and to raise questions about Piccadilly. Shortly, you will partition the system by events and build event-response data models. (If you are unfamiliar with these terms and models, your trail will introduce them before you need them for the case study.)

Why do you need several methods? Because we have found in our own projects the enormous benefit of building a first-cut data model of the users' business. This model is an invaluable vehicle for starting to understand the business policy. However, the first-cut model is too large to support very detailed thinking. Later, you will use event-response data models to partition the data model into head-sized pieces so that you can confirm all the details and assumptions. The result will be a data model that is descriptive and reliable because you have used the policy in the event-response models to verify it.

✚ Ski Patrol

Your data model should be substantially the same as ours. Naturally, there are sure to be some differences in interpretation of the meaning of data and in your choice of names. This is expected when building a first-cut model. Your reason for building the model is to discover potential misunderstandings and arrive at a consensus. If, after reviewing the answer, you feel you've accomplished the objective of the exercise, proceed directly to the Trail Guide for further directions.

If you had some problems—if perhaps the whole exercise of building a data model had no meaning for you—then Chapter 2.4 *Data Viewpoint* will give you the reason for needing a data model. Similarly, if you feel it to be overkill to build both a data model and a data flow model of the same system, we remind you that the purpose is to analyze and specify the complete system. Your specification is more rigorous and easier to build if you have both data and process models. This will be reinforced when you get to the essential modeling chapters.

Some commonly encountered problems in data modeling concern the questions of which attributes make up an entity, when to make a relationship, and how to name relationships. Chapter 2.5 *Data Models* will answer these questions, and the exercises there will give you some more practice. Also, remember that your data model is only a first cut, based on your current fragmentary knowledge of Piccadilly. As you do more of the Project, your increased knowledge will add to and improve your data model.

Trail Guide

This Trail Guide duplicates the one at the end of Chapter 1.3 *What About the Business Data?*

● Easiest: Go to Chapter 2.6 *More on Data Flow Diagrams.* You will leave data models for the moment to expand your knowledge of data flow models. You will rejoin the Piccadilly Project shortly.

■ More Difficult: Now is the time to consider an appropriate viewpoint for this stage of the Piccadilly Project. Go to Chapter 2.8 *Current Physical Viewpoint.*

◆ Most Difficult: Continue on with the Piccadilly Project, when you will now get some more background into the company. Go to Chapter 1.4 *The Piccadilly Organization.*

❋ Promenade: This wasn't intended to be part of your journey. However, if you have already been through the data modeling chapters (2.4 *Data Viewpoint* and 2.5 *Data Models*), look at the data model in Chapter 3.2 (Figure 3.2.8), then pick up your trail in Chapter 2.10 *Essential Viewpoint.*

REVIEW: THE PICCADILLY ORGANIZATION 3.3

Before You Reached Here ...

You have drawn Diagram 0 and updated your context and data models, as presented in Chapter 1.4 *The Piccadilly Organization.*

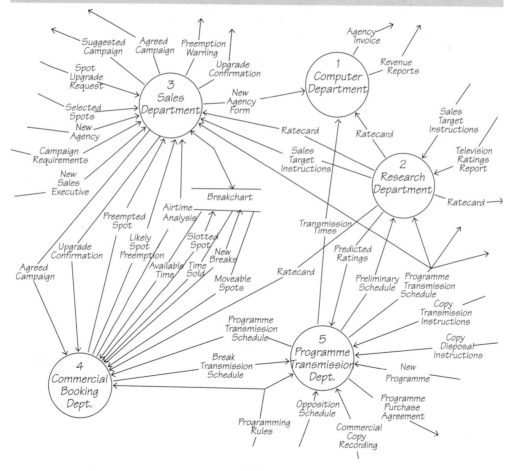

Figure 3.3.1: Diagram 0. This model is partitioned to highlight Piccadilly's current organization.

Sample Model of Piccadilly

The partitioning of this current physical model mirrors the business that you see when you visit Piccadilly's offices. The model is useful to you at the moment, as you do not yet have a complete understanding of the television business. The model also highlights some problems. The largest number of interfaces is between the Sales and Commercial Booking departments. This heavy traffic indicates that the work done by these two departments is closely connected. Perhaps there is some functionality being handled in the wrong department, or perhaps there is some overlapping or duplicate processing. It is too early yet to know. These data flows simply tell you that you must analyze both departments before you can see the complete picture.

High-level current physical models such as this one can be used to explain your project to other people. In the real Piccadilly Project, we pinned this model to a prominent wall and used it whenever we discussed the Project with the users or with others on the analysis team. We also found it an extremely good starting point for introducing new people into the Project.

Verifying Your Context

The model you have just built, Diagram 0, is a more detailed version of the context diagram. The functionality of the system is now decomposed into five lower-level bubbles. These five bubbles must interface with each other, which means you must introduce new data flows into the model. Note that the boundary data flows from the context diagram must also appear in the lower-level diagram.

Compare the data flows in your context diagram with the flows that enter or leave your Diagram 0. They should match, with the exception of any new flows that were discovered through the lower-level description. The flows NEW AGENCY, NEW SALES EXECUTIVE, OPPOSITION SCHEDULE, and COPY DISPOSAL INSTRUCTIONS do not appear in the context diagram (Figure 3.1.1). The business description mentions OTHER TV CHANNELS and PERSONNEL DEPARTMENT. Since you have no way of changing other television companies, they are outside your context; show them as a terminator. In a real-world project, you'd have to get a ruling from user management whether the Personnel Department is inside or outside the context before attempting to correct your models. In this Project, let's say that Piccadilly management has decided that the Personnel Department is outside the context. So we'll correct the context by adding PERSONNEL DEPARTMENT as a terminator, along with the other new terminator and the new flows. The updated context diagram appears in Figure 3.3.2.

As you learn about the details of a system, you will usually find that some information has been omitted from the higher-level models. That's because people forget to tell you something, or they assume that you already know something, or, as happens occasionally, they just don't know enough to give you the complete picture. People also tend to contradict one another.

These inconsistencies are irritating, but not debilitating. The reason that you are modeling the system is to come up with a precise definition, and it's far easier and cheaper to find the inconsistencies during analysis than at some later stage.

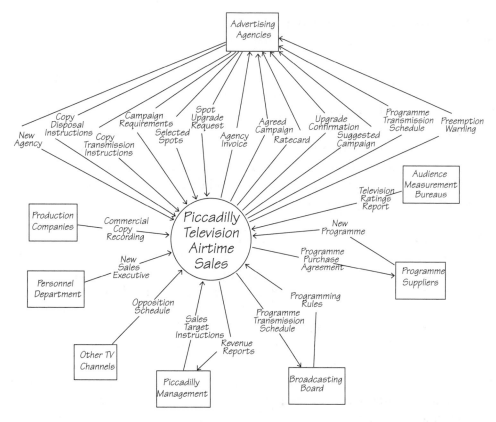

Figure 3.3.2: Updated context diagram. The data flows NEW AGENCY, NEW SALES EXECUTIVE, OPPOSITION SCHEDULE, and COPY DISPOSAL INSTRUCTIONS, plus the terminators PERSONNEL DEPARTMENT and OTHER TV CHANNELS are added to the context diagram to make it balance with Diagram 0.

Adding to Your Data Model

From the business description, you know that Perry Vale in the Programme Transmission Department needs to record the opposition's programme schedules. He does this so that he can schedule high-rating programmes whenever the oppo-

371

sition is planning a blockbuster. He cannot afford to let Piccadilly give away too big a share of the audience.

Therefore, the system has to remember the opposition's programming schedules, which means that the data model also changes. Figure 3.3.3 is an updated version of the first-cut data model we showed you in Figure 3.2.8.

Some Analysts' Questions

Suppose I didn't have a written description of Piccadilly. How would I model the organization?

There are other possible sources of information. As the users describe their work and workplace to you in person, draw data flow diagrams as they talk. They can help if you get stuck. Another source of information is reports and documents used within the company. Start by drawing a bubble for each department in the organization chart, and then treat each report or document as a data flow. Some of these will flow between departments, and some will be boundary data flows to or from terminators. Then, complete the model by asking the users, and looking for data flows that travel by telephone or voice.

Will the Piccadilly users think my model is too complicated?

The users won't find this picture nearly as complicated as you do. After all, it is a model of something they know very well: their own organization. The data flow names are terms they use every day. You can enhance the model by noting in each bubble the name of the manager or the key people in each department. Your objective is not to deny the complexity, but to control it.

One simplification that we have chosen to use in this model is the single unnamed data flow between the Sales Department and the breakchart. As we are not yet certain of the exact data content in this case, we use this flow as a summary of all the flows. Later, when we study the Sales Department in detail, we will identify the individual flows in the lower-level model. We were much more certain about the communication between the Commercial Booking Department and the breakchart. In this case, we felt confident enough to show separately named flows in the model, and these will be confirmed by our detailed study of the department.

When reviewing your models with the users, don't try to explain what the symbols mean; they'll either pick up the meaning from the context or ignore it. Instead, talk about the business you are modeling. Use your models to focus on the topics and areas you need clarified. When you point to a data flow, don't call it a data flow, call it by its business name. Talk through the model and make it come alive.

For instance, if you were talking to Dagenham Heathway in the Research Department, you could point to the data flow SALES TARGET INSTRUCTIONS and

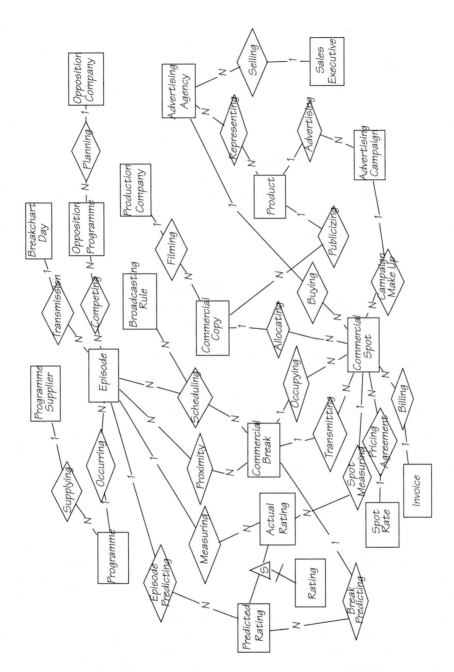

Figure 3.3.3: The first-cut data model is updated to reflect the additional information about the opposition companies and their programming. There are two new entities: OPPOSITION COMPANY *and* OPPOSITION PROGRAMME. *A new relationship keeps track of which company is* PLANNING *to schedule which opposition programme. Another new relationship keeps track of all the opposition programmes that are* COMPETING *with a Piccadilly programme.*

say, "When you get the sales target instructions from management, you use the programme transmission schedule and the television ratings report to do some financial modeling and to set the ratecard for the next quarter. Have I understood that correctly? Are there any other data you use that I don't know about?" Talking through the diagram leads Dagenham to verify your understanding and tell you about anything that is missing. If he corrects your model, thank him. Remember that it is not an interruption to your work if somebody makes changes to your model—it is the purpose of your work.

How long does it take to build a current physical model?
Sometimes, you can sketch a current physical model on a whiteboard in less than half an hour. In other cases, it can take weeks. We have found the time depends on three factors:

- how well you know the people involved
- how much user knowledge you have yourself
- how available are the people you need to see

Remember that the current physical model is not the specification for your future system. This model's purpose is to give you an understanding of the business. Instead of asking "How long does it take?" perhaps a better question is "How much time should I spend?" The answer is as much time as you need to understand the business, and no more. Later, after discussing event modeling, we can define more precisely how much knowledge you'll need. Also, as you progress through the Piccadilly Project, you will see the point at which we suggest that you stop physical modeling and move on to another view.

Are there any dangers in building current physical views of a system?
Yes. Many projects have wasted a great deal of time by building overly detailed current physical models. Some projects have built models that specify every detail right down to the mini specifications. That is not the purpose of the current physical model. Shortly before it was canceled, one project we know of had produced three thousand mini specifications detailing the current system. This was tragic, because the analysts could have captured enough knowledge with one or two levels of data flow diagrams, some data dictionary definitions, and *no* mini specifications.

The reason this happened was that the people building the models were not sure why they were doing it. Without a clear plan of how much physical modeling is needed, analysts find it is all too easy to continue doing it past the point of being useful.

Physical modeling is also fun to do. Most analysts enjoy the experience of talking to users, modeling what they said, verifying the model, and moving down to the next level of detail. There is also the problem that when analysts are talking to the users, they are providing details of their business. Many analysts feel that they have to model all the details so that they don't lose any of them and have to ask for the information again. However, at this stage of the Piccadilly Project, you don't know what details are needed, and what will become superseded by the future system. We urge you to procrastinate. You can always postpone details and then go back to pick up the details when you know what you need. It is much more difficult to get rid of lovingly crafted models even after they have proved to be unnecessary.

Your goal is to specify a system, and to do that, you have to know how to *selectively* build your physical models.

Why do you recommend starting the Piccadilly Project with a current physical view?
We recommend this approach because you had no prior knowledge of Piccadilly's business, and the current physical model helps you to understand its business. This model is a convenient vehicle to collect information about the system.

A current physical model is not wasted effort, and you've used it to accomplish two things. First, the model helps to ensure that you are beginning the project with the correct context. Second, it helps you get to know the key users. You've demonstrated your interest in what they told you by building a model of it. You've discovered most of the data that flows around and is stored by the company. You also know, at a high level, what processes exist to treat those data. While it is early yet, you have already traveled quite a distance.

Once you have built up enough knowledge of Piccadilly's business, it will be time to move on to other views of the system.

✚ Ski Patrol

We have mentioned that a possible pitfall with the current physical model is the temptation to show a bubble for each process mentioned in the text, thus resulting in a model with far too many bubbles. If this happened to you, a quick revisit to Chapter 2.7 *Leveled Data Flow Diagrams* would be useful for you to redo your model.

The data flow names should be substantially the same as those used in the text. You are free to change names, but you should always have a solid reason for doing so. Remember that you have to confirm this diagram with the users before proceeding, so they must be able to recognize the name.

Your diagram should have balanced with the context. If it didn't, or if you didn't do the balancing check, or if you didn't add the new boundary flows to your context diagram, visit Chapter 2.7 *Leveled Data Flow Diagrams* to read about keeping the leveled models in balance with each other. Balancing is a critical part of your modeling effort.

You also should have discovered the additions to the data model. You should be happy with your data model at this stage as a reflection of your growing knowledge of the business. It does not have to look exactly like ours. In later modeling efforts in the Project, you will refine your model, and at the same time prove or disprove your ideas. However, we want you to feel comfortable with the idea and notation of data models before proceeding. Look in Chapters 2.4 *Data Viewpoint* and 2.5 *Data Models* for help on data models.

Trail Guide

● Easiest: The next step in the Piccadilly Project is to define the data flows and stores. To find out how to do this, go to Chapter 2.9 *Data Dictionary.*

■ More Difficult: If you already know about data dictionaries, proceed to Chapter 1.5 *Building the Data Dictionary.* If you need to brush up on some rusty skills, may we suggest a quick detour through Chapter 2.9 *Data Dictionary.*

◆ Most Difficult: Put your skill with data dictionaries into practice by going to Chapter 1.5 *Building the Data Dictionary.*

✳ Promenade: This chapter is not required reading for you. Turn to Chapter 2.8 *Current Physical Viewpoint,* where you can pick up your trail.

REVIEW: BUILDING THE DATA DICTIONARY 3.4

Before You Reached Here ...
You have written some data dictionary definitions and probably updated your data model, as assigned in Chapter 1.5 *Building the Data Dictionary*.

Defining Piccadilly Entries

Your task was to define the information that Piccadilly receives from its programme suppliers. This is the minimum that Perry Vale needs to make his programme buying decisions. The entries are listed alphabetically.

Director Name = * Data element. Identifies the director of a programme. *

New Programme = * Data flow. Describes a programme offered by an English
        or overseas programme supplier. *
        Programme Name + Programme Type + Programme Description
        + Programme Duration + Programme Price
        + ?Programme Episodes
        + {Performer Name}
        + ?(Producer Name + Director Name)
        + ?Supplier Name

Performer Name = * Data element. Identifier of an actor or actress appear-
        ing in a programme. *

Producer Name = * Data element. Identifies the producer of a programme. *

Programme Description = * Data element. Synopsis of the contents of a pro-
        gramme. *

Programme Duration = * Data element. Running time of a programme. Units: hrs/mins/secs. *

Programme Episodes = * Data element. The number of episodes of a programme covered by an agreement with a supplier or by Piccadilly's internal production plans. *

Programme Name = * Data element. The name that uniquely identifies this programme, for example, News at Ten, Brideshead Revisited, Coronation Street. *

Programme Price = * Data element. Price paid to the supplier of a programme. Units: pounds sterling. *

Programme Type = * Data element *
[First-Run Film | Sporting Event | Documentary | Talk Show | Old Movie]
* The types of programmes that Piccadilly transmits. Note that there are other programme types to add to these. *

Supplier Name = * Data element. Identification for a programme supplier. *

Although not specifically mentioned, a supplier would likely tell Piccadilly who he is when the new programme information is sent, since Perry must inform the supplier of his buying decision. Because the information was not specifically mentioned by the user, a question mark is used to indicate that an assumption is being made. A question mark can be used anywhere you're not certain of what you are writing. Definitions and data flows highlighted by this notation can then be clarified by the users. There is nothing wrong with showing what you don't know, but there is everything wrong with hiding it.

The supplier's address is not listed, as it is likely to be on file and can be retrieved using the name. Perry can confirm this when you go back with questions.

You were given the information "Some programmes include the names of the producer and director." When names appear in the data flow, does it mean that both the producer's and the director's names are there? Or does it mean that one or the other may be present? The English language doesn't have any precedence rules for "and," but the data dictionary does. You could have defined (PRODUCER NAME + DIRECTOR NAME) to mean that if any names appear, then both do; or (PRODUC-

ER NAME) + (DIRECTOR NAME) to say that one or both or none may be present. Alternatively, if it is possible to have more than one producer and more than one director, then {PRODUCER NAME} + {DIRECTOR NAME} would be used. Again, a question mark is used in the definition as you cannot be sure about the correct meaning until you have checked with Perry Vale.

Although the user did not mention it, the analyst realized that there might be several episodes of a programme. The ?PROGRAMME EPISODES has been included in the definition so that the analyst remembers to ask the question.

In the description, you were told that there are other types of programmes. The comment in the data dictionary entry for PROGRAMME TYPE reminds you that some more values need to be defined.

Defining Ratecard

This definition was derived from a sample page of a ratecard presented in Figure 1.5.2.

Ratecard = * Data flow. Prices, moveability, and preemption rules of time
            available for sale. *
            Rate From Date
            + ?{Rate Spot Duration
            + {Rate Segment Day
            + {Rate Segment Start + Rate Segment End
            + {Rate Moveability + Spot Price}}}}

Rate From Date = * Data element. The commencement date for a new rate-
            card period. Format: Day/Month/Year. *

Rate Moveability = * Data element. Rate moveability as defined in the rate-
            card. *
            [Fixed: fixed on a nominated day and break | Broad: moveable
            within a specified segment on a nominated day | ROD: run-of-
            day, moveable to any similarly priced segment on a nominated
            day | ROW: run-of-week, moveable to any similarly priced seg-
            ment during a week]

Rate Segment Day = * Data element. Day(s) of week on which this segment
            of time occurs. *
            [Weekday | Saturday | Sunday]

379

Rate Segment End = * Data element. The end time for a ratecard segment. *

Rate Segment Start = * Data element. The start time for a ratecard segment. *

Rate Spot Duration = * Data element. Duration of spots as defined in the
ratecard. Units: seconds. *
[10 | 20 | 30 | 40 | 50 | 60]

Spot Price = * Data element. Ratecard price for a spot rate. Units: pounds
sterling. *

Your names may be different from those that we have used, but their meanings should be substantially the same. Take a moment to reconcile your names with ours. Also, make sure that you understand ours because, naturally enough, we will be using them for the rest of the Project.

Our definition of RATECARD has a ? before RATE SPOT DURATION. We have assumed that when a new ratecard is issued for a quarter, the rates for all durations are changed. So for one RATE FROM DATE, there are a number of RATE SPOT DURATIONs. The brace indicates that there are several RATE SPOT DURATIONs, and the question mark says that we are not certain.

Make sure that you have your pairs of braces in the right places. If you specify the repeating groups incorrectly, it will mislead the file designers later when they receive your specification.

The ratecard is made up of a number of components; these are defined in turn. When the component is a data element, it has the comment * Data element *. For example, RATE SPOT DURATION is a data element, and its definition shows all the possible values it can have. RATE FROM DATE is another data element, but has an almost infinite number of possible values. There is no point attempting to list them in the dictionary, and dates are commonly understood. So the definition simply has a comment, and points out that the format is day/month/year.

It is not possible to define all possible values of SPOT PRICE by looking at the sample ratecard page in Figure 1.5.2. There are other pages of the ratecard, for spots of different durations. You were told there are similar ratecard pages for 10-, 20-, 40-, 50-, and 60-second spots. Besides, over the life of the system, there will be an almost infinite number of values for this item. To find more information about this data element, you would look through the other pages of the ratecard. You'll find that the cheapest price is £150 for a 10-second ROW spot in the last segment of the day. The highest-priced spot, a 60-second fixed in the prime segment, sells for £25,000. You could enhance your definition in the data dictionary with

Spot Price = * Data element. Ratecard price for a spot rate. Units: pounds
           sterling. Value range: >= 150 and <= 25000. *

Adding to Your Data Model

You were also asked to study your definitions of the data flows to see if they revealed any new or updated entities. Then you added any new information you found to the data model. Figure 3.4.1 is an updated version of the data model in Figure 3.3.3.

When you built your first-cut data model, you included an entity called PRO-GRAMME. At that stage, you didn't have enough information to write a detailed definition of it. However, you now know that most of the data content of PRO-GRAMME comes from the data flow NEW PROGRAMME. Now that you have defined the content of the flow, you can begin to define the entity PROGRAMME.

Let's say that the users have accepted our version of the definition for a new programme:

New Programme = * Data flow. Describes a programme offered by an English or
           overseas programme supplier. *
           Programme Name + Programme Type + Programme Description
           + Programme Duration + Programme Price
           + ?Programme Episodes
           + {Performer Name}
           + ?(Producer Name + Director Name)
           + ?Supplier Name

The first five items are data elements, and, as they describe a programme and are remembered by the system, you can say they are attributes of the entity PRO-GRAMME. Entities are defined by listing their components just as if they were data flows (a sample appears below).

PROGRAMME EPISODES has a ? beside it in the definition of the data flow. The reason is because a question occurred to the analyst: "Do the suppliers tell Piccadilly if a new programme has multiple episodes?" If the answer to this question is yes, the question mark will be removed. Otherwise the element will be removed from the definition. Let us say that in this case the users agreed that PRO-GRAMME EPISODES should be part of the definition. So we can also attribute it to

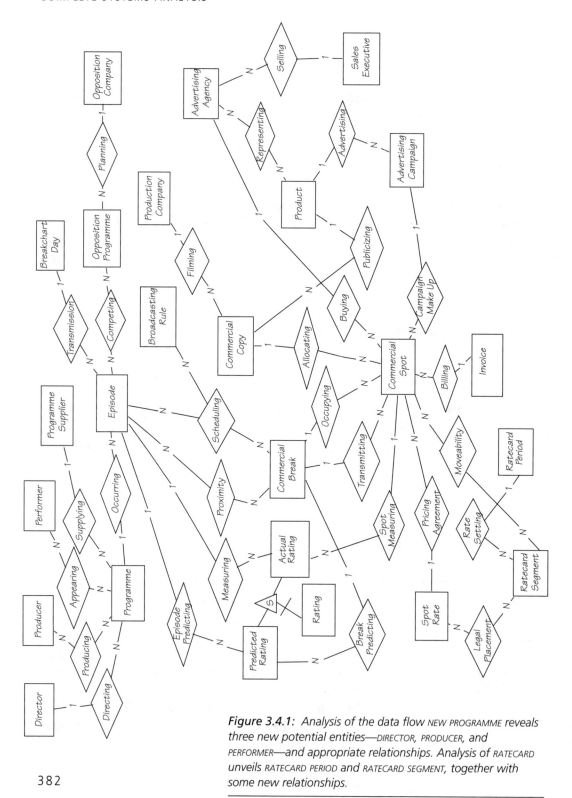

Figure 3.4.1: *Analysis of the data flow* NEW PROGRAMME *reveals three new potential entities—*DIRECTOR, PRODUCER, *and* PERFORMER*—and appropriate relationships. Analysis of* RATECARD *unveils* RATECARD PERIOD *and* RATECARD SEGMENT, *together with some new relationships.*

the entity PROGRAMME. Your dictionary can now support your data model with this definition:

Programme = * Entity. A television programme made by Piccadilly or bought
from a programme supplier. *
Programme Name + Programme Type + Programme Description
+ Programme Duration + Programme Price
+ Programme Episodes + Programme Purchase Date

We added the * *Entity* * comment to differentiate between the types of definitions in the data dictionary. You will find this approach a useful aid in finding your way around a large data dictionary.

Did you include {PERFORMER NAME} in the definition of PROGRAMME? We hope not. While {PERFORMER NAME} lists the main performers in the programme, you cannot say that a performer's name describes a programme; it describes a performer. Instead of attributing this data item to PROGRAMME, it should be attributed to a new entity called PERFORMER. As usual, you support your data model with a data dictionary definition:

Performer = * Entity. Actor or actress appearing in a programme. *
Performer Name

Two other entities derived from the data in NEW PROGRAMME are DIRECTOR and PRODUCER. Adding these two entities to the data model, you define them:

Director = * Entity. The director of a programme. *
Director Name

Producer = * Entity. The producer of a programme. *
Producer Name

The NEW PROGRAMME data flow carries data about a number of different entities that are grouped in one data flow because there is a need for Piccadilly to know which director has directed a programme. When the data are stored as entities, there is a need to remember the relationship between them. The name DIRECTING given to the relationship in Figure 3.4.1 describes the reason for the link. Similarly, the data flow reveals a relationship PRODUCING between the PROGRAMME and its PRODUCER.

Think of these additions to the data model as potential entities and relationships. For example, we are not yet sure, within this context of study, whether Piccadilly thinks of DIRECTOR as an entity or whether the DIRECTOR NAME is an attribute of the PROGRAMME entity. We will verify this when we do a detailed analysis of the use of the data. Until then, you need to show all potential entities and relationships in the data model because by doing so, you make the questions obvious.

In the first version of the data model that you built (Figure 3.2.8), an entity called SPOT RATE is related to many commercial spots. At that stage of your analysis, Piccadilly's pricing looked quite straightforward. Since analyzing the sample rate-card, you now find that the pricing is more complex than it appeared when you did the first-cut data model. Your detailed study of the ratecard has resulted in this definition:

Ratecard = * Data flow. Prices, moveability, and preemption rules of time
          available for sale. *
          Rate From Date
          + ?{Rate Spot Duration
          + {Rate Segment Day
          + {Rate Segment Start + Rate Segment End
          + {Rate Moveability + Spot Price}}}}

The data model should reflect the reality of the business. So instead of the single entity called SPOT RATE, you can break it down into separate entities, each one describing something that is familiar and important to the business:

Ratecard Period = * Entity. Period during which given rates apply. *
          Rate From Date

Ratecard Segment = * Entity. Continuous band of time defined on ratecard. *
          Rate Segment Day + Rate Segment Start + Rate Segment End

Spot Rate = * Entity *
          Rate Spot Duration + Spot Price + Rate Moveability

Relationships

Like entities, relationships must reflect the policy of the business. The LEGAL PLACE-MENT relationship between RATECARD SEGMENT and SPOT RATE is there to define the segment placement offered by a particular set of moveability rules. This relationship is introduced because the price paid for a spot depends in part on the segments in which it can be transmitted. The relationship is many to many, as a rate can apply to several segments, while the one segment can have many rates applicable to it.

The MOVEABILITY relationship between COMMERCIAL SPOT and RATECARD SEGMENT is established when the agency agrees to buy a spot with certain moveability conditions. (Broad spots can be moved within a segment; run-of-day spots moved to similarly priced segments on the same day; and run-of-week spots moved to similarly priced segments over the week.) The relationship exists to link a commercial spot with the segments in which it may be broadcast.

Eventually, you'll verify each relationship and define each of these relationships in the data dictionary, and you can do this when you acquire more knowledge of the processes that create and use the relationships.

✚ Ski Patrol

Here's some first aid with the data dictionary definitions. At this stage, we want you to feel secure about defining data. We trust that the reason for writing definitions was revealed by the exercise, namely to better understand the data and the system that uses the data. If you think you can model a system without defining the data, please rethink. If you are not happy with the meaning of the operators, we can only suggest reviewing them in Chapter 2.9 *Data Dictionary.* If you were not happy with our interpretation of the data, take a few moments to reread the problem statement in Chapter 1.5 *Building the Data Dictionary.* This time, have the sample answers beside the problem statement. Note how the nouns that are important to Piccadilly's business make up the data dictionary definitions.

We suspect the data model was your biggest problem. The most important thing we have to say about the data model is, "Don't panic!" Data models don't come easily to most people, so you are not alone. Data modeling requires knowledge of the subject matter and the ability to view that subject at a high level of abstraction. The Piccadilly Project will give you plenty of practice in developing these skills, and you will progressively improve your data model. If you attempted to do the data modeling part of the exercise without reading and doing the exercises in Chapters 2.4 *Data Viewpoint* and 2.5 *Data Models,* we suggest a detour through those chapters.

So far, we have approached the data model somewhat piecemeal. We have attempted to build a data model from a general description of the business, and have made some small alterations to it by analyzing data flows. This approach could be classified as a "fuzzy top-down approach." Its strength lies in providing a way of getting started with a complex problem.

There is more to come: When we work through the chapters on essential event-response models, we'll look at another way to model the system's stored data. This approach works by partitioning the data model into small logical chunks based on the need to support one event (this is explained in Chapter 2.11 *Event-Response Models*). By the time you have experienced both approaches, we're sure that data modeling will be much clearer to you. And, as is necessary in practice, you will be in a position to blend both approaches.

Finish comparing your data dictionary and data model upgrades with the samples. Make sure that you can reconcile any differences between your answers and ours.

Trail Guide

● Easiest, ■ More Difficult, and ◆ Most Difficult: Go to Chapter 1.6 *Selling the Airtime*. There you will expand on your model of the Piccadilly system.

❋ Promenade: The most appropriate destination for you is to rejoin your trail in Chapter 2.8 *Current Physical Viewpoint*.

REVIEW: SELLING THE AIRTIME 3.5

The Data Flow Diagram As a Recording Device

During your early interviews with the users, the data flow diagram can serve as a note-taking device. As people talk, draw processes and data flows. Your diagrams won't be great art, but they nevertheless have a use. You capture what the users said, and immediately afterward confirm that you have the correct interpretation of what they said.

For example, the model in Figure 3.5.1 shows that bubble 3.4 SET SALES POLICY RATES updates the BREAKCHART with BREAK MINIMUM RATE. You show this model to Stamford Brook, sales manager, and ask him, "Is this right? Is it true that when you set the sales policy, you update the minimum rates on the breakchart?" If this isn't true, if he said something he didn't mean or you misunderstood what he said, now is the time to find out. You have no ego invested in this model (nobody can love a diagram with fourteen bubbles), and you cannot mind changing it. After all, its purpose is to provide you with an accurate and unambiguous statement of what happens in the Sales Department. Changes now will save time later on.

The model should be comprehensible to Stamford. If he were available, you would walk through the model with him and raise all the questions you could. Maybe he would disagree with your model, or contradict something he said, or remember something he forgot to tell you. What you would be doing is communicating with him in a language he could understand.

Don't worry if initially there are no data flows joining all the processes in your diagram; you'll find these when you go over the model with the users. Also don't be too concerned if your model breaks some of the rules discussed in this book. You can clean it up later.

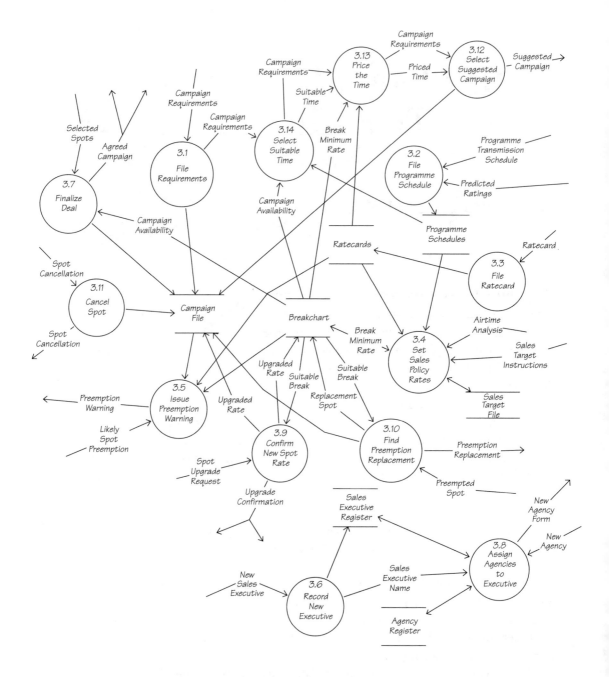

Figure 3.5.1: *Preliminary version of Diagram 3 Sales Department, reflecting what Stamford told you. There are too many processes in this diagram, but this can be fixed later by leveling upward. At this stage, the model is used to confirm Stamford's description of the business.*

Leveling Upward to Reduce Complexity

The first thing that strikes you about this model is that there are more bubbles in it than we recommend in Chapter 2.7 *Leveled Data Flow Diagrams*. When you use a data flow diagram as a note-taking device, this often happens. Sometimes, it results in models even more complex than this one. As an information-gathering tool, that's fine, but you cannot leave the model in this state forever.

To control the complexity and reduce the number of bubbles, group processes to create a high-level summary. Before making this higher level, though, you have to identify which processes can be logically grouped together.

There are two guidelines. The first is to group processes so that you minimize the number of interfaces at the highest level. This can be done by trial and error, but it is probably faster to look for groupings of bubbles that are closely connected to each other but loosely connected to anything else. The second guideline is to look for groups of bubbles with similar functionality so that the group can be given a functional name that is meaningful to the business being studied.

Figure 3.5.2 is the result of leveling upward. Bubbles 3.5, 3.9, 3.10, and 3.11 (from Figure 3.5.1) were functionally related and were consolidated into the honestly named process MAKE CHANGES TO SPOTS. We also grouped bubbles 3.1, 3.12, 3.13, and 3.14, which had data flow connections, into one high-level bubble called PLAN CAMPAIGN.

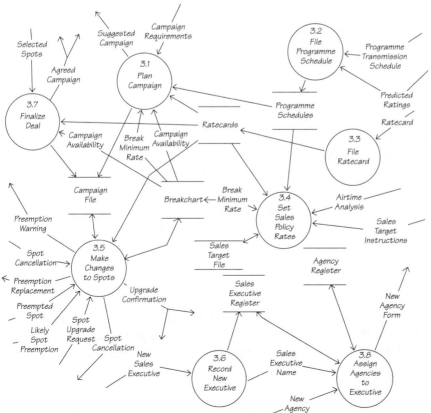

Figure 3.5.2: Second version of Diagram 3 Sales Department, combining some of the bubbles of the previous version. It was leveled upward to make the model less complex. Notice how many of the processes have interfaces through the use of common physical data stores.

389

The model shown in Figure 3.5.2 is more presentable. It shows all of the functionality of the Sales Department, but does it in only eight bubbles. This model would be used to present an overview of the department, or to reassure Stamford if he became apprehensive over the complexity of the previous version. The details of the department are not lost, and they are now shown in two lower-level diagrams 3.1 and 3.5. These appear in Figures 3.5.3 and 3.5.4, respectively.

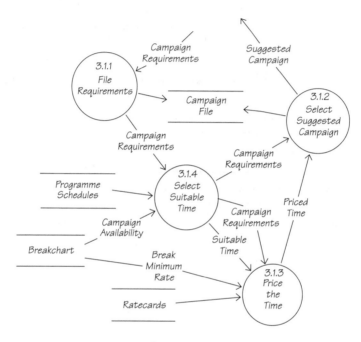

Figure 3.5.3: Diagram 3.1 Plan Campaign.

The Big Picture

Your work in the Sales Department has identified two new data flows: SPOT CANCELLATION, which comes from the agency; and PREEMPTION REPLACEMENT, which goes to the agency.

To keep your models in balance, you have to add these data flows to Diagram 0 (Figure 3.3.1) and the context diagram (Figure 3.3.2). Figures 3.5.5 and 3.5.6 are updated versions of these diagrams.

✚ Ski Patrol

A common error with lower-level data flow models is to omit the processes that store data. If a data flow that originates outside the context of your diagram is to be stored, there must be a process to store it. For example, some of Piccadilly's data

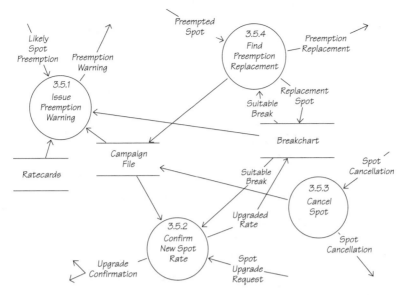

Figure 3.5.4: Diagram 3.5 Make Changes to Spots. This diagram shows the processes that were summarized in the higher-level diagram.

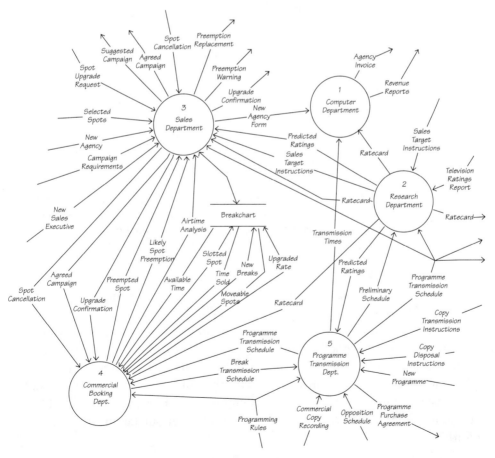

Figure 3.5.5: Updated Diagram 0 contains the two new data flows identified during your work in the Sales Department.

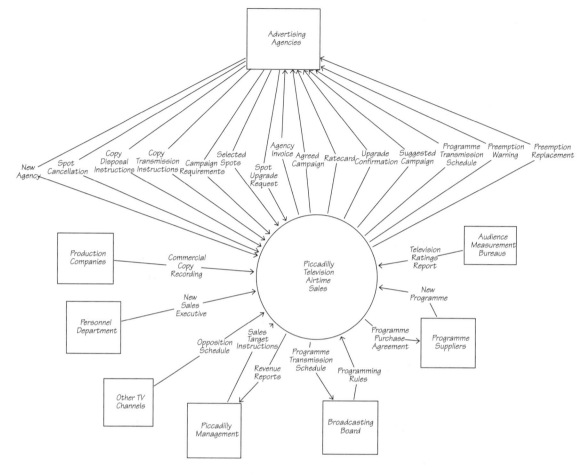

Figure 3.5.6: Updated context diagram. The two additional data flows were both interfaces with terminators, so they must be shown in the context diagram.

originate in the advertising agencies, but they have no mechanism to store data inside Piccadilly's system. Besides, it is not desirable for outsiders to be able to change data. Similarly, if the data originate in a bubble that is not part of your diagram, you still must have a process to store it. Data cannot come from out of nowhere and go directly into your data stores. You must control the storage.

It is not necessary that you have exactly the same diagrams as these samples. You may have partitioned the problem a different way. However, you should have one or more models that you would feel confident about talking through with Stamford Brook.

By this stage of the Project, we want you to feel comfortable with drawing leveled data flow diagrams, and maintaining the correct balance between levels. Remember that having balancing abnormalities does not always mean that the lower-level diagram is wrong. Sometimes, the parent diagram needs to be changed because it fails to show a flow or store that appears in the more detailed child diagram.

If you have any questions about the symbology used in data flow diagrams, review Chapters 2.2 *Data Flow Diagrams* and 2.6 *More on Data Flow Diagrams.* Chapter 2.7 *Leveled Data Flow Diagrams* provides reference material on how leveling works. If you are having trouble with why you want to build a model of the current system, try a detour through Chapter 2.8 *Current Physical Viewpoint.* We also want you to be confident with your ability to record data dictionary entries. A refresher on the notation can be found in Chapter 2.9 *Data Dictionary.*

What to Do

Just so you don't have to spend all your time building current physical models, we've built the lower-level models for the other Piccadilly departments. They are packaged in Chapter 3.6. You will be using the current physical model as the basis for your future work, so take some time now to find your way around it.

Trail Guide

You'll need to remember your next destination before leaving here because you are going to Chapter 3.6, the complete physical model for Piccadilly. You will visit Chapter 3.6 many times from many different locations, so we are not able to provide a trail guide out of there.

● Easiest, ■ More Difficult, and ◆ Most Difficult: Go to Chapter 1.7 *Strategy: Focusing on the Essentials.* Now you'll leave the physical model to move on to a more logical view of the system.

❈ Promenade: You may rejoin your trail in Chapter 2.8 *Current Physical Viewpoint,* where you will learn more about the change in direction the Project is about to take.

3.6 COMPLETE CURRENT PHYSICAL MODEL

Before You Reached Here ...

You could have been at any of a number of locations, since many of the Project chapters refer to the models here. Don't forget where you came from, as we have no way to direct you back there.

This chapter contains the complete current physical model for Piccadilly Television. Since you would spend too much of your time if you were to build the complete model, we have completed the parts that you've not done. The model is for reference purposes. Look through it so that you are familiar with its contents.

Contents

This complete current physical model is made up of a collection of leveled data flow diagrams, a data model, and a supporting data dictionary, specifically:

1. A package of data flow diagrams:

   - Context diagram (Piccadilly Project at the highest level)
   - Diagram 0 (Piccadilly's current implementation)
   - Diagram 1 Computer Department
   - Diagram 2 Research Department
   - Diagram 3 Sales Department, version 2
   - Diagram 3.1 Plan Campaign
   - Diagram 3.5 Make Changes to Spots
   - Diagram 4 Commercial Booking Department
   - Diagram 5 Programme Transmission Department

2. Updated version of the data model.

3. Data dictionary, with the definitions of all terms in the data flow diagrams and data model. The entries are listed in alphabetical order. Each entry has a comment that classifies its type as one of the following:

* Data element *
* Data element grouping *
* Data flow *
* Data store *
* Entity *
* Material store *
* Relationship *

Sometimes, the definition of an entry is followed by one or more comments that define the purpose, define the values, or point to other sources of knowledge about the entry.

✚ Ski Patrol

Your patroller cannot know why you are reading this model; you could have reached this chapter from a number of others, so the only help the ✚ Ski Patrol can give is to refer you to the appropriate chapter for each of the model's components:

- data flow diagrams discussed in Chapters 2.2 and 2.6
- leveled data flow diagrams in Chapter 2.7
- data dictionary described in Chapter 2.9
- data models in Chapter 2.5

Trail Guide

● ■ ◆ ❊ All trails: This chapter is used as reference material for many other chapters. As we have no idea where you came from, our advice is to return whence you came.

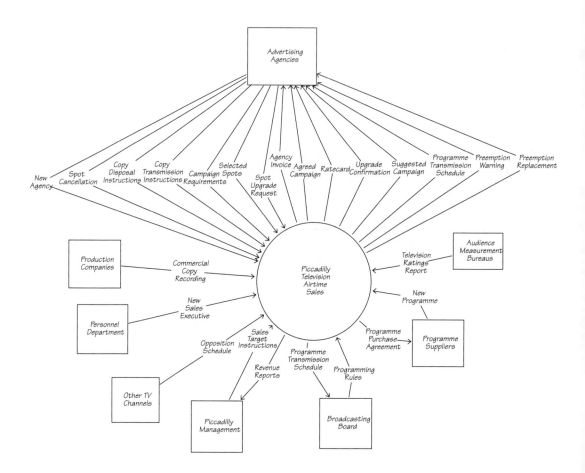

Figure 3.6.1: Context diagram, which defines the scope of the current Piccadilly system.

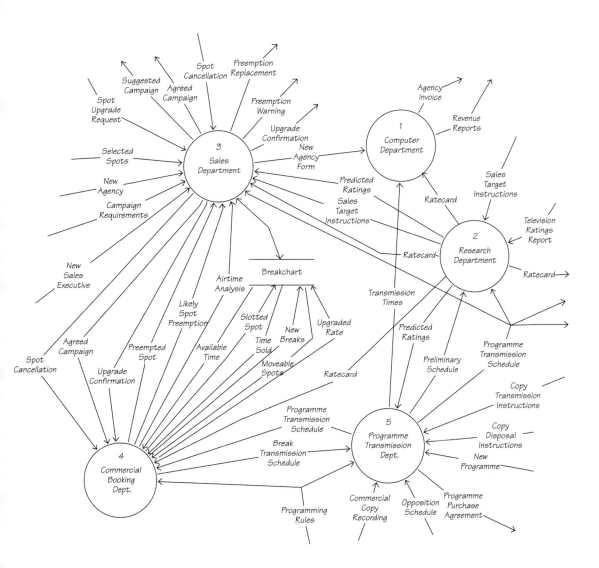

Figure 3.6.2: Diagram 0, the first breakdown of the context diagram. This is intended to be a readily recognizable physical model, so the departments have been chosen as the partitioning theme.

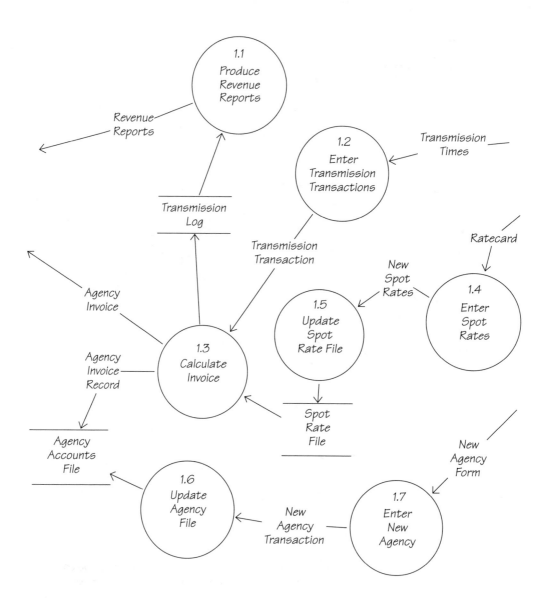

Figure 3.6.3: Diagram 1 Computer Department.

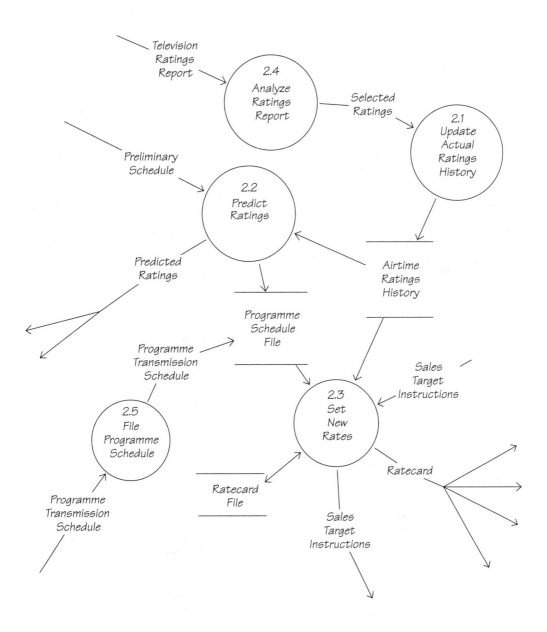

Figure 3.6.4: *Diagram 2 Research Department.*

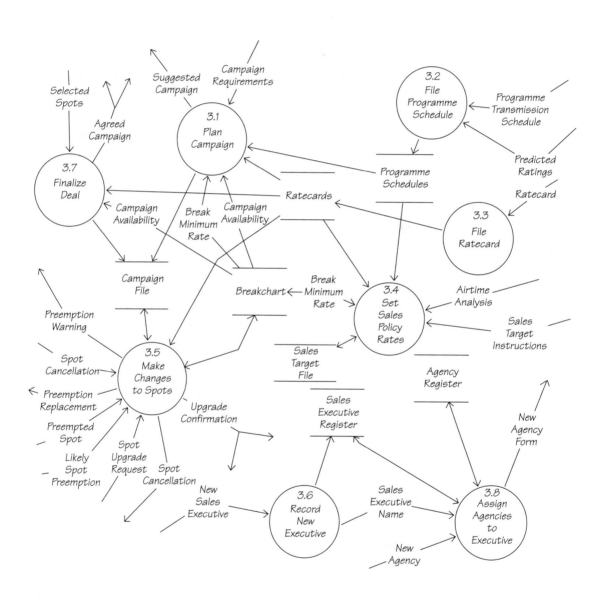

Figure 3.6.5: Diagram 3 Sales Department, version 2.

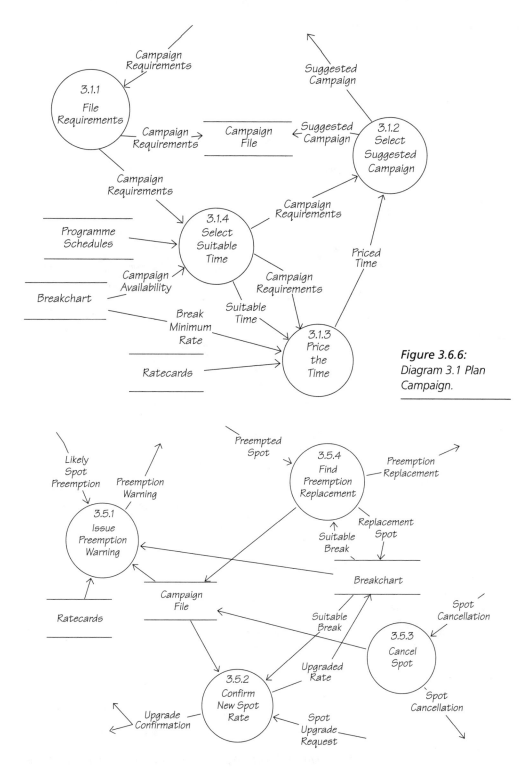

Figure 3.6.6: Diagram 3.1 Plan Campaign.

Figure 3.6.7: Diagram 3.5 Make Changes to Spots.

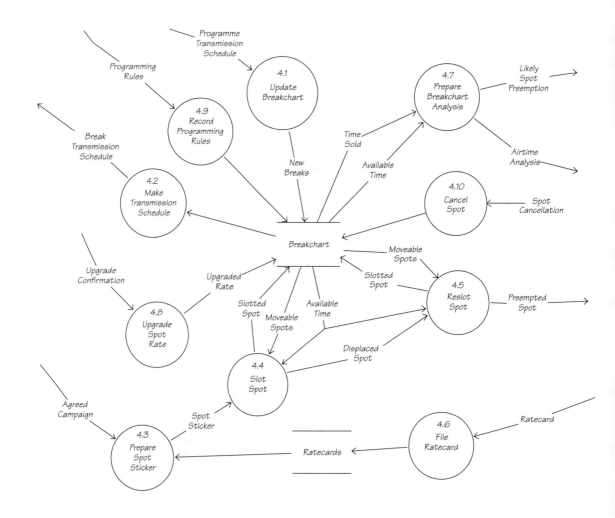

Figure 3.6.8: Diagram 4 Commercial Booking Department.

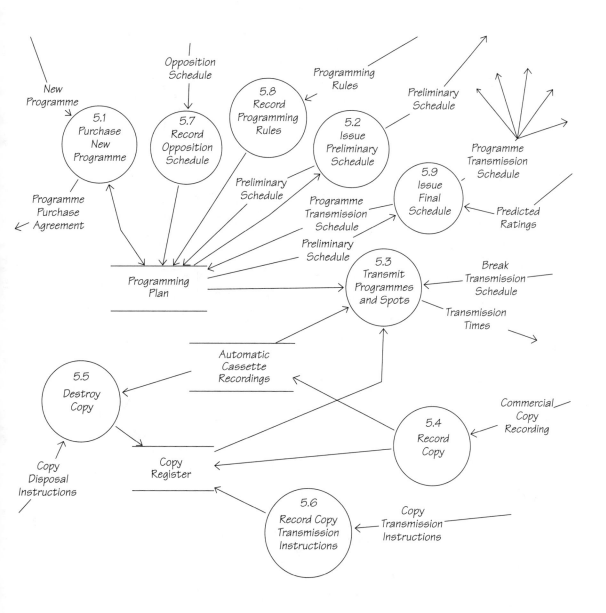

Figure 3.6.9: Diagram 5 Programme Transmission Department.

403

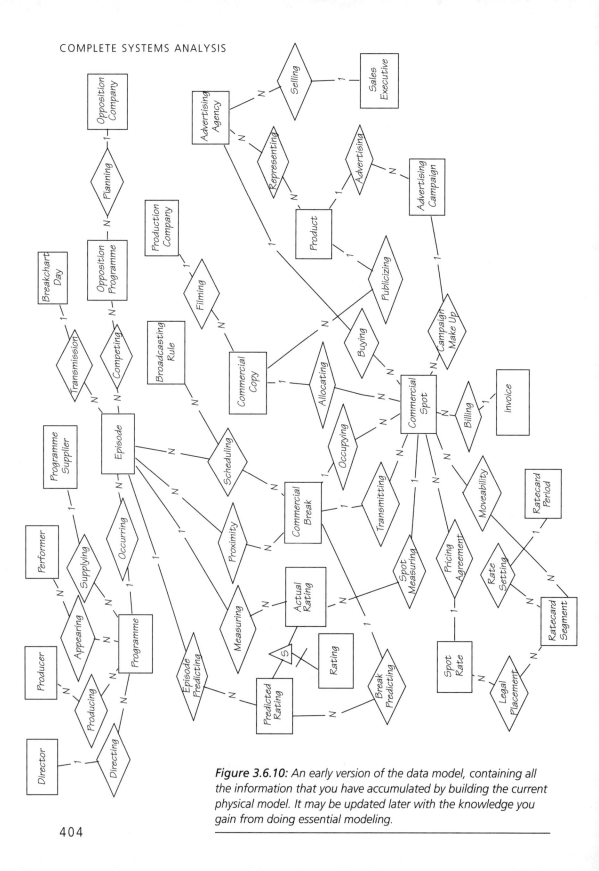

Figure 3.6.10: *An early version of the data model, containing all the information that you have accumulated by building the current physical model. It may be updated later with the knowledge you gain from doing essential modeling.*

Data Dictionary

This data dictionary defines the terms used in the current physical model and in the first-cut data model. You will notice that some of the definitions are not complete and that there are some questions in the dictionary. For example, some of the entities and relationships don't have a note defining their purpose, which indicates there are still unanswered questions about the relevancy and use of the data. Some of the entities have their unique identifiers underlined; those that don't need composite identifiers, which will be specified in the mini specifications. When you build your essential requirements models, you will get answers to such outstanding questions as you make changes, additions, and deletions to the definitions in the data dictionary.

Actual Rating = * Entity. A rating measurement taken during transmission. *
      Actual Rating Percentage + Rating Time

Actual Rating Percentage = * Data element. The percentage of a given audience type watching Piccadilly Television at the time this rating was measured. *

Advertising = * Relationship. Keeps track of which products are involved in advertising campaigns. Cardinality: for each Product, there are many Advertising Campaigns; for each Advertising Campaign, there is one Product. Participation: Product optional, Advertising Campaign mandatory. *

Advertising Agency = * Entity. Buyer of commercial spots for advertising campaigns. *
      Agency Name + Agency Address + Agency Phone Number

Advertising Campaign = * Entity. Records the conditions and aims for a campaign to advertise a product. *
      Campaign Number + Campaign Start Date
      + Campaign End Date + Target Audience
      + Target Rating Percentage + Campaign Predicted Rating
      + Campaign Budget Total + Piccadilly Budget Amount
      + Campaign Duration + {Required Spot Duration}
      * Work necessary to remove or justify the repeating group *

Agency Accounts File = * Data store. Computerized current system accounts file. *
{Agency Name + Agency Address + Agency Phone Number + {Agency Invoice} + {Agency Payment}}

Agency Address = * Data element *

Agency Invoice = * Data flow *
Agency Name + Agency Address + Invoice Number
+ Invoice Date + {Campaign Number + {Spot Number
+ Spot Duration + Rate Moveability + Spot Transmitted Time
+ Spot Transmitted Date + Spot Price }} + Invoice Total

Agency Invoice Record = * Data flow. A transaction in the current computer system. *
Agency Invoice

Agency Name = * Data element. Identifier for an advertising agency. *

Agency Phone Number = * Data element *

Agency Register = * Data store. File of agencies and responsible sales executives kept by the Sales Department. *
{Agency Name + Agency Address + Agency Phone Number
+ Sales Executive Name + Servicing Start Date}

Agreed Campaign = * Data flow. Spots and rates agreed between Piccadilly and an advertising agency. *
Agency Name + Agency Address + Campaign Number
+ Product Name + Campaign Start Date + Campaign End Date
+ {Spot Number + Spot Duration + Spot Price
+ Rate Moveability + Spot Booking Agreement
+ ([Breakchart Date + Break Start Time | {Breakchart Date}])}
+ {Unavailable Slot}

Airtime Analysis = * Data flow *
        {Breakchart Date + {Break Start Time + Break Sold Value
        + Break Unsold Value + {Seconds Available
        + (Rate Moveability)}}} + Total Sold Value + Total Unsold Value

Airtime Ratings History = * Data store. Television ratings supplied by audi-
        ence measurement bureaus. *
        {Television Ratings Report}

Allocating = * Relationship. Keeps track of which commercial copy should be
        transmitted for a commercial spot. Cardinality: for each
        Commercial Spot, there is one Commercial Copy; for each
        Commercial Copy, there are many Commercial Spots.
        Participation: Commercial Copy optional, Commercial Spot
        optional. *

Appearing = * Relationship. Keeps track of which performers appear in pro-
        grammes because this can affect which products can be adver-
        tised during a programme. Cardinality: for each Performer,
        there are many Programmes; for each Programme, there are
        many Performers. Participation: Programme optional, Performer
        mandatory. *

Audience Type = * Data element. Used to classify ratings figures. *
        [Homes | Homemakers | Adults | Men | Women | Children]

Automated Cassette Recording = * An actual cassette containing a record-
        ing of a television commercial. This technology is used by the
        current system. *

Automatic Cassette Recordings = * Data store and material store *
        {Commercial Copy Number + Automated Cassette Recording
        + Production Company Name + Agency Name + Product Name}

Available Time = * Data flow *
        {Breakchart Date + {Break Start Time + {Seconds Available
        + (Rate Moveability)}}}

Average Break Predicted Rating = * Data element. Derivable: average of all
        Predicted Ratings for a Target Audience for all Episodes with a
        Proximity relationship to a given Commercial Break. Integer. *

Billing = * Relationship. Keeps track of which spots have been invoiced.
        Cardinality: for each Invoice, there are many Commercial Spots;
        for each Commercial Spot, there is one Invoice. Participation:
        Invoice mandatory, Commercial Spot optional. *

Breakchart = * Data store. Board containing hanging files plus breaksheets,
        used to record available time and sold time. *
        {Breakchart Date + {Programme Name + Episode Start Time
        + Episode End Time} + {Break Start Time + Break End Time
        + Break Minimum Rate + {Spot Number + Product Name
        + Spot Duration + Spot Price + Rate Moveability
        + Spot Booking Agreement}}} + {Programming Rule}

Breakchart Date = * Data element. A date on which Piccadilly will transmit
        programmes and commercials. *

Breakchart Day = * Entity *
        <u>Breakchart Date</u> + Breakchart Day Start
        + Breakchart Day End + Daily Revenue

Breakchart Day End = * Data element *

Breakchart Day Start = * Data element *

Break Duration = * Data element. Duration of a commercial break. Units:
        mins/secs. *
        ["1 min" | "1 min 30 secs" | "2 mins" | "2 mins 30 secs" | "3 mins"]

Break End Time = * Data element *

Break Minimum Rate = * Data element. Minimum rate to be charged for a
        break. *

Break Predicting = * Relationship. Cardinality: for each Commercial Break,
        there are many Predicted Ratings; for each Predicted Rating,
        there is one Commercial Break. Participation: Commercial
        Break optional, Predicted Rating mandatory. *

Break Sold Value = * Data element. Derivable: total of Spot Prices for each
        Commercial Spot that has an Occupying relationship with a
        given Commercial Break. *

Break Start Time = * Data element *

Break Transmission Schedule = * Data flow. Commercial Booking Department's
        notification of the scheduled transmission time and the spots
        that will occupy the next day's breaks. *
        Breakchart Date + {Break Start Time + {Spot Number
        + Campaign Number + Spot Duration + Product Name}}

Break Unsold Value = * Data element. Derivable:
        ((Break Duration x Break Minimum Rate) − Break Sold Value). *

Broadcasting Rule = * Entity *
        <u>Programming Rule + Rule Effective Date</u>

Budgeting = * Relationship. Keeps track of the sales target for a number of
        breakchart days. Cardinality: for each Sales Target, there are
        many Breakchart Days; for each Breakchart Day, there is one
        Sales Target. Participation: Sales Target mandatory,
        Breakchart Day optional. *

Buying = * Relationship. Cardinality: for each Advertising Agency, there are
        many Commercial Spots; for each Commercial Spot, there is
        one Advertising Agency. Participation: Advertising Agency
        optional, Commercial Spot mandatory. *

Campaign Availability = * Data flow. Airtime available for a specific
        campaign.*
        {Breakchart Date + {Break Start Time + {Seconds Available
        + (Rate Moveability)}}}

Campaign Budget Total = * Data element. The total campaign budget that the agency intends to spend on television advertising. Units: pounds sterling. *

Campaign Duration = * Data element. Number of weeks during which spots for a campaign will be transmitted. Integer. Value range: >= 2 and <= 52. *

Campaign End Date = * Data element. Date marking the last day spots will be transmitted for a campaign. *

Campaign File = * Data store. File kept in the Sales Department. *
{Campaign Requirements + Suggested Campaign}

Campaign Make Up = * Relationship. Keeps track of all the spots that make up a campaign plan to achieve the desired target ratings. Cardinality: for each Advertising Campaign, there are many Commercial Spots; for each Commercial Spot, there is one Advertising Campaign. Participation: Advertising Campaign optional, Commercial Spot mandatory. *

Campaign Number = * Data element. Identifier for an advertising campaign. *

Campaign Predicted Rating = * Data element. Derivable: total of the Predicted Ratings for Commercial Breaks that have an Occupying relationship with Commercial Spots in a given Advertising Campaign. *

Campaign Requirements = * Data flow. An agency's description of require-ments for an advertising campaign. *
Agency Name + Product Name + Campaign Budget Total
+ Piccadilly Budget Amount + Target Audience
+ Target Rating Percentage + Campaign Duration
+ Campaign Start Date + Campaign End Date
+ {Required Spot Duration}

Campaign Start Date = * Data element. Date marking the first day spots for a campaign will be transmitted. *

Commercial Break = * Entity. Each commercial break represents a number of
        seconds that can be sold for the transmission of advertising
        copy. *
        Break Start Time + Break End Time + Break Duration
        + Break Sold Value + Break Unsold Value + Break Minimum Rate

Commercial Copy = * Entity. The material that is transmitted during the time
        occupied by a commercial spot. *
        <u>Commercial Copy Number</u> + Physical Copy
        + {Copy Transmission Date} + Disposal Date
        * Note that this contains a repeating group so is not yet fully
        normalize. *

Commercial Copy Number = * Data element. Unique identifier for commercial
        copy. *

Commercial Copy Recording = * Data flow plus physical cassette. Commercial
        copy recording sent by the production company to Piccadilly's
        Programme Transmission Department. *
        Commercial Copy Number + Automated Cassette Recording
        + Production Company Name + Agency Name + Product Name

Commercial Spot = * Entity. Represents time that has been sold to an adver-
        tising agency. *
        <u>Spot Number</u> + Spot Duration + Spot Booking Agreement

Competing = * Relationship. Keeps track of which opposition programmes are
        in competition with the episodes that Piccadilly plans to trans-
        mit. Cardinality: for each Episode, there are many Opposition
        Programmes; for each Opposition Programme, there are many
        Episodes. Participation: Episode optional, Opposition
        Programme optional. *

Containing = * Relationship. Keeps track of which commercial breaks are
        planned for a breakchart day. Cardinality: for each Commercial
        Break, there is one Breakchart Day; for each Breakchart Day,
        there are many Commercial Breaks. Participation: Breakchart
        Day optional, Commercial Break mandatory. *

Copy Disposal Instructions = * Data flow. Instruction from an agency to dispose of an outdated commercial copy recording. *
Commercial Copy Number + Product Name + Disposal Date

Copy Register = * Data store *
{Commercial Copy Number + Product Name + Campaign Number + {Copy Transmission Date} + (Disposal Date)}

Copy Transmission Date = * Data element. The date that a piece of copy is scheduled to be transmitted. *

Copy Transmission Instructions = * Data flow. Details of when particular commercial copy is to be transmitted. *
Agency Name + Product Name + Commercial Copy Number + {Copy Transmission Date}

Current Rate = * Data element. Derivable: the average rate currently being paid for a 30-second spot within a particular break. Units: pounds sterling. Integer. *

Daily Revenue = * Data element. Revenue made by Piccadilly for all the commercial spots transmitted on one breakchart day. Derivable: total of all Spot Prices for every Commercial Spot having a Transmitting relationship with the Commercial Breaks on today's date. *

Directing = * Relationship. Cardinality: for each Director, there are many Programmes; for each Programme, there is one Director. Participation: Director mandatory, Programme optional. *

Director = * Entity. The director of a programme. *
<u>Director Name</u>

Director Name = * Data element. Identifies the director of a programme. *

Displaced Spot = * Data flow *
Spot Number + Spot Price + Rate Moveability + Spot Booking Agreement

Disposal Copy Material = * Data flow *
Commercial Copy Number + Physical Copy

Disposal Date = * Data element. Date by which a piece of commercial copy
must be destroyed. *

Episode = * Entity. An episode is one occurrence of a programme. *
Episode Number + Episode Scheduled Date
+ Episode Start Time + Episode End Time

Episode End Time = * Data element *

Episode Number = * Data element *

Episode Predicting = * Relationship. Cardinality: for each Episode, there are
many Predicted Ratings; for each Predicted Rating, there is one
Episode. Participation: Episode optional, Predicted Rating
mandatory. *

Episode Scheduled Date = * Data element *

Episode Start Time = * Data element *

Filming = * Relationship. Cardinality: for each Production Company, there are
many Commercial Copy(s); for each Commercial Copy, there is
one Production Company. Participation: Production Company
mandatory, Commercial Copy mandatory. *

Invoice = * Entity. A record of money owed by an advertising agency. *
Invoice Number + Invoice Total + Invoice Date

Invoice Date = * Data element *

Invoice Number = * Data element *

Invoice Total = * Data element. Derivable: sum of the Spot Price related to
each Commercial Spot related to a given Invoice. *

413

Legal Placement = * Relationship. Keeps track of the ratecard-selling conditions concerning the price of a spot and the segments of time
that it can occupy. Cardinality: for each Spot Rate, there are
many Ratecard Segments; for each Ratecard Segment, there
are many Spot Rates. Participation: Spot Rate mandatory,
Ratecard Segment mandatory. *

Likely Spot Preemption = * Data flow. Trends in airtime sales indicate that
this spot may be preempted. *
Spot Number + Product Name + Break Start Time
+ Spot Duration + Rate Moveability + Spot Price
+ Recommended Rate

Measuring = * Relationship. Keeps track of the actual ratings for an episode.
Cardinality: for each Episode, there are many Actual Ratings;
for each Actual Rating, there is an Episode. Participation:
Actual Rating mandatory, Episode optional. *

Moveability = * Relationship. Keeps track of the ratecard segments in which a
particular commercial spot may be transmitted. Cardinality:
for each Commercial Spot, there are many Ratecard Segments;
for each Ratecard Segment, there are many Commercial Spots.
Participation: Commercial Spot mandatory, Ratecard Segment
optional. *

Moveable Spots = * Data flow *
Break Start Time + {Spot Number + Spot Duration + Spot
Price + Rate Moveability + Spot Booking Agreement}

New Agency = * Data flow *
Agency Name + Agency Address + Agency Phone Number

New Agency Form = * Data flow *
Agency Name + Agency Address + Agency Phone Number

New Agency Transaction = * Data flow *
Agency Name + Agency Address + Agency Phone Number

New Breaks = ✳ *Data flow. Breaks progressively added to the breakchart for three months in the future.* ✳
{Breakchart Date + Break Start Time + Break End Time
+ Break Duration}

New Programme = ✳ *Data flow. Describes a programme offered by an English or overseas programme supplier.* ✳
Programme Name + Programme Type + Programme Description
+ Programme Duration + Programme Price
+ Programme Episodes + {Performer Name} + (Producer Name
+ Director Name) + Supplier Name

New Sales Executive = ✳ *Data flow* ✳
Sales Executive Name + Sales Executive Address
+ Sales Executive Start Date

New Spot Rates = ✳ *Data flow. A transaction in the current computer system.*✳
Ratecard

Occupying = ✳ *Relationship. Keeps track of which spots are occupying a break so that the spot manipulators can make decisions about moving and slotting spots. Cardinality: for each Commercial Spot, there is one Commercial Break; for each Commercial Break, there are many Commercial Spots. Participation: Commercial Spot optional, Commercial Break optional.* ✳

Occurring = ✳ *Relationship. Keeps track of the episodes of a programme. Cardinality: for each Programme, there are many Episodes; for each Episode, there is one Programme. Participation: Programme optional, Episode mandatory.* ✳

Opposition Company = ✳ *Entity. Piccadilly competes with the government channels for viewing audiences. Piccadilly also competes with other commercial channels for the advertiser's money.* ✳
<u>Television Company Name</u>

Opposition Predicted Rating = ✳ *Data element. The rating predicted for one of the opposition company's programmes.* ✳

415

Opposition Programme = * Entity. Piccadilly's programme purchasing and
             scheduling is influenced by the programmes planned for trans-
             mission by other commercial channels and by the government
             channels. *
             <u>Opposition Transmission Date + Opposition Transmission Time</u>
             <u>+ Opposition Programme Name</u> + Opposition Predicted Rating

Opposition Programme Name = * Data element. The name of a programme
             transmitted by an opposition company. *

Opposition Schedule = * Data flow *
             Television Company Name + {Opposition Transmission Date
             + Opposition Transmission Time + Opposition Programme Name
             + (Opposition Predicted Rating)}

Opposition Transmission Date = * Data element. The date an opposition pro-
             gramme is scheduled for transmission. *

Opposition Transmission Time = * Data element. The time an opposition pro-
             gramme is scheduled for transmission. *

Performer = * Entity. Actor or actress appearing in a programme. *
             <u>Performer Name</u>

Performer Name = * Data element. Identifier of an actor or actress appearing
             in a programme. *

Physical Copy = * A physical copy of the material to be transmitted for a
             commercial spot. Current system uses ACR. *

Piccadilly Budget Amount = * Data element. The portion of the campaign
             budget total that the agency intends to spend with Piccadilly.
             Units: pounds sterling. *

Planning = * Relationship. Cardinality: for each Opposition Company, there are
             many Opposition Programmes; for each Opposition Programme,
             there is one Opposition Company. Participation: Opposition
             Company optional, Opposition Programme mandatory. *

Predicted Rating = * Entity. Predicted ratings are used to plan advertising
campaigns and programme scheduling. *
Predicted Rating Percentage + Predicted Rating Date

Predicted Rating Date = * Data element. Date on which a predicted rating
was made. *

Predicted Rating Percentage = * Data element. Percentage of an audience
type predicted to watch a given programme episode. *

Predicted Ratings = * Data flow. The Research Department's predictions of
programme ratings. *
{Breakchart Date + {Programme Name + Episode Number
+ {Audience Type + Predicted Rating Percentage
+ Predicted Rating Date}}}

Preempted Spot = * Data flow. A spot that has been preempted by another
spot and dropped from the breakchart. *
Spot Number + Spot Duration + Rate Moveability

Preemption Replacement = * Data flow. Details of a spot that has been
booked to replace a preempted spot. *
Agency Name + Product Name + Campaign Number
+ Preempted Spot + [Replacement Spot | Upgrade Unavailable]

Preemption Warning = * Data flow. A sales executive warning to an agency
that a spot is in danger of being preempted . *
Agency Name + Product Name + Campaign Number
+ Spot Number + Breakchart Date + Break Start Time
+ Spot Duration + Recommended Rate

Preliminary Schedule = * Data flow. The Programme Transmission Department's
first cut of a programme transmission schedule. *
{Programme Transmission Date + {Programme Name
+ Episode Start Time + Episode End Time}}

Priced Time =∗ Data flow ∗
          Agency Name + Product Name + Campaign Start Date
          + Campaign End Date + {Breakchart Date + {Break Start Time
          + Seconds Available + Total Available Price + Current Rate
          + Audience Type + Average Break Predicted Rating}}

Pricing Agreement = ∗ Relationship. Keeps track of ratecard pricing conditions
          under which a spot is sold. Cardinality: for each Commercial
          Spot, there is one Spot Rate; for each Spot Rate, there are
          many Commercial Spots. Participation: Commercial Spot
          mandatory, Spot Rate optional. ∗

Producer = ∗ Entity. The producer of a programme. ∗
          <u>Producer Name</u>

Producer Name = ∗ Data element. Identifies the producer of a programme. ∗

Producing = ∗ Relationship. Cardinality: for each Programme, there are many
          Producers; for each Producer, there are many Programmes.
          Participation: Programme optional, Producer mandatory. ∗

Product = ∗ Entity. Piccadilly needs to record which products are being adver-
          tised because some of the broadcasting rules govern the place-
          ment of commercials depending on the product. ∗
          <u>Product Name</u>

Production Company = ∗ Entity. Producer of television commercials. ∗
          <u>Production Company Name</u>

Production Company Name = ∗ Data element ∗

Product Name = ∗ Data element. Identifier for one of the products represent-
          ed by an advertising agency. ∗

Product Number = ∗ Data element. Piccadilly's unique identifier for a
          product. ∗

Programme = * Entity. A television programme made by Piccadilly or bought
    from a programme supplier. *
    <u>Programme Name</u> + Programme Type + Programme Description
    + Programme Duration + Programme Price
    + Programme Episodes + Programme Purchase Date

Programme Description = * Data element. Synopsis of the contents of a pro-
    gramme. *

Programme Duration = * Data element. Running time of a programme.
    Units: hrs/mins/secs. *

Programme Episodes = * Data element. The number of episodes of a pro-
    gramme covered by an agreement with a supplier or by
    Piccadilly's internal production plans. *

Programme Name = * Data element. The name that uniquely identifies this
    programme, for example, News at Ten, Brideshead Revisited,
    Coronation Street. *

Programme Price = * Data element. Price paid to the supplier of a programme.
    Units: pounds sterling. *

Programme Purchase Agreement = * Data flow. Terms negotiated with an
    external supplier. *
    Programme Name + Supplier Name + Programme Price

Programme Purchase Date = * Data element. Date of purchase from a pro-
    gramme supplier. *

Programme Schedule File = * Data store. File kept by the Research
    Department. *
    {Programme Transmission Schedule + Preliminary Schedule
    + Predicted Ratings}

Programme Schedules = * Data store. File kept in the Sales Department. *
  {Programme Transmission Schedule} + {Predicted Ratings}

Programme Supplier = * Entity *
  <u>Supplier Name</u>

Programme Transmission Date = * Data element. The actual transmission
  date of a particular episode. *

Programme Transmission Schedule = * Data flow. Piccadilly's planned trans-
  mission for the next quarter. *
  {Programme Transmission Date + {Episode Number
  + Episode Start Time + Episode End Time + Programme Name
  + Programme Description + Programme Type
  + Predicted Rating} + {Break Start Time + Break End Time}}

Programme Transmission Time = * Data element. The actual transmission
  time of an episode. *

Programme Type = * Data element *
  [First-Run Film | Sporting Event | Documentary | Talk Show |
  Old Movie]
  * The types of programmes that Piccadilly transmits. Note
  that there are other programme types to add to these. *

Programming Plan = * Data store. File in the Programme Transmission
  Department. *
  {Programme Name + Programme Type + Programme Description
  + Programme Duration + Programme Price
  + Programme Purchase Date + {Performer Name}
  + (Producer Name) + (Director Name) + Supplier Name
  + {Programme Transmission Date}} + Preliminary Schedule
  + Programme Transmission Schedule + Opposition Schedule
  + Programming Rules

Programming Rule = * Data element. A rule set by the Broadcasting Board.
  Each rule addresses some aspect of the mixture, content, and
  placement of programmes and commercial breaks. *

Programming Rules = * Data flow. File in the Programme Transmission
Department. *
{Programming Rule}

Proximity =* Relationship. Keeps track of which breaks are within one hour of
an episode so that the spot manipulators can apply the appro-
priate programming rules when moving and slotting spots.
Cardinality: for each Commercial Break, there are many
Episodes; for each Episode, there are many Commercial Breaks.
Participation: Commercial Break mandatory, Episode optional. *

Publicizing = * Relationship. Keeps track of which product is advertised by a
particular piece of commercial copy. Cardinality: for each
Commercial Copy, there is one Product; for each Product, there
are many Commercial Copy(s). Participation: Commercial Copy
mandatory, Product optional. *

Purchase Decision = * Data flow *
Programme Name + Programme Type + Programme Description
+ Programme Duration + Programme Price
+ Programme Episodes + {Opposition Programme}
+ {Performer Name} + {Producer Name} + (Director Name)
+ Supplier Name + {Episode Scheduled Date
+ Episode Start Time + Episode End Time}

Ratecard = * Data flow. Prices, moveability, and preemption rules of time
available for sale. *
Rate From Date + {Rate Spot Duration + {Rate Segment Day
+ {Rate Segment Start + Rate Segment End
+ {Rate Moveability + Spot Price}}}}

Ratecard File = * Data store. File kept by the Research Department. *
{Ratecard}

Ratecard Period = * Entity. Period during which given rates apply. *
<u>Rate From Date</u>

Ratecards = * Data store. File kept in the Sales Department. There is a duplicate file in the Commercial Booking Department. *
{Ratecard}

Ratecard Segment = * Entity. Continuous band of time defined on ratecard. *
        Rate Segment Day + Rate Segment Start + Rate Segment End

Rate From Date = * Data element. The commencement date for a ratecard
        period. Format: Day/Month/Year. *

Rate Moveability = * Data element. Rate moveability as defined in the rate-
        card. *
        [Fixed: fixed on a nominated day | Broad: moveable within a speci-
        fied segment on a nominated day | ROD: run-of-day, moveable to
        any similarly priced segment on a nominated day | ROW: run-of-
        week, moveable to any similarly priced segment during a week]

Rate Segment Day = * Data element. Day(s) of week on which this segment of
        time occurs. *
        [Weekday | Saturday | Sunday]

Rate Segment End = * Data element. The end time for a ratecard segment. *

Rate Segment Start = * Data element. The start time for a ratecard seg-
        ment. *

Rate Setting = * Relationship. Keeps track of the ratecard segments applic-
        able to a ratecard period. Cardinality: for each Ratecard
        Segment, there is one Ratecard Period; for each Ratecard
        Period, there are many Ratecard Segments. Participation:
        Ratecard Segment mandatory, Ratecard Period mandatory. *

Rate Spot Duration = * Data element. Duration of spots as defined in the
        ratecard. Units: seconds. *
        [10 | 20 | 30 | 40 | 50 | 60]

Rating = * Entity *
        Rating Date + Audience Type

Rating Date = * Data element. The date to which a rating refers. *

Rating Time = * Data element. Time to which an actual rating refers. Ratings are taken every minute. Units: hrs/mins/secs. *

Recommended Rate = * Data flow. The rate recommended for a spot that is in danger of preemption. *
Rate Moveability + Spot Price

Replacement Spot = * Data flow. A spot to replace another spot that has been preempted. *
Product Name + Spot Number + Spot Duration + Spot Price + Rate Moveability

Representation End Date = * Data element. Date an agency ends representing a product. *

Representation Start Date = * Data element. Date an agency starts representing a product. *

Representing = * Relationship. Keeps track of which agency is responsible for a product. Keeps track of historical relationships between agencies and products for collecting bad debts. Cardinality: for each Advertising Agency, there are many Products; for each Product, there are many Advertising Agency(s). Participation: Advertising Agency mandatory, Product mandatory. *
Representation Start Date + Representation End Date

Required Spot Duration = * Data element. Required length of commercial spots for a campaign. See Spot Duration for values. *

Revenue Reports = * Data flow. Computer reports used by management to help set sales targets. *
Breakchart Date + Daily Revenue + {Spot Number + Spot Rate + Product Name + Spot Transmitted Time + Spot Price}

Rule Effective Date = * Data element. The date that a new rule from the Broadcasting Board must be applied to all Piccadilly's programme and commercial transmissions. *

Sales Executive = * Entity *
> <u>Sales Executive Name</u> + Sales Executive Address
> + Sales Executive Start Date

Sales Executive Address = * Data element *

Sales Executive Name = * Data element *

Sales Executive Register = * Data store. File of sales executives' responsibili-
> ties kept by the Sales Department. *
> {Sales Executive Name + {Agency Name + Servicing Start Date
> + Servicing End Date}}

Sales Executive Start Date = * Data element. The date a sales executive
> starts working for Piccadilly. *

Sales Target = * Entity. The revenue that Piccadilly aims to make within a
> specified period. *
> <u>Sales Target From + Sales Target To</u> + Sales Target Amount

Sales Target Amount = * Data element. The amount of a sales target. Units:
> pounds sterling rounded up to the nearest thousand. *

Sales Target File = * Data store *
> {Sales Target Instructions}

Sales Target From = * Data element. The start date for a sales target. *

Sales Target Instructions = * Data flow. Set by Piccadilly management. *
> Sales Target From + Sales Target To + Sales Target Amount

Sales Target To = * Data element. The end date of a sales target. *

Scheduling = * Relationship. Keeps track of the scheduling decisions made concerning episodes, commercial breaks, and broadcasting rules. Cardinality: for each instance of one Episode and one Broadcasting Rule, there are many Commercial Breaks; for each instance of one Commercial Break and one Broadcasting Rule, there are many Episodes; for each instance of one Commercial Break and one Episode, there are many Broadcasting Rules. Participation: Broadcasting Rule optional, Commercial Break optional, Episode optional. *

Seconds Available = * Data element. Derivable: Break Duration minus total of Spot Durations for all Commercial Spots that have an Occupying relationship with this Commercial Break. Units: seconds. Value range: 0 to 180. *

Seconds Sold = * Data element. Derivable: sum of Spot Durations for all Commercial Spots that have an Occupying relationship with a Commercial Break. *

Selected Ratings = * Data flow. Selected records from the television ratings report that are stored by the Research Department. *
Television Ratings Report

Selected Spots = * Data flow. Spots selected by an agency as part of a campaign. *
Agency Name + Campaign Number + Product Name
+ Campaign Start Date + Campaign End Date + {Spot Number
+ Spot Duration + Spot Price + Rate Moveability
+ ([Breakchart Date + Break Start Time | {Breakchart Date}])}

Selling = * Relationship. Keeps track of which sales executive is responsible for selling to an advertising agency. Cardinality: for each Advertising Agency, there are many Sales Executives; for each Sales Executive, there are many Advertising Agency(s). Participation: Advertising Agency mandatory, Sales Executive mandatory. *
Servicing Start Date + (Servicing End Date)

Servicing End Date = * Data element. The date a sales executive stops servicing an advertising agency. *

Servicing Start Date = * Data element. The date on which a sales executive starts servicing an advertising agency. *

Slotted Spot = * Data flow. Relocation of a spot due to its place being taken by another spot at a higher rate. *
Breakchart Date + Break Start Time + Spot Number
+ Spot Price + Rate Moveability + Spot Booking Agreement

Spot Booking Agreement = * Data element grouping. The conditions under which a spot is sold. *
[Spot Date Agreed + Spot Break Agreed | Spot Date Agreed
+ Spot Segment Agreed | Spot Date Agreed | Spot Week
Agreed + {Spot Segment Agreed}]

Spot Break Agreed = * Data element. Agreed break in which a spot must be transmitted. *

Spot Cancellation = * Data flow *
Agency Name + Product Name + Campaign Number
+ Spot Number

Spot Date Agreed = * Data element. Agreed date on which a spot must be transmitted. *

Spot Duration = * Data element. Duration of a commercial spot. Units: seconds. *
[10 | 20 | 30 | 40 | 50 | 60]

Spot Measuring = * Relationship. Keeps track of the actual ratings for a commercial spot. Cardinality: for each Commercial Spot, there are many Actual Ratings; for each Actual Rating, there is one Commercial Spot. Participation: Actual Rating mandatory, Commercial Spot optional. *

Spot Number = * Data element. Identifier for a commercial spot. *

Spot Price = * Data element. Ratecard price for a spot rate. Units: pounds
            sterling. Value range: >= 150 and <= 25000. *

Spot Rate = * Entity *
            Rate Spot Duration + Spot Price + Rate Moveability

Spot Rate File = * Data store. File in the current computer system. *
            {Ratecard}

Spot Segment Agreed = * Data element. Agreed segment in which a spot will
            be transmitted .*

Spot Sticker = * Data flow *
            Spot Number + Spot Duration + Spot Price + Rate Moveability
            + Spot Booking Agreement + Product Name
            + Campaign Number

Spot Transmitted Date = * Data element. The breakchart day on which a
            commercial spot is transmitted. *

Spot Transmitted Time = * Data element. The time when a commercial spot is
            transmitted. Units: hrs/mins/secs. *

Spot Upgrade Request = * Data flow. Increase in rate to avoid preemption. *
            Agency Name + Product Name + Campaign Number
            + {Spot Number + Spot Duration + Rate Moveability}

Spot Week Agreed = * Data element. Agreed week during which a spot must
            be transmitted. *

Spots Transmitted = * Data flow *
            Spot Transmitted Date + {Agency Name + {Campaign Number
            + {Spot Number + Spot Transmitted Time}}}

Suggested Campaign = * Data flow. Suggestions to an agency about the
makeup of an advertising campaign. *
Agency Name + Product Name + Campaign Number
+ Campaign Start Date + Campaign End Date
+ {Required Spot Duration + Spot Price + Rate Moveability
+ ([Breakchart Date + Break Start Time | {Breakchart Date}])}
+ Target Rating Percentage + Campaign Predicted Rating

Suitable Break = * Data flow. Same definition as Moveable Spots. *

Suitable Time = * Data flow *
Agency Name + Product Name + Campaign Start Date
+ Campaign End Date + {Breakchart Date + Break Start Time
+ Seconds Available + Current Rate + Audience Type
+ Average Break Predicted Rating}

Supplier Name = * Data element. Identification for a programme supplier. *

Supplying = * Relationship. Keeps track of the programmes that are supplied
by a programme supplier. Cardinality: for each Programme
Supplier, there are many Programmes; for each Programme,
there is one Programme Supplier. Participation: Programme
mandatory, Programme Supplier optional. *

Target Audience = * Data element. The audience at which a campaign is
aimed. See Audience Type for values. *

Target Rating Percentage = * Data element. Percentage rating aimed for by a
campaign. Integer. Value range: > 0 and <= 100. *

Television Company Name = * Data element *

Television Ratings Report = * Data flow. Statistical analysis of programmes
and types of viewers showing ratings that are measured every
5 minutes during the viewing day. *
{Rating Date + {Programme Name + Rating Time
+ {Audience Type + Actual Rating Percentage}}}

Time Sold = * Data flow *
> {Breakchart Date + {Break Start Time + Break Sold Value
> + {Seconds Sold + Rate Moveability}}} + Total Sold Value

Total Available Price = * Data element. Derivable: ((Available Seconds/30) x
> Spot Price for fixed spot in Ratecard Segment corresponding
> to Commercial with Spot Duration equal to 30 secs). *

Total Sold Value = * Data element. Derivable: sum of Break Sold Value for all
> the Commercial Breaks within a specific period. *

Transmission = * Relationship. Keeps track of when an episode of a pro-
> gramme is transmitted. Cardinality: for each Episode, there is
> one Breakchart Day; for each Breakchart Day, there are many
> Episodes. Participation: Episode optional, Breakchart Day
> optional. *
> Programme Transmission Time + Programme Transmission Date

Transmission Log = * Data store *
> Spot Transmitted Date + {Spot Number
> + Spot Booking Agreement + Spot Price + Rate Moveability
> + Product Name + Spot Transmitted Time}

Transmission Times = * Data flow *
> Spot Transmitted Date + {Spot Number
> + Spot Booking Agreement + Spot Price + Rate Moveability
> + Product Name + Spot Transmitted Time}

Transmission Transaction = * Data flow *
> Spot Transmitted Date + {Spot Number
> + Spot Booking Agreement + Spot Price + Rate Moveability
> + Product Name + Spot Transmitted Time}

Transmitting = * Relationship. Keeps track of when a spot is transmitted.
      Cardinality: for each Commercial Spot, there is one Commercial
      Break; for each Commercial Break, there are many Commercial
      Spots. Participation: Commercial Spot optional, Commercial
      Break optional. *
      Spot Transmitted Time + Spot Transmitted Date

Unavailable Campaign = * Data flow. Piccadilly has no time available to suit
      the campaign requirements. *
      Agency Name + Product Name + Target Audience
      + Target Rating Percentage + Campaign Duration
      + Campaign Start Date + Campaign End Date
      + Required Spot Duration

Unavailable Slot = * Data element grouping. A suitable slot cannot be found
      for this spot at this price. The recommendation is to upgrade
      the spot to the next higher price. *
      Spot Number + Spot Price + Spot Duration

Upgrade Confirmation = * Data flow. Agreement with an agency to upgrade
      the rate of a spot. *
      Agency Name + Campaign Number + Product Name
      + Spot Number + Spot Duration + Spot Price
      + Rate Moveability

Upgraded Rate = * Data flow *
      Spot Booking Agreement + Spot Price + Rate Moveability

Upgrade Unavailable = * Data flow. Message sent to an agency when it is not
      possible to upgrade a preempted spot. *
      Spot Number + Spot Duration

REVIEW: IDENTIFYING EVENTS 3.7

Before You Reached Here ...
You have identified and listed all the events for the Piccadilly system. This exercise was given in Chapter 1.8 *Identifying Events.*

The Context Is Your Guide

The problem statement in Chapter 1.8 suggested that the Piccadilly context diagram from the current physical model (Figure 3.6.1) is the primary input to building the event list. Figure 3.7.1 repeats that diagram except that each boundary data flow is connected to an event by its number in the event list in Figure 3.7.2.

Piccadilly Event List

By going through the remainder of the context diagram and referring to the background material on Piccadilly as well as the lower-level diagrams, we came up with the event list in Figure 3.7.2.

Your task was to name all the events. Let's start with the data flow CAMPAIGN REQUIREMENTS, which is tagged to event 1. Since the flow comes from the terminator ADVERTISING AGENCIES, it's an external event. Something happens in the agencies to cause this data flow to enter the Piccadilly context. We suggested that descriptive names for external events consist of the terminator's name, and the reason that the terminator sends the system the data.

Following our suggestion and using the singular form, you'd name the first part of the event name "Agency." Next consider the reason an agency sends CAMPAIGN REQUIREMENTS to Piccadilly. Remember, Stamford Brook told you, "Our clients are the advertising agencies, which are hired by companies that want to run advertising campaigns for their products." So the reason that agencies send CAMPAIGN REQUIREMENTS to Piccadilly is they want to run a campaign. This means the event should be called *Agency wants to run a campaign.* So much for the first one. You should give the other events similarly descriptive names. (Ours are in the list in Figure 3.7.2.)

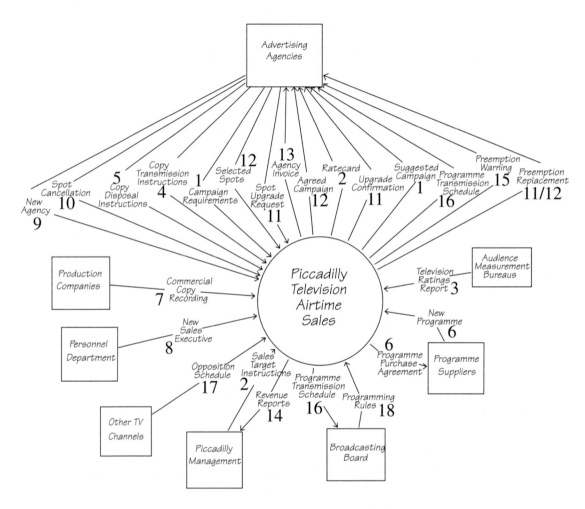

Figure 3.7.1: Piccadilly context diagram with tagged data flows. The numbers correspond to those in the event list in Figure 3.7.2.

Event Name	Associated Data Flows
1. Agency wants to run a campaign	CAMPAIGN REQUIREMENTS (IN) SUGGESTED CAMPAIGN (OUT)
2. Management sets a sales target	SALES TARGET INSTRUCTIONS (IN) RATECARD (OUT)
3. Bureau prepares TV ratings	TELEVISION RATINGS REPORT (IN)
4. Agency decides the transmission instructions for a commercial	COPY TRANSMISSION INSTRUCTIONS (IN)
5. Agency decides a commercial is outdated	COPY DISPOSAL INSTRUCTIONS (IN)
6. Supplier wants to sell a new programme	NEW PROGRAMME (IN) PROGRAMME PURCHASE AGREEMENT (OUT)
7. Production company makes a commercial	COMMERCIAL COPY RECORDING (IN)
8. Personnel hires a sales executive	NEW SALES EXECUTIVE (IN)
9. New agency wants to do business	NEW AGENCY (IN)
10. Agency cancels a spot	SPOT CANCELLATION (IN)
11. Agency wants to upgrade a spot	SPOT UPGRADE REQUEST (IN) UPGRADE CONFIRMATION (OUT) PREEMPTION REPLACEMENT (OUT)
12. Agency chooses spots for a campaign	SELECTED SPOTS (IN) AGREED CAMPAIGN (OUT) PREEMPTION REPLACEMENT (OUT)
13. Spots are transmitted	AGENCY INVOICE (OUT)
14. Time to analyze revenue	REVENUE REPORT (OUT)
15. Time to analyze the breakchart	PREEMPTION WARNING (OUT)
16. Time to finalize new programme schedule	2:2 {PROGRAMME TRANSMISSION SCHEDULE} (OUT)
17. Another channel sets a schedule	OPPOSITION SCHEDULE (IN)
18. Broadcasting Board makes rules	PROGRAMMING RULES (IN)

Figure 3.7.2: Event list for Piccadilly.

You were also asked to annotate your list with all the data flows connected with each event response. The current physical model for the Sales Department (Figure 3.6.5) reveals that SUGGESTED CAMPAIGN is output from this event response. That gives us

1. Agency wants to run a campaign   CAMPAIGN REQUIREMENTS (IN)
                                               SUGGESTED CAMPAIGN (OUT)

Temporal events take place because the system has a contract with a terminator to provide information at a certain time. You begin temporal event names with "Time to," and you add whatever the system is expected to do. For example, the information about the Programme Transmission Department in Chapter 1.4 *The Piccadilly Organization* told you, "Every quarter, finalized programme transmission schedules are sent to the Sales, Commercial Booking, and Research departments, as well as to the Broadcasting Board for its review. The version of the schedule that Perry sends to all of the agencies highlights the new high-rating programmes in the hope that it will encourage them to book their spots early."

This description says that an event occurs whenever it is *Time to finalize new programme schedule,* and the associated output data flow is, naturally enough, PROGRAMME TRANSMISSION SCHEDULE. This is in our list as event 16. Notice the annotation 2:2 {PROGRAMME TRANSMISSION SCHEDULE} to indicate that two copies of the schedule leave the context of the system as a result of event 16. Your context diagram shows that one copy is sent to the advertising agencies and another is sent to the Broadcasting Board. We know there are three other copies of the schedule that are sent to internal Piccadilly departments, but you are not concerned with these internal flows when making your event list. For the moment you are focusing only on the context flows for each event because this will provide you with minimally connected subsystems, one for each event. Later, when you model the details of each event response, you will deal with these internal flows.

Note that we are only dealing with the events that are caused by the flows in the context diagram. These flows concentrate on activities that are fundamental to the system. If your work in event partitioning caused you to identify other potential context flows and events that are concerned with maintaining the data, congratulations! You have exceeded requirements. We'll come back later to discuss these other events.

✚ Ski Patrol

We anticipate you might have had problems with a few things: first, the names. The event names you choose should be descriptive enough to indicate the likely response that the system makes. If the name is too general, or it only describes the data flow's arrival, your users may well say, "So what?" and be unable to confirm the information. On the other hand, if you have good, descriptive names, the next part of the Project—building event-response models—will be a lot easier. Take a few moments now to review your event names and to satisfy yourself that they cannot be improved.

A difficult part of this exercise is deciding exactly what is an event, and what is part of an event. Think of an event response as a chain reaction that continues until all the resulting data flows have reached data stores or terminators.

The response may end before you expect. Take, for example, event 1 *Agency wants to run a campaign*. The incoming flow CAMPAIGN REQUIREMENTS triggers the sales executive into action. When the response generates the flow SUGGESTED CAMPAIGN, the action finishes. Because SUGGESTED CAMPAIGN goes to a terminator, and the system has to wait for the terminator's action, that's the end of the response. The executive now has to wait until the agency decides what it's going to do. When the terminator does act, it is a separate event, and has its own response. The agency contact may call back soon to say that the campaign will run, or the agency may take a few days to call back. (When the agency responds, it is event 12.) The agency may never call back.

A response may generate some data flows that you didn't expect. We show event 12 *Agency chooses spots for a campaign* with these data flows:

SELECTED SPOTS (IN)
AGREED CAMPAIGN (OUT)
PREEMPTION REPLACEMENT (OUT)

Why is PREEMPTION REPLACEMENT included as part of the response to this event? Why isn't it the response to a separate event *Spot is preempted*? To answer this, let's look at what happens when the flow SELECTED SPOTS arrives at Piccadilly (see Figure 3.7.3).

The agency tells the sales executive which spots are wanted for a campaign. The executive tells the Commercial Booking people about the agreed campaign so that they can put the spots on the breakchart. Sometimes, placing new spots on the breakchart causes spots from another campaign to be preempted. When this happens, the Commercial Booking people tell the appropriate sales executives, who each tell the agency concerned about the preemption and recommend a preemption replacement. In this case, PREEMPTION REPLACEMENT is part of the chain reaction when the system responds to an agency choosing spots for a campaign.

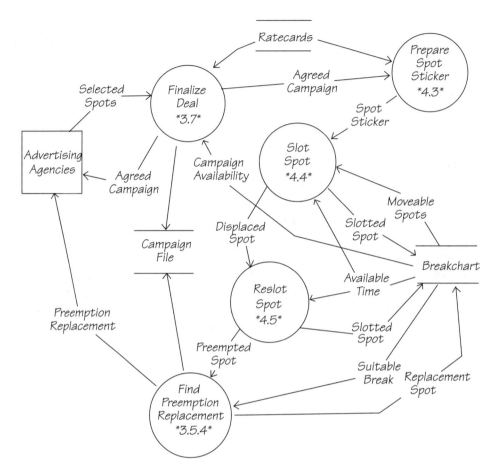

Figure 3.7.3: Event-response model for event 12 **Agency chooses spots for a campaign.** *The numbers in the bubbles are those of the process in the current physical model. They are included to make it easier for you to relate this model to the diagrams in the current physical model.*

Notice that the response to event 11 *Agency wants to upgrade a spot* can also produce the data flow PREEMPTION REPLACEMENT. When a spot is upgraded, it may displace another spot, which in turn may displace another, and so on. If there is no more room on the breakchart for the last displaced spot, it is preempted, and the chain reaction results in a PREEMPTION REPLACEMENT.

Don't be tempted to group similar processes and call it one event. For example, all of the agency's communications about spots might be lumped together under the name *Agency changes its spots.* This event would cover events 10, 11, and

12. Even if the agency sent the same type of message (say it used the same printed form for all three events), the response the system would make is different in each case. If the events require different responses by the system, they are different events. The result of following these guidelines is manageably sized subsystems, one for each event. Later on in the Project, you will model each event response in detail and you will model the inter-event dependencies.

When you are satisfied with your efforts, proceed to your next assignment.

Trail Guide

● ■ ◆ ✳ All trails: Return to Chapter 1.8 *Identifying Events,* to "A Strategic Point" heading, where there is some more information you'll need before proceeding.

3.8 REVIEW: MODELING AN EVENT RESPONSE

Before You Reached Here ...
You have built an event-response model for event 9, as described in Chapter 1.9 *Modeling an Event Response*.

Sample Event-Response Model
Compare your event-response process model for event 9 *New agency wants to do business* with Figure 3.8.1.

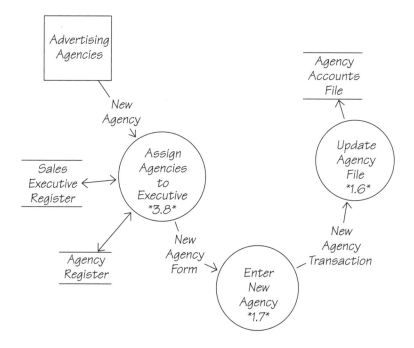

Figure 3.8.1: These are the data stores and processes that currently respond to event 9. The numbers enclosed in asterisks are those of the process in the current physical model. They are included to make it easier for you to relate this model to the diagrams in the current physical model.

Look at process 3.8 ASSIGN AGENCIES TO EXECUTIVE. Every time a new agency applies to do business with Piccadilly, a Sales Department clerk checks the register to find a sales executive who is available to deal with the new agency, and writes the new agency's name next to the executive's name. The clerk adds the new agency to the AGENCY REGISTER, generates a NEW AGENCY FORM, and sends it to the Computer Department.

In the Computer Department, process 1.7 is carried out when the operator enters information about the new agency into the computer system. Process 1.6 UPDATE AGENCY FILE is part of a computer program. It records the new agency in the AGENCY ACCOUNTS FILE.

Your event-response model brings together all the pieces by following the data flows through the system and identifying all the processes, data flows, and data stores that respond to an event. Make sure that your model is substantially the same as the sample. The next stage is to refine the model so that it contains only essential processes and data.

✚ Ski Patrol

If you had trouble isolating the responses to this event, keep in mind that the context of the response is bounded by terminators and data stores. This particular event is triggered by the data flow NEW AGENCY. Once it arrives at the context of the system, you model the system's response by following the data flows until they are either stored or sent to a terminator. In this event response, data flows are stored in SALES EXECUTIVE REGISTER and AGENCY REGISTER. That means the end of the line in that direction. However, the flow NEW AGENCY FORM went off for more processing, so you had to follow the trail to the Computer Department. Eventually, the chain of processes ended when the data flow was stored in AGENCY ACCOUNTS FILE. There are no flows going back to the terminator, so there is nothing left for you to follow.

If you missed some processes, you need to remember that the response to an event continues until all the data flows have been resolved. It doesn't matter that the processes take place in different geographical locations, nor that the implementation may delay this continuous processing. The computer operator may decide to wait until after lunch to enter the new agency information, but there is no essential policy to contradict the notion of continuous processing until the response has finished.

Note that the data flow SALES EXECUTIVE NAME from bubble 3.6 into 3.8 is part of the response to another event, and so is not shown in Figure 3.8.1. Although event 9 does use the data element SALES EXECUTIVE NAME, it gets it from the data store called SALES EXECUTIVE REGISTER, which contains the names of all the sales executives employed by Piccadilly. The physical partitioning of the current system means that parts of two event responses have been bundled together into one process. When you model the response to event 8 *Personnel hires a sales executive*, you will discover that the data flow SALES EXECUTIVE NAME will be included as part of the response to that event.

We took the easy way and built this model in two stages. If you built a model that included only essential stored data and no implementation-dependent processes, you should compare your answer with the one in Chapter 3.9.

If the ✚ Ski Patrol has not provided all the help you need, we suggest you work through Chapter 2.11 *Event-Response Models* before returning to the Piccadilly Project.

Trail Guide

● Easiest, ■ More Difficult, and ◆ Most Difficult: Go to Chapter 1.10 *Refining an Event Response,* where you will remove the physical characteristics of the model you just built.

✻ Promenade: Either go with the others to Chapter 1.10 *Refining an Event Response*, or resume the ✻ Promenade Trail in Chapter 2.10 *Essential Viewpoint*.

REVIEW: REFINING AN EVENT RESPONSE 3.9

Before You Reached Here ...
You refined your event-response model, as assigned in Chapter 1.10 *Refining an Event Response.*

The Event-Response Model

The event-response model collects all of the processes that make up the system's reaction to an event. Once these processes are together, it is easier to determine which of them play a role in the essential policy of the system, and which are there only because of arbitrary implementation choices of the past.

The event-response model that you have built in Chapter 1.9 *Modeling an Event Response* is a collection of processes in the current system that reacted to the event. Now that you have brought the processes together, you can eliminate anything that does not contribute to the essence of the system.

We'll start by eliminating the implementation processes, and then consider the data. You may feel more comfortable starting with the data model, and then determining which processes are necessary to support that data. If so, read our treatment of the data model before you look at the discussion of the processes.

Ignoring the Implementation

A process is implementation dependent if it exists only to support the technology that the business uses. In Figure 3.8.1, the bubble UPDATE AGENCY FILE *1.6* is there because the AGENCY ACCOUNTS FILE is in a computer. The data held in this file are very similar to the data of AGENCY REGISTER that is updated by the process ASSIGN AGENCIES TO EXECUTIVE *3.8*. The reason for the duplication is that one file is a computer file, and the other a manual file in the Sales Department. The sales people want their own file because they cannot get fast enough access to the computer file. The computer people need to hold the information in a database. While there is an essential requirement to remember data about agencies, there is no essential need to remember it in two places using two different technologies. So, for the moment, connect AGENCY ACCOUNTS FILE directly to bubble ASSIGN AGENCIES TO EXECUTIVE *3.8* using the data flow NEW AGENCY FORM. Once you have done that, you can eliminate bubble UPDATE AGENCY FILE *1.6*.

Now let's run bubble ENTER NEW AGENCY *1.7* through the Techn-O-Filter and see if any essential policy remains. The computer operator enters the NEW AGENCY FORM into the computer system as a NEW AGENCY TRANSACTION. The data dictionary definitions for NEW AGENCY FORM and NEW AGENCY TRANSACTION are identical, and this process does not have any other inputs or outputs. So the bubble does nothing more than transport data from one technology, the clerk, to another, the computer system. There is no essential policy here, and as you have already eliminated the process that received the NEW AGENCY TRANSACTION, you can do the same to ENTER NEW AGENCY *1.7*.

You are left with bubble ASSIGN AGENCIES TO EXECUTIVE *3.8*. At the moment, it is recording the new agency, as well as assigning the executive. You can, if you wish, split this functionality into two bubbles. We are leaving it as one as we think that the process is simple enough to specify. We can always break it up later should the need arise. The last physical detail we can remove from the bubble is the number *3.8* that ties it to the current physical viewpoint. The essential number of the bubble is 9 to tie it to event 9. The resulting model is in Figure 3.9.1.

Congratulate yourself if you raised the question "Should there be a data flow to the agency to tell who the assigned sales executive is?" This is a business policy question and can only be answered by the users. The analyst's job is to raise the question. Detailed event-response modeling will help you to think of questions that may not have occurred to you before. To keep things simple, we'll act the part of the users and answer that there is no need for a data flow back to the advertising agency.

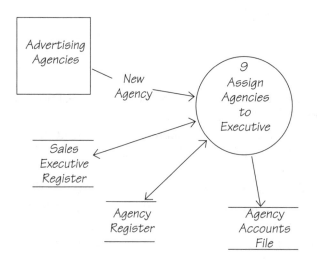

Figure 3.9.1: The more essential event-response process model for event 9 **New agency wants to do business**. *The implementation-dependent processes have been eliminated, but the current system's data stores are still there.*

Connecting to the Data Model

Now we must investigate the model's stored data, with the objective of replacing each access to a physical data store with the essential data entities and relationships used by that access. The bubble that assigns the new agency to a sales executive has four accesses:

1. It records the agency's name, address, and phone number in the AGENCY ACCOUNTS FILE.

2. It looks at the SALES EXECUTIVE REGISTER to find the sales executive most suitable for selling to the new agency.

3. It records the agency's name, address, and phone number, and the sales executive's name in the AGENCY REGISTER.

4. It records the agency's name and the servicing start date in the SALES EXECUTIVE REGISTER.

You'll notice some duplication in the data storage, since in the current physical world, technological boundaries sometimes make this necessary. In the essential world, each piece of data needs to be remembered only once. To eliminate the duplication, partition the essential data into data entities and relationships. Look at the data elements that are stored or retrieved by the process, and attribute them to the appropriate data entity or relationship.

Accesses 1 and 3 from the above list give you a data entity

Advertising Agency = Agency Name + Agency Address + Agency Phone Number

Accesses 2 and 3 yield an entity

Sales Executive = Sales Executive Name

and a data–bearing relationship

Selling = Servicing Start Date

Why is this a relationship? Access 4 tells you that the agency name is recorded against the selected sales executive in the SALES EXECUTIVE REGISTER because Piccadilly needs to know which executive services which agency. However, the SALES EXECUTIVE entity cannot hold data about an agency. (Remember that all the attributes of an entity describe only that entity.) To remember the link between an executive and the agency, we must establish a relationship between the two.

The SERVICING START DATE is an attribute of the relationship. It doesn't describe the executive, and it doesn't describe the agency. However, it does describe the relationship. In fact, if there was no relationship between the entities, there would be no reason to remember this date.

Figure 3.9.2 is the essential data model for the response to the event 9 *New agency wants to do business*. The C and R annotation is the start of a CRUD check, which will eventually be used to verify all the events and to ensure that no process nor data has been missed. For instance, the R in the SALES EXECUTIVE data entity means that this event references that entity. It does this to get the name of the sales executive chosen to deal with the account. Similarly, the SELLING relationship is referenced to assess each sales executive's work load. When an executive is selected, the event response creates a selling relationship between that executive and the newly created agency.

If you have not been using an early version of the data model that you built in the Project (Figure 3.2.8 or Figure 3.4.1), now is a good time to refer to it. There are data entities called ADVERTISING AGENCY and SALES EXECUTIVE and a SELLING relationship between them. You added these to the model because your knowledge of the business led you to suspect that you needed them. It seemed like a reasonable assumption, although at the time you had no details or proof. Now the event-response model confirms your choice. Moreover, it has established some of the attributes for the data entities and the relationship.

Figure 3.9.2: The essential data model that satisfies the requirement to relate an advertising agency with the sales executive responsible for selling to it. This models the event response from the point of view of the essential data.

The next step is to replace the physical files in the event-response model with the essential entities and relationships.

The model in Figure 3.9.3 shows the relationship SELLING as a data store because it contains data, the SERVICING START DATE. When relationships do not contain data, you should omit them from the event-response process model.

That almost completes the first event-response model for the Project. Now update your data dictionary with the preliminary definitions of the new entities and relationship. The next step is to write a mini specification for the process.

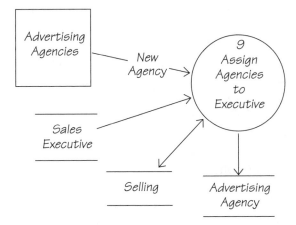

Figure 3.9.3: *The essential event response process model for event 9. Notice that the names used for the entities and relationship in the essential data model are identical to the names of the stores in the essential process model.*

We've constructed this event-response model in a very detailed, methodical way. Our intention was to give you a complete explanation of the procedure for building these models. Now that you understand it, you can speed up a little. For example, you need not draw two versions of the model with the implementation-dependent processes removed. Nor do you need to draw the current physical response in the first place. Once you feel you understand the actions and data of the event response, you can skip the first two steps and model only the essentials. If you find this doesn't work for you, you can always revert to the step-by-step approach. Although there are benefits in having separate event-response data models, you may feel happier modeling the response's data in the system data model.

Use the remainder of the Project to find a way of working that is most productive for you without sacrificing the ability to accurately capture all the details.

✚ Ski Patrol

First, don't feel bad if this exercise gave you problems. Differentiating between implementation processes and essential policy is not easy.

You need to separate the method used from the work being done. For example, recording a new agency is the work. Recording it in a book or in a computer file is the method. Note that because some bubbles carry out both essential policy and implementation processing, you of course need to retain the essential part when eliminating the implementation.

Go back over the event-response model you built in Chapter 1.9 *Modeling an Event Response*. For every process and data element, ask whether each has to do with the work or with the method used to do the work. Read carefully through our explanation of why some parts of the model are essential, and why some exist because of the implementation.

It may be useful, now that you have attempted a Piccadilly example, to revisit Chapter 2.10 *Essential Viewpoint* for its definitions of what is essential and what is not. It may also be worth revisiting Chapter 2.11 *Event-Response Models* for the material on event-response data models.

If you have already done that, stick with the Project. Although the pace is going to speed up, there are many more opportunities for you to improve your skills with the Piccadilly work and the discussions about each chapter.

Here we go.

Trail Guide

This duplicates the Trail Guide in Chapter 1.10 *Refining an Event Response*.

● Easiest: Now that you have an essential event response, it's time to describe its processes. Go to Chapter 2.12 *Mini Specifications*.

■ More Difficult: Since you already know how to write mini specifications, you can proceed directly to Chapter 1.11 *Writing Mini Specifications*.

◆ Most Difficult: Proceed straight to Chapter 1.11 *Writing Mini Specifications*.

✻ Promenade: Writing mini specifications is probably a bit out of your line, but you should know at least what they are. Go to Chapter 2.12 *Mini Specifications* to read about them.

REVIEW: WRITING MINI SPECIFICATIONS 3.10

Before You Reached Here …

You have written a mini specification for the process ASSIGN AGENCY TO EXECUTIVE. This problem was defined in Chapter 1.11 *Writing Mini Specifications.*

Reviewing the Specification

We chose structured language for our first version of the specification, and translated Stamford Brook's information like this:

Mini specification 9 Assign Agency to Executive

For a NEW AGENCY

   Find SALES EXECUTIVE(s) who has fewer than 10 SELLING relationships with ADVERTISING AGENCY(s)

   Select the executive with the least number of SELLING relationships

   If more than one SALES EXECUTIVE has the least number

      Select an executive with the earliest SERVICING START DATE for the most recent SELLING relationship

   Create an occurrence of ADVERTISING AGENCY

   Create a SELLING relationship between ADVERTISING AGENCY and the selected SALES EXECUTIVE

   Set the SERVICING START DATE to today's date

Figure 3.10.1: The mini specification for process 9 ASSIGN AGENCY TO EXECUTIVE.

The second line of the mini specification raises the first questions for the users: "What if there are no executives with fewer than ten agencies? Do I assign just

447

anyone? Do I send an alert to sales management?" Unfortunately, Stamford Brook is away from the office this week, and cannot resolve these questions immediately. In the meantime, you can move on to other issues.

Combining the Models

We suggested that you needed the event–response data model in Chapter 3.9 for this part of the Project. Have a look at it now. Figure 3.9.2 shows which entities and relationships are involved in this particular event response. For instance, the C inside the ADVERTISING AGENCY entity indicates that this event response may create a new instance of that entity. So your mini specification must mention the creation of this entity. Similarly, the creation and referencing of the SELLING relationship and the referencing of the SALES EXECUTIVE must be defined in the specification.

While you can annotate the event–response data model to show the specific entities and relationships that are created, referenced, updated, and deleted, you need the mini specification to define the rules for these operations.

The mini specification also has a strong connection to the data dictionary. When the specification says "Create an instance of ADVERTISING AGENCY," the meaning of ADVERTISING AGENCY has already been defined in the data dictionary (see Chapter 3.6 *Complete Current Physical Model*). This is true of all the relationships and data flows mentioned in the specification.

Using an Alternative Mini Specification

There is an alternative mini specification (see Figure 3.10.2) that is less procedural and consequently much briefer than the previous example. The data model represents the condition of the stored data after the system has responded to the event. It is included in the specification along with a minimal specification of the decisions to be made. The CRUD notation in the data model shows the possible actions of this event response on the data.

This alternative specification method is just as complete as the first version, but by including the event–response data model, it is less wordy.

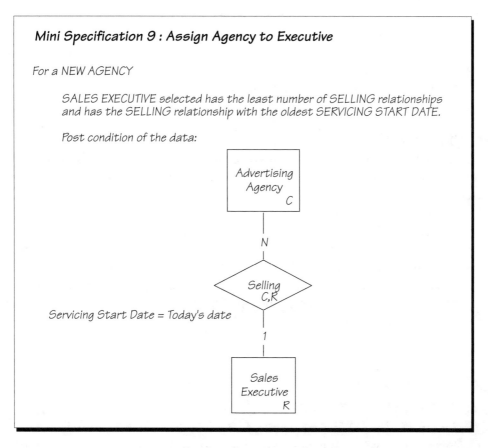

Figure 3.10.2: This alternative mini specification combines the data model with structured language.

More Questions for the User

Mini specifications are usually the final component of the essential requirements model to be developed. When you write them, you confirm your understanding of the system's business policy. Look for any inconsistencies and omissions remaining in your model. The detailed mini specification sometimes raises new questions, for example:

1. "Should the system send an acknowledgment to the agency telling them which sales executive has been assigned?"

2. "Should there be a data flow going to sales management, and/or the executive, telling them who has been assigned?"

3. "Who is chosen if all the sales executives are already selling to ten advertising agencies?"

That completes one mini specification. By the time you finish the Project, you will have specified all the processes in the essential model in a similar manner.

✚ Ski Patrol

You didn't have to use one of the two techniques that we demonstrated. Other techniques are quite acceptable, provided they convey the same policy we described.

If you did not write a satisfactory specification, we suggest some remedial work in Chapter 2.12 *Mini Specifications*. Pay particular attention to the exercises there. If you had trouble with the data model, a trip through the event-response data model in Chapter 2.11 *Event-Response Models* may be in order.

Regardless of the trouble you may be having, keep going. You've come too far with the Project to consider giving up. There are many more opportunities to practice and improve your skills.

Trail Guide

● Easiest, ■ More Difficult, and ◆ Most Difficult: You all go straight on with the Piccadilly Project. Your next assignment is waiting in Chapter 1.12 *Another Event Response*.

❋ Promenade: If you want to see more event-response modeling, go with the others to Chapter 1.12 *Another Event Response*; otherwise, resume the ❋ Promenade Trail in Chapter 2.13 *Modeling New Requirements*.

REVIEW: ANOTHER EVENT RESPONSE 3.11

Sample Event–Response Model for Event 1

Our physical event-response model is shown in Figure 3.11.1, which you should compare with your own.

The response to event 1 involves four processes and touches several data stores before delivering its final output data flow to the terminator ADVERTISING AGENCIES. If you didn't have all these processes in your model, remember the rule that the processing for an event response is continuous until the data flows concerned with that event reach either a terminator or a data store.

The model as it appears here is quite physical. It shows, as much as possible, the method that sales executives use to plan campaigns, and it's what you'd see if you were standing in the Piccadilly offices. This is a cautious approach. First, build a physical view of the event response and then, having confirmed it with the users, advance to the essential view.

If you're satisfied with your model, and have resolved all differences with ours, reread "Your Strategy" in Chapter 1.12 *Another Event Response* for advice on building the essential view of this event response. If your model is already in an essential state (because you started with an essential viewpoint and used essential data instead of the current physical files), jump over the ✚ Ski Patrol section to the next section.

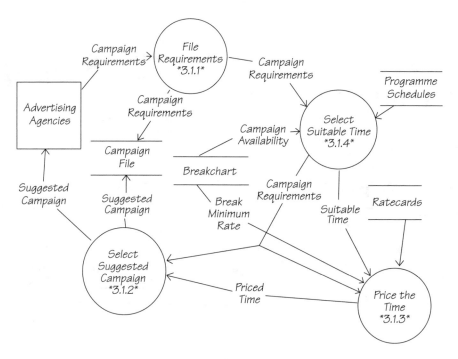

Figure 3.11.1: *The current physical response to the event* **Agency wants to run a campaign.** *The data stores in this model are physical data stores in the current Piccadilly system. Eventually, you will refine the model and replace these physical stores with their essential equivalents.*

✚ Ski Patrol

One of the main problems with building event-response models is knowing when to stop. Think of the event response as being surrounded by terminators and data stores. When the boundary data flow enters, the system spreads out like an octopus—the data flow tentacles reach out to entangle processes and are stopped only by the barriers of data stores and terminators. Read through the interview in Chapter 1.12 again and this time whenever Dollis mentions a process, ask yourself, "Does this process send any data flows to other processes?" If the answer to the question is yes, you know the response is not yet over. Follow each flow to the next process and ask the question again.

If a data flow leads to a store or terminator, you have finished tracking that part of the event response. Go back over the event-response model, and the interviews, and the current physical models, until you understand why these processes and data stores make up this event response. After doing that, check out the idea of linking the processes in Chapter 2.11 *Event-Response Models*.

In any case, make sure that you are comfortable that the sample answer is an accurate representation of what happens inside Piccadilly when an agency wants to run a new campaign. The next step is making the essential version of the event-response model. If you aren't feeling confident about your essential skills, review Chapter 2.10 *Essential Viewpoint* before returning to the Project.

Ignoring Physical Details

We suggested that a way to turn the physical model into an essential one is to replace all the current physical files with essential entities from the preliminary data model. First, you need to look at the content of the data flows to know what data are being accessed. Figure 3.11.2 shows the physical processes accessing stores of the equivalent essential data entities from Figure 3.6.10.

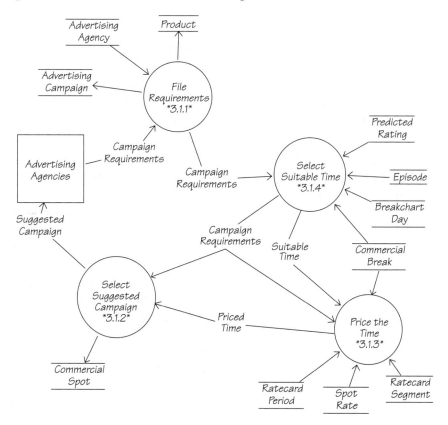

Figure 3.11.2: The first modification to the current physical event-response model. The files have been replaced with entities from the data model.

Be aware that as we look more closely at the processing for this event response, we may need to make corrections to the preliminary data model. Your first question may be, "How did we select these entities to replace the physical files?" The event response needed to react to the incoming data flow that triggered the processing. The data dictionary definition for this flow gave us

Campaign Requirements = * Data flow. An agency's description of requirements for an advertising campaign. *
Agency Name + Product Name + Campaign Budget Total
+ Piccadilly Budget Amount + Target Audience
+ Target Rating Percentage + Campaign Duration
+ Campaign Start Date + Campaign End Date
+ {Required Spot Duration}

Each of these elements must be stored or used. For each of the processes, we determined what data they needed. We then looked for a suitable entity in the data model. If one was not available, we created one and noted its attributes. The first bubble was fairly easy: AGENCY NAME from the data flow was used to reference AGENCY NAME from the ADVERTISING AGENCY entity, PRODUCT NAME was stored in the PRODUCT entity, and the remainder of the flow was stored in ADVERTISING CAMPAIGN.

The process SELECT SUITABLE TIME needed to match the CAMPAIGN REQUIREMENTS with the entities that are the logical equivalent of BREAKCHART and PROGRAMME SCHEDULES. The stores that we show in Figure 3.11.2 are the entities from the preliminary data model that provided the needed data. If this seems as if it is being done with smoke and mirrors, go through the dictionary definition of these entities, and check for yourself that they reconcile with all the elements of the data flow.

Before going on, make sure that you can justify the selection of the entities that we have used to replace the files. Once the essential data are incorporated into the model, the processes can be examined to determine if they are essential or are dependent on the implementation.

In Figure 3.11.2, consider why the sales executive records the campaign requirements in the file before selecting and pricing the time. The reason is that it takes the executive a period of time, sometimes several days, before he finishes planning the campaign and then advises the agency of his suggestions. Since every executive is busy and is interrupted frequently and doesn't want to risk forgetting the agency's requirements, he writes them in the campaign file. But isn't that to do with the implementation, which is a relatively slow, fallible human? If the system doesn't have to be concerned with the limitations of a processor, things would be different.

Also consider that if there were no available airtime or if the agency's budget were insufficient, there would be no campaign and no need to record the campaign requirements. So, essentially, the campaign requirements do not have to be recorded until a campaign has been arranged. Take the FILE REQUIREMENTS bubble off the model for the moment. You can reintroduce the necessary process after the campaign is selected.

The physical model shows us that the campaign requirements are used to select the suitable time. What happens if no time is available? If the executive cannot find enough time for the campaign, there must be some communication with the agency. After all, the executive can't just walk away and forget the campaign. So there must be some rejection flow; let's call it UNAVAILABLE CAMPAIGN and show it from SELECT SUITABLE TIME to ADVERTISING AGENCIES.

The current physical model misses this flow. Dollis Hill forgot to tell you what happens when things don't go as planned. This kind of thing happens all the time in systems analysis. If you missed it here, you would have picked it up when you wrote the mini specification.

The process PRICE THE TIME is essential. Before the final suggestions of the campaign can be formulated, the system must be able to append a price to each of the suitable time slots. Although you could have an essential model that combined SELECT SUITABLE TIME and PRICE THE TIME into one bubble, the complexity of the piece will help you make this decision. If the resulting mini specification is too complex, you must break the process into two.

What about the requirement to record the campaign requirements? The answer must now be that the requirements will be committed to the system's memory only when the system is certain that it is able to offer the agency a campaign. So it is appropriate to record the campaign requirements along with the suggested campaign as part of the final process of the event response. We call this process CREATE SUGGESTED CAMPAIGN.

The resulting model is shown in Figure 3.11.3. The data flow CAMPAIGN REQUIREMENTS flows to 1.1 SELECT SUITABLE TIME and 1.2 CREATE SUGGESTED CAMPAIGN because all the data in the flow are needed by both processes. However, the data are not needed by 1.3 PRICE THE TIME. The physical model showed CAMPAIGN REQUIREMENTS traveling to every process because the executives carried the agency's letter around with them. In the essential model, each process obeys the Rule of Data Conservation and only receives the data that are necessary for it to carry out its policy. In this case, 1.3 PRICE THE TIME only needs the SUITABLE TIME to do its essential task.

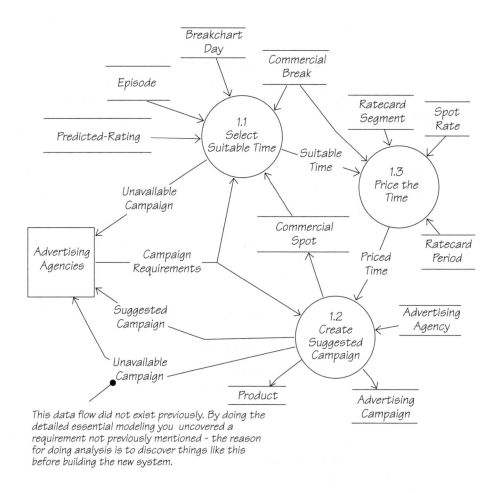

Figure 3.11.3: *This refined version of the event-response process model shows only essential processes and essential data. Note that the three bubbles have been renumbered to indicate they are part of the response to event 1.*

Note the new data flow UNAVAILABLE CAMPAIGN being generated by two bubbles. It can occur first when there is no available time for the campaign, and later when the suggested campaign is being created if the executive discovers that the budgets and targets cannot be met by the price and ratings of the time Piccadilly can make available.

The flows to and from the entity data stores are unnamed as a way of minimizing the complexity of the model. The informative, subject-specific names that you have given to the entity stores make the model's purpose understandable. The mini specifications will specify exactly which data elements are used by each process.

Make sure you can reconcile all the data stores and flows before proceeding.

Connecting to the Data Model

In this refined model (Figure 3.11.3), each access to a physical file was replaced by the essential data entities and relationships used by that access. You can see the essential data in a different way.

In the unrefined event-response model in Figure 3.11.1, the process SELECT SUITABLE TIME is using two physical files: PROGRAMME SCHEDULES and BREAKCHART. Does the bubble really need all the data in both those files to do its job? No, but the current design of the files provides all the data, whether the process needs it or not. Essentially, the bubble is looking at each EPISODE that is associated with a BREAKCHART DAY within the proposed campaign period. The PREDICTED RATINGS for the EPISODE are used to decide whether the COMMERCIAL BREAKs during the programme suit the CAMPAIGN REQUIREMENTS. The bubble determines how much time is available in each COMMERCIAL BREAK by deducting the duration of each COMMERCIAL SPOT that is currently OCCUPYING the break.

By looking outward from the process to the stored data, we can determine all of the essential data for the process. To do it this way, you need to have the data already rationalized into entities and to know the attributes of the entities. We used the data model and data dictionary from Chapter 3.6 to help us with our task. If you had not built a first-cut data model, you would start from scratch and build a data model that is solely concerned with this event's data. Figure 3.11.4 is the event-response data model for this event.

If you haven't already done so, complete your essential event-response process model by ensuring that each of its data stores is equivalent to a data entity or a data-bearing relationship in the event-response data model.

The event-response data model is an accurate representation of only some of the system's essential stored data. This means that you must now reexamine your system data model to ensure that it harmonizes with the smaller event-response version. If you find a difference between them, the event-response model is the correct one. Eventually the combination of all your event-response data models will provide an accurate system data model.

Conversely, the system data model is an aid in building your event-response data models. Use it to get an idea of what data the event needs, and then refine your ideas by asking detailed questions about the real policy of the system. With the distraction of the implementation out of the way, you can concentrate on the essential processes and data needed by the system.

457

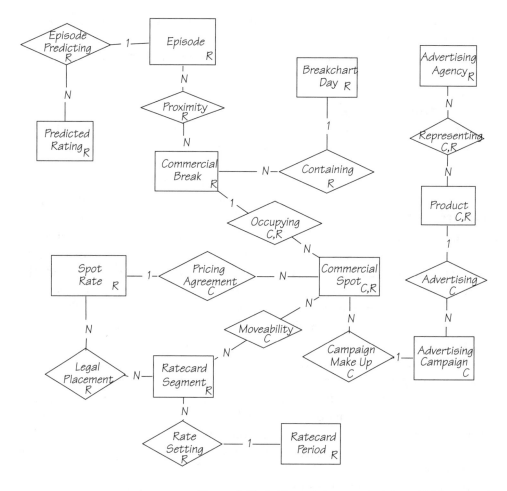

Figure 3.11.4: The event-response data model for event 1. Due to the size of most system data models, we find it easier to focus on the details if we build a separate data model for each event response. Later, you can combine all the event-response data models to form a system data model.

Defining the Essential Activity

An essential activity model is a one-bubble event-response process model. Figure 3.11.5 shows the essential activity model for event 1.

Analysts use essential activity models because they are more manageable, and are quicker to build than a detailed event-response model. You will find that you can often specify event responses with one process. Try starting each of your event-response models by drawing this one-bubble model. It should show all the essential data needed by the activity, but the details of the processing may remain hidden for a while. You can level downward as you need to.

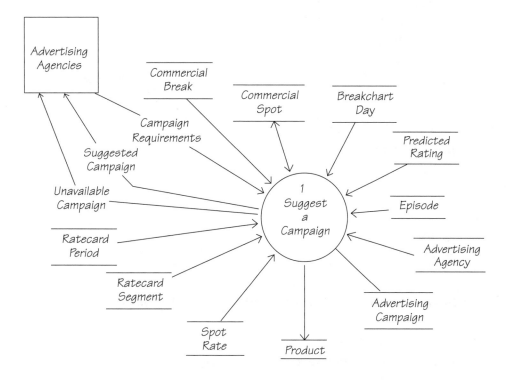

Figure 3.11.5: The essential activity model for event 1. This model was produced by leveling the event-response process model upward and showing it as one essential activity. The numbering of the process ties this model to event 1.

This *top-down approach* lets you see an overview of the event response before plunging into the details. When you have an essential activity model, consider if it is possible to write a suitable mini specification for it. What looks like a complex physical model, may, in fact, turn out to be a simple essential process and can be specified without any further partitioning.

Speaking of mini specifications, we asked you to write a mini specification for the processes in your essential event-response model. Here are the ones we wrote.

Mini specification 1.1 Select Suitable Time

Input: CAMPAIGN REQUIREMENTS
With reference to the event-response data model

Find the COMMERCIAL BREAKs with a CONTAINING relationship to BREAKCHART
DAYs that lie within the CAMPAIGN START DATE and the CAMPAIGN END DATE
>   For each COMMERCIAL BREAK
>   Derive AVAILABLE SECONDS
>   If AVAILABLE SECONDS > 0
>   Derive AVERAGE BREAK PREDICTED RATING
>   If AVERAGE BREAK PREDICTED RATING >= TARGET RATING
>   PERCENTAGE
>   Add defined data to SUITABLE TIME
If > 1 break has been added to SUITABLE TIME
>   Issue SUITABLE TIME
Otherwise
>   Issue UNAVAILABLE CAMPAIGN

Output: SUITABLE TIME, UNAVAILABLE CAMPAIGN

Mini specification 1.2 Create Suggested Campaign

Input: CAMPAIGN REQUIREMENTS, PRICED TIME
With reference to the event-response data model

Select SUGGESTED CAMPAIGN from PRICED TIME according to CAMPAIGN REQUIRE-
MENTS * This job relies on the skill and experience of the sales executive *
If a SUGGESTED CAMPAIGN can be built
>   Create an instance of ADVERTISING CAMPAIGN
>   If PRODUCT NAME does not match an existing PRODUCT
>   Create an instance of PRODUCT
>   Create a REPRESENTING relationship between PRODUCT and
>   ADVERTISING AGENCY
>   Create an ADVERTISING relationship between PRODUCT and ADVER-
>   TISING CAMPAIGN
For each SPOT DURATION added to the SUGGESTED CAMPAIGN
>   Create an instance of COMMERCIAL SPOT
>   Create a CAMPAIGN MAKE UP relationship between COMMERCIAL SPOT and
>   ADVERTISING CAMPAIGN

Create a PRICING AGREEMENT relationship between COMMERCIAL SPOT and
each RATECARD SEGMENT that matches the RATE MOVEABILITY of that
COMMERCIAL SPOT
Issue SUGGESTED CAMPAIGN
Otherwise
Issue UNAVAILABLE CAMPAIGN

Output: SUGGESTED CAMPAIGN, UNAVAILABLE CAMPAIGN

Mini specification 1.3 Price the Time

Input: SUITABLE TIME
With reference to the event-response data model

For each BREAK START TIME
Match the BREAKCHART DATE and BREAK START TIME and CURRENT RATE
with the RATECARD PERIOD and RATECARD SEGMENT and SPOT RATE
Derive TOTAL AVAILABLE PRICE for CURRENT RATE

Output: PRICED TIME

✚ Ski Patrol

The hardest part of this exercise is recognizing what is essential, and what is implementation dependent. This is also probably the hardest thing to do in analysis, and where analysts make most of their errors. Time and experience will be valuable in helping you to avoid errors. After all, essential modeling is not something that you can learn in a day. Also keep in mind that when you are doing this on your own projects, you have the opportunity to discuss the essence with your fellow analysts and users. (Don't forget the users. They are often very good at separating essence from implementation.)

We've given you a detailed description of how we derived the essence of this event. At this stage, if you are not satisfied with your ability to identify the essential processes and data, review the models and the description. Evaluate each stage of the refinement, with the aim of understanding why each component is essential. We hesitate to send you elsewhere, but Chapter 2.10 *Essential Viewpoint* may be of some help here.

If data models are bothering you, try some remedial action, returning to Chapters 2.4 *Data Viewpoint* and 2.5 *Data Models* for a quick review. Since you are no longer bothered by *why* you want them, now you are learning *how* to build them.

Whatever you do, hang in there. There will be many more opportunities to practice working with this kind of model, and lots more examples to learn from.

Trail Guide

● Easiest, ■ More Difficult, and ◆ Most Difficult: Go to Chapter 1.13 *More Events,* where you will model the remaining events for the Piccadilly system.

❋ Promenade: The other trails lead to a lot of work. If you are a genuine promenader, you have just come here for a stroll through the subject, and are probably not interested in doing that much work. As you have already seen the essential models, Chapter 2.13 *Modeling New Requirements* is an appropriate destination.

REVIEW: MORE EVENTS 3.12

Before You Reached Here ... '

You have done an exercise in Chapter 1.13 *More Events* to model the sixteen remaining responses in the event list.

How to Use This Chapter

This chapter contains event-response process models, event-response data models, and mini specifications for each of these sixteen events in the Piccadilly system:

2. *Management sets a sales target*
3. *Bureau prepares TV ratings*
4. *Agency decides the transmission instructions for a commercial*
5. *Agency decides a commercial is outdated*
6. *Supplier wants to sell a new programme*
7. *Production company makes a commercial*
8. *Personnel hires a sales executive*
10. *Agency cancels a spot*
11. *Agency wants to upgrade a spot*
12. *Agency chooses spots for a campaign*
13. *Spots are transmitted*
14. *Time to analyze revenue*
15. *Time to analyze the breakchart*
16. *Time to finalize new programme schedule*
17. *Another channel sets a schedule*
18. *Broadcasting Board makes rules*

The data dictionary definitions that support these models are packaged in Chapter 3.6. The combined data model is included in this chapter as well.

Compare each model that you have built with the corresponding one in this chapter. If you are happy with the result, return to Chapter 1.13 *More Events* to find the next event response you want to model. Keep doing this until you have modeled all the events. If you are dissatisfied with the results you're getting, consult the ✚ Ski Patrol in Chapter 1.13 *More Events* for help.

Modeling the Piccadilly Essential Event Responses

As you build your event-response models, you will discover more about the data. An updated version of the system data model for the Piccadilly Project is included here, as well as the individual event-response data models. Building such individual data models is an effective way to work, for the size and complexity of the system data model make it difficult to focus on the parts that are relevant to one event. Later, you'll consolidate these individual data models and use them to do a CRUD check of the entire Piccadilly system. This check verifies that you have not missed any events.

Some of the sample event-response models contain questions about the business policy. These questions must be answered by the users before a new system can be built. When you build the essential event-response models, you examine the system in far more detail, so you'll probably discover missing or undetermined pieces of business policy. Any questions that you raise for the users at this stage are far more valuable than questions later in the Project because the essential models are far easier to correct than the installed system. If it bothers you to have to change a paper specification, think how much more bother it would be to change installed software.

Event 2 Management sets a sales target

Management decides how much revenue Piccadilly can reasonably generate for the upcoming period. This is not quite as simple as it sounds. Management has to estimate the expected revenue for the whole of the independent network, and then estimate how much Piccadilly's share should be. If the target is set too high, the rates will be too high, and the advertisers will look for better value elsewhere. If the target is set too low, Piccadilly will lose income.

The main part of the response to this event is setting the new ratecard. Figure 3.12.1 shows the system's current response.

You can now start to refine the response. The new ratecard is duplicated and sent to three locations. Usually, duplication like this indicates that the processes that receive the duplicated data are themselves duplicates. For example, bubbles *3.3* and *4.6* are identical because they are storing exactly the same data. Some inves-

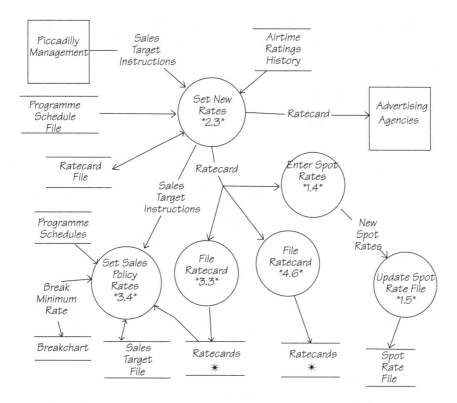

Figure 3.12.1: The current physical model for event 2. The numbers in the bubbles indicate that the response happens in different departments.

tigation in the data dictionary proves that bubbles *2.3* and *1.5* are also recording the same data. There is an essential reason for the system to remember the ratecard, but not multiple times. The reason that the current system looks like this is because the four processes all keep their own copies. Essentially, you only need one. The process SET NEW RATES stores the ratecard information, so both the bubbles called FILE RATECARD (*3.3* and *4.6*) and the bubble called UPDATE SPOT RATE FILES (*1.5*) are redundant (see Figure 3.12.2).

The process ENTER SPOT RATES is part of the current system for technological reasons; the ratecard must be entered into the computer. Essentially, all of the system is executed by a single processor, so there is no need to move data from one processor to another. Let's eliminate this bubble.

UPDATE SPOT RATE FILE is another duplicate. It does the same thing as the bubbles called FILE RATECARD. The file it updates, SPOT RATE FILE, has a data dictionary definition of {RATECARD}. The definition of RATECARD FILE that is updat-

465

ed by the process SET NEW RATES is {RATECARD}. Clearly, the process UPDATE SPOT RATE FILE is repeating some of the function of another bubble. There is no reason for keeping it in the essential model.

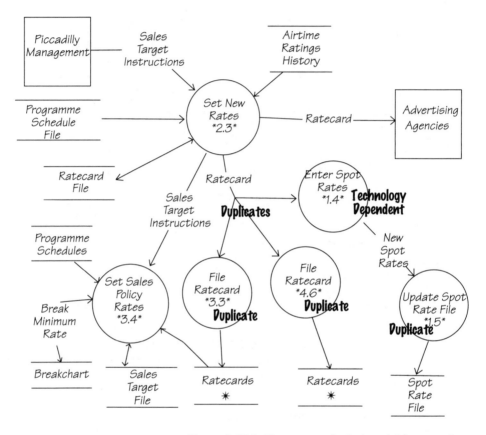

Figure 3.12.2: The current physical model for event 2, annotated to show the nonessential processes and data.

You will often find that a single physical bubble is involved in several events. This happens when a person or a processor is involved in multiple tasks, typically because the tasks are part of somebody's job or because they have related subject matter. The partitioning of the physical model results in a bubble that forms part of the response to several events. However, the tasks that were grouped because of a physical design decision now must be shown as fragments of the different events they respond to.

In the Sales Department, process 3.4 manages the sales policy. Fragments of this process respond to two different events. In event 2, the one we are concerned with here, the bubble SET SALES POLICY RATES represents the fragment of the event response that records the sales target.

However, the task of recording the sales target is simple enough to be amalgamated with the remainder of the essential process. When we eliminate the duplication from the model in Figure 3.12.2 and amalgamate the sales target recording, we are left with one essential process. The physical files can now be replaced with the essential entities, and the resulting essential model is shown in Figure 3.12.3.

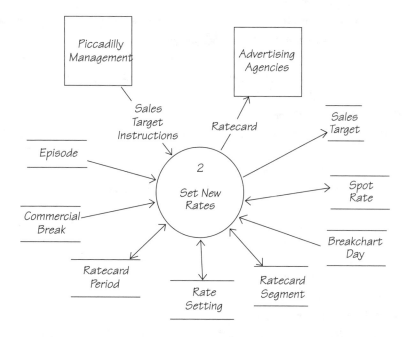

Figure 3.12.3: The essential event response for event 2. The data stores are entities from the data model.

Now let's consider the mini specification for this process. The fundamental idea is for the process to set new rates such that they will generate enough revenue to meet the sales target. You were not given all the information you needed, so you may have a different specification. Here's what we came up with:

Mini specification 2 Set New Rates

Input: SALES TARGET INSTRUCTIONS
With reference to the event-response data model

✳ First see if a SPOT PRICE on the current ratecard will satisfy the sales target ✳

For each date within the SALES TARGET FROM and SALES TARGET TO

    For each EPISODE with that EPISODE SCHEDULED DATE

        For each COMMERCIAL BREAK that has a PROXIMITY relationship

            Using the latest RATECARD PERIOD

            Find the RATECARD SEGMENT corresponding with the BREAK START TIME

            Find related SPOT RATE with RATE MOVEABILITY = "BROAD" and with RATE SPOT DURATION = 30 seconds

            ESTIMATED BREAK REVENUE =(BREAK DURATION / 30) x (SPOT PRICE)

            Add ESTIMATED BREAK REVENUE to TOTAL PREDICTED

If TOTAL PREDICTED < SALES TARGET AMOUNT

    PERCENTAGE INCREASE =

    (SALES TARGET AMOUNT – TOTAL PREDICTED) / SALES TARGET AMOUNT) x 100

Otherwise

    PERCENTAGE INCREASE = 0

    ✳ Create entities and relationships for new ratecard ✳

    Create SALES TARGET

    Create BUDGETING relationships between SALES TARGET and all BREAKCHART DAYS within the SALES TARGET FROM and SALES TARGET TO

    Create RATECARD PERIOD

    RATE FROM DATE = SALES TARGET FROM

    Per the previous ratecard

    Create RATECARD SEGMENT entities per the previous RATECARD PERIOD

    Create RATE SETTING relationships with RATECARD SEGMENT entities

        For each RATE MOVEABILITY

            For each RATE SPOT DURATION

                Create SPOT RATE entity

                    If PERCENTAGE INCREASE > 0

                    SPOT PRICE = current SPOT PRICE ((current SPOT PRICE x PERCENTAGE INCREASE) / 100)

                    Otherwise

                    SPOT PRICE = current SPOT PRICE

                    Create LEGAL PLACEMENT relationship with RATECARD SEGMENTS

Output: RATECARD

The first part of the specification establishes whether the current ratecard can deliver the required revenue. For the ratecard algorithms, Piccadilly management considers the thirty-second broad price to be the average for all spots sold over a period. In other words, Piccadilly makes its prediction by calculating the total revenue for all available airtime at thirty-second broad rates. If this prediction falls short of the requirement, management increases the rates by a percentage that is sufficient to meet the target.

Can the percentage increase be too big, and the advertising agencies refuse to buy Piccadilly's airtime? Not if management has made reasonable predictions about the revenue to expect as its share of the total advertising money to be spent over the whole network.

Is it safe to estimate the revenue based on the thirty-second broad rate? This is the most common rate used. Additionally, sales management knows how to fine tune the revenue by adjusting the minimum rates for each break.

The event-response data model for event 2 is shown in Figure 3.12.4.

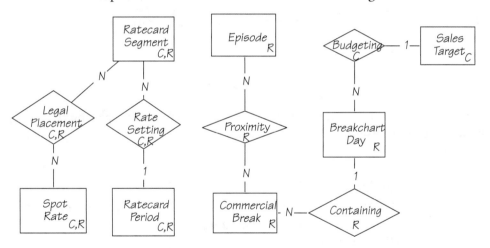

Figure 3.12.4: The event-response data model for event 2.

Event 3 Bureau prepares TV ratings

The processing for this event response is relatively simple. Figure 3.12.5 shows the current response, which is to select the applicable ratings and to record them in the store AIRTIME RATINGS HISTORY.

There is something rather suspicious about this model. The process ANALYZE RATINGS REPORT doesn't access any stored data. So you must ask, "How does it make a decision to select a rating or ratings?" Either the process uses stored data not shown in the model, or the ratings are selected by personal whim. The data dictionary tells you that the flow SELECTED RATINGS is identical to TELEVISION RATINGS REPORT, so you may conclude that SELECTED RATINGS are just those applicable to

Piccadilly. For the essential model, this selection process can be merged with the process that stores the data.

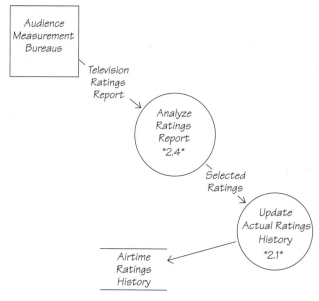

Figure 3.12.5: The current physical model for event 3.

Now let's turn our attention to the data, with the intention of replacing the current physical file with essential data. The data dictionary gives this definition of the incoming data flow:

Television Ratings Report = * Data flow. Statistical analysis of programmes and types of viewers showing ratings that are measured every 5 minutes during the viewing day. *
{Rating Date
+ {Programme Name + Rating Time
+ {Audience Type + Actual Rating Percentage}}}

Why does Piccadilly need these data? Piccadilly needs to know the ratings because the breaks in the more popular programmes are sold at higher rates. The essential task here is to attach the appropriate rating to the programme and the commercial breaks.

We can begin to write a mini specification for the essential process:

For each Piccadilly PROGRAMME NAME in the TELEVISION RATINGS REPORT
Find the EPISODE whose EPISODE START TIME
and EPISODE END TIME contain the RATING TIME

This involves the entities PROGRAMME to match the PROGRAMME NAME, and EPISODE to match the times. These entities should be part of the event-response data model. In theory, there could be a rating time without matching episode times. However, this suggests that people are watching when no programme is being broadcast, and we do not think it a serious possibility.

Ratings are taken at frequent intervals, but there may be no commercial break happening at the rating time. So this part of the specification will read

Find the BREAKCHART DAY that matches the RATING DATE
If RATING TIME is within a COMMERCIAL BREAK
    For each COMMERCIAL SPOT that has an OCCUPYING relationship with the COMMERCIAL BREAK
        Create a SPOT MEASURING relationship with the ACTUAL RATING

The ACTUAL RATING entity mentioned here is where the system stores the rating information. ACTUAL RATING PERCENTAGE is the key piece of data here, for it refers to the percentage of Piccadilly viewers. This number has to be recorded and related to the spot and the episode.

As you know, the Research Department at Piccadilly predicts ratings for programmes and their breaks. These predicted ratings are recorded before the event takes place. Because they have different meanings, the predicted rating and the actual rating must be modeled separately. We have modeled RATING as a supertype of ACTUAL RATING and PREDICTED RATING (see Figure 3.6.10). However, PREDICTED RATING plays no part in this event so it does not appear in the event-response data model. (If you have forgotten supertypes, reread Chapter 2.5 *Data Models*.)

Using the description above, look at the data model in Figure 3.12.6. The CRUD operators describe what this event response is doing.

The essential process model for this event is in Figure 3.12.7, with all the physical files replaced by the essential entities. The complete mini specification is now the most interesting item because it specifies all the details of the event response.

Mini specification 3 Record Ratings History

Input: TELEVISION RATINGS REPORT
With reference to the event-response data model

For each PROGRAMME NAME in the TELEVISION RATINGS REPORT
    Find the EPISODE whose EPISODE START TIME and EPISODE END TIME contain the RATING TIME

> For each AUDIENCE TYPE
>> Create an instance of RATING
>> Create an instance of ACTUAL RATING
>> Create an instance of PROGRAMME MEASURING relationship
> Find the BREAKCHART DAY that matches the RATING DATE
> If RATING TIME is within a COMMERCIAL BREAK
>> For each COMMERCIAL SPOT that has an OCCUPYING relationship with the COMMERCIAL BREAK
>>> Create a SPOT MEASURING relationship with the ACTUAL RATING

Note that the event response finds the appropriate episode and commercial break and relates the rating to it. Why doesn't the system simply record the rating information and relate it later when analyzing the ratings?

The answer lies in the difference between essential and physical. Physical is what Piccadilly does now, but because the rating carries a time, RATING is logically related to the episode and the breaks. Piccadilly needs to know the ratings for its programmes and spots, so as soon as the data flow enters the system, it becomes part of the system's data. If there is a logical link between items of data, they are related by the essential process that receives the data flow.

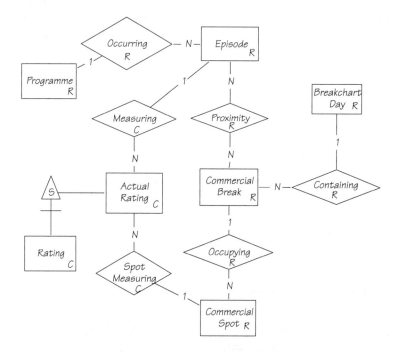

Figure 3.12.6: The event-response data model for event 3.

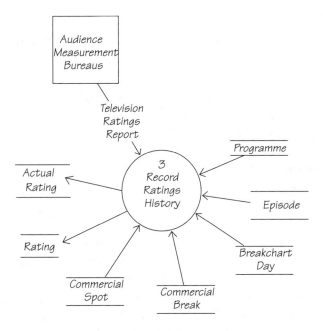

Figure 3.12.7: The essential process model for event 3.

Event 4 Agency decides the transmission instructions for a commercial

For event 4, the response again is to record the incoming information. The current physical model (Figure 3.12.8) shows the data being stored in the COPY REGISTER physical data store.

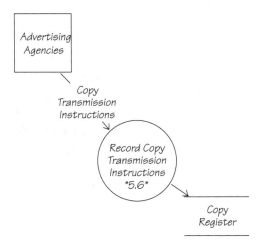

Figure 3.12.8: The current physical model for event 4.

Let's examine the dictionary definitions of the data for this response:

*Copy Register = * Data store *
                 {Commercial Copy Number + Product Name + Campaign Number
                 + {Copy Transmission Date}
                 + (Disposal Date)}*

*Copy Transmission Instructions = * Data flow. Details of when particular com-
                 mercial copy is to be transmitted. *
                 Agency Name + Product Name + Commercial Copy Number
                 + {Copy Transmission Date}*

The COPY TRANSMISSION INSTRUCTIONS show both the agency's name and the product's name. This is an example of nonessential data coming into the system. As a product may be represented by only one agency at a time, and as the product is already known to the system, there is no essential need for the agency name to be stored. Another event has already created the relationship between agency and product. The COMMERCIAL COPY NUMBER uniquely identifies the copy, so that is an essential piece of data. As it is a unique identifier, there must be an entity for it to identify. The natural choice is a COMMERCIAL COPY entity.

The COPY REGISTER tells us that COMMERCIAL COPY NUMBER and PRODUCT NAME are recorded together, thus indicating a relationship between COMMERCIAL COPY and PRODUCT. Sometimes, there are several different copies for one product (as we learned in Chapter 1.4 *The Piccadilly Organization*), so it is necessary to have a many-to-one relationship between COMMERCIAL COPY and PRODUCT. (That the COPY REGISTER repeats the PRODUCT NAME for each piece of copy is because of the way that the current technology implements the relationship.)

The COPY REGISTER tells us that each piece of copy is logged and the dates on which it is to be played are recorded. We have included {COPY TRANSMISSION DATE} as attributes of COMMERCIAL COPY. The repeating group makes us suspicious of this policy. Further questions might result in changes to the policy and the creation of another entity type and relationship in the data model.

With this description of the data, the task of the essential process becomes that of creating a COMMERCIAL COPY entity and relating it to an existing PRODUCT entity. The relevant part of the data model is shown in Figure 3.12.9, and the essential process model is in Figure 3.12.10.

Figure 3.12.9: The essential data model for event 4.

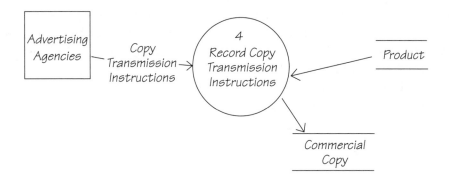

Figure 3.12.10: The essential process model for event 4.

The mini specification has to describe how the process makes use of the data that are already known to the system, and how the new data are recorded. We wrote this:

Mini specification 4 Record Copy Transmission Instructions

Input: COPY TRANSMISSION INSTRUCTIONS

Match the COMMERCIAL COPY NUMBER *in the* COPY TRANSMISSION INSTRUCTIONS
with the COMMERCIAL COPY NUMBER *on a* COMMERCIAL COPY *entity*
If a match is found
　　　Update COMMERCIAL COPY *with {*COPY TRANSMISSION DATE*}*
Otherwise
　　　Match PRODUCT NAME *on* COPY TRANSMISSION INSTRUCTIONS *with*
　　　PRODUCT NAME *on a* PRODUCT *entity*
　　　Create an instance of COMMERCIAL COPY *entity*
　　　Create an instance of PUBLICIZING *relationship with* PRODUCT

This mini spec raises a question. Is it correct to create a new COMMERCIAL COPY entity if one does not already exist? Yes, this is typically the first communication from the agencies about copy. We have not been told what happens if the automated cassette recording containing the commercial copy does not already exist in the library. There should be another event to check that all scheduled automated cassette recordings exist sometime prior to transmission time. None of this was apparent before we wrote the mini spec.

Event 5 Agency decides a commercial is outdated

The current response to this event takes place in the Programme Transmission Department. The process that receives the triggering data flow is insulated from other processes by data stores, so the response does not continue. The current physical model is shown in Figure 3.12.11. The response locates the physical cassette and records its deletion in the register. The essential version will be similar, except that the current physical data stores will be replaced by essential entities and relationships.

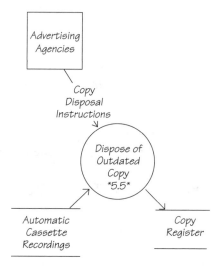

Figure 3.12.11: The current physical model for event 5. Note that the response involves a single process.

The definition for the incoming data shows

Copy Disposal Instructions = * Data flow. Instruction from an agency to dispose of an outdated commercial copy recording. *
Commercial Copy Number + Product Name + Disposal Date

The product name in this flow is nonessential. It is used only by people to check that they have found the correct ACR. Essentially, the system can match by COMMERCIAL COPY NUMBER alone. When the copy is located, the ACR must be removed from the library, and the copy marked as deleted.

The existing data model shows a COMMERCIAL COPY entity, which is the most likely replacement for the physical files. The data dictionary defines

*Commercial Copy = * Entity. The material that is transmitted during the time*
*occupied by a commercial spot. **
Commercial Copy Number + Physical Copy
+ {Copy Transmission Date} + Disposal Date
** Note that this contains an internal repeating group so is not*
*yet fully normalized **

By matching to this entity, the system can remove the copy. Our essential process model is shown in Figure 3.12.12. The mini specification reads

Mini specification 5 Dispose of Outdated Copy

Find the COMMERCIAL COPY *that matches the* COMMERCIAL COPY NUMBER *in the*
COPY DISPOSAL INSTRUCTIONS
DISPOSAL DATE *on the* COMMERCIAL COPY = DISPOSAL DATE *from the*
COPY DISPOSAL INSTRUCTIONS
Destroy the physical medium of the copy and record the DISPOSAL DATE

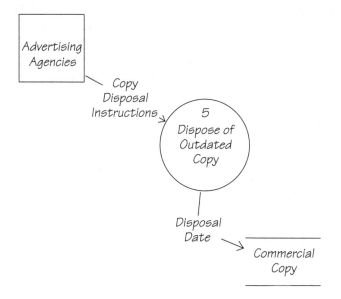

Figure 3.12.12: The essential process model for event 5. As there is only one entity involved in this event response, you don't need to draw a data model.

A question raised by this response is what happens if the copy is deleted and there is no other copy available for a campaign that is still running? Does the system refuse to delete the copy (against the wishes of the agency that has said that the copy is not to be used)? Or does the system delete the copy and issue a warning? Perhaps it might even suspend any bookings for the campaign.

Event 6 Supplier wants to sell a new programme

Again, we visit the Programme Transmission Department for the response to event 6. The current response involving one physical data store is shown in Figure 3.12.13.

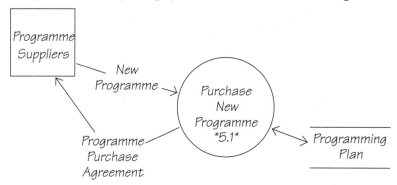

Figure 3.12.13: The current physical model for event 6. The response involves a single process.

The task is to determine if Piccadilly wants the NEW PROGRAMME and, if so, to buy it by issuing the PROGRAMME PURCHASE AGREEMENT. We started our essential model by examining the incoming and outgoing data flows to determine the necessary stored data. We used entities and relationships from the current data model to help here. Our version of the essential model is shown in Figure 3.12.14.

Let's look at the data the system uses to assess the offered programmes:

New Programme = * Data flow. Describes a programme offered by an English
          or overseas programme supplier. *
          Programme Name + Programme Type
          + Programme Description + Programme Duration
          + Programme Price + Programme Episodes
          + {Performer Name}
          + (Producer Name + Director Name)
          + Supplier Name

The outgoing flow is defined like so:

Programme Purchase Agreement = * Data flow. Terms negotiated with an
          external supplier. *
          Programme Name + Supplier Name + Programme Price

The following mini specifications explain the processes we chose. Although the organization of your specifications may be different, they should contain all the elements of the incoming data flow. If an element is disregarded, you should have a reason for doing so.

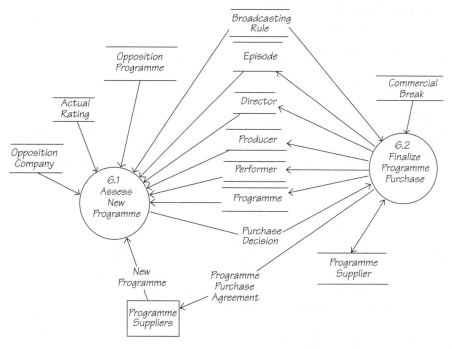

Figure 3.12.14: The essential process model for event 6. It uses two processes. The data stores are essential entities from the data model.

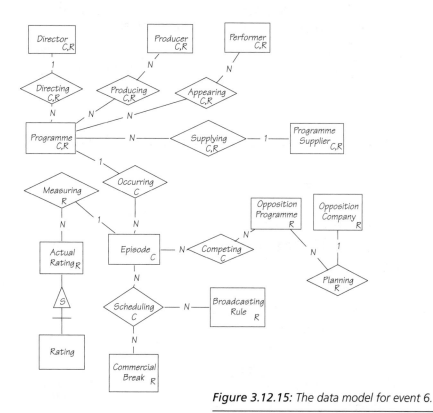

Figure 3.12.15: The data model for event 6.

Mini specification 6.1 Assess New Programme

Input: NEW PROGRAMME

∗ *Deciding which new programmes to buy requires experience in programme scheduling. Rather than attempting to define the decision-making process carried out by the buyer, this specification defines the data used to make the decision. We have used this approach because we've frequently uncovered data conservation problems with this kind of subjective process. Understanding the data brings us closer to understanding the skills used by the programme buyer.* ∗

Compare NEW PROGRAMME with all PROGRAMMEs of the same PROGRAMME TYPE
Ensure we do not already have the programme
Check that we have had successful ACTUAL RATINGs for this PROGRAMME TYPE
Consider the DIRECTOR, {PRODUCER}, and {PERFORMER}
Consider whether the NEW PROGRAMME satisfies the BROADCASTING RULES
Compare the NEW PROGRAMME with all OPPOSITION PROGRAMMEs of the same
PROGRAMME TYPE
If a decision is made to buy the NEW PROGRAMME
      Include NEW PROGRAMME in PURCHASE DECISION
If a decision has been made about when to transmit the NEW PROGRAMME
      For each EPISODE
          Include the EPISODE SCHEDULED DATE, EPISODE START TIME,
          and EPISODE END TIME in PURCHASE DECISION

Issue PURCHASE DECISION

Output: PURCHASE DECISION

Mini specification 6.2 Finalize Programme Purchase

Input: PURCHASE DECISION

Create an instance of PROGRAMME
Find a DIRECTOR entity that matches the DIRECTOR NAME in the
PURCHASE DECISION
If no match is made
      Create an instance of DIRECTOR
      Create an instance of DIRECTING

For each PRODUCER NAME in the PURCHASE DECISION
>   Find a matching PRODUCER entity
>   If no match is found
>>   Create an instance of PRODUCER
>>   Create an instance of PRODUCING

For each PERFORMER NAME in the PURCHASE DECISION
>   Find a matching PERFORMER entity
>   If no match is found
>>   Create an instance of PERFORMER
>>   Create an instance of APPEARING

Find a PROGRAMME SUPPLIER entity that matches SUPPLIER NAME
in the PURCHASE DECISION
If no match is found
>   Create an instance of PROGRAMME SUPPLIER
>   Create an instance of SUPPLYING
For each EPISODE SCHEDULED DATE in the PURCHASE DECISION
>   Create an instance of EPISODE
>   Create an OCCURRING relationship with PROGRAMME
For each OPPOSITION PROGRAMME in the PURCHASE DECISION
>   Create a COMPETING relationship with EPISODE
For each COMMERCIAL BREAK that lies between EPISODE START TIME and EPISODE
END TIME with reference to the BROADCASTING RULES
>   Create a SCHEDULING relationship
Issue PROGRAMME PURCHASE AGREEMENT

Output: PROGRAMME PURCHASE AGREEMENT

Event 7 Production company makes a commercial
We're still in the Programme Transmission Department for event 7. Again, the response termination points are at data stores. Figure 3.12.16 shows the current approach to receiving commercial copy.

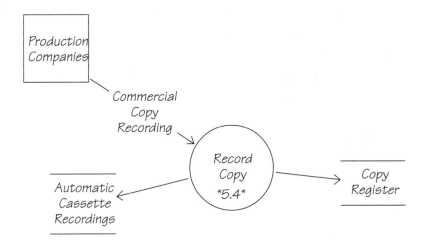

Figure 3.12.16: The current physical model for event 7.

Examine the stored data for the current model. There is some overlap:

Copy Register = * Data store *
            {Commercial Copy Number + Product Name + Campaign Number
            + {Copy Transmission Date}
            + (Disposal Date)}

Automatic Cassette Recordings = * Data store and material store *
            {Commercial Copy Number + Automated Cassette Recording
            + Production Company Name + Agency Name + Product Name}

The store AUTOMATIC CASSETTE RECORDINGS holds the physical cassette (ACR) along with the information that is written on the case of the cassette. The copy register is the paper copy of the information. Suppose that you merged the two data stores into one and eliminated the redundant data. Let's call this working definition INFORMATION ABOUT COPY:

Information About Copy = Commercial Copy Number + Product Name
            + Campaign Number + Production Company Name
            + Agency Name + Automated Cassette Recording
            + {Copy Transmission Date}
            + (Disposal Date)}

How much of these data can be gained from the incoming flow? The definition is

Commercial Copy Recording = * Data flow plus physical cassette. Commercial
*   copy recording sent by the production company to Piccadilly's*
*   Programme Transmission Department. **
*   Commercial Copy Number + Automated Cassette Recording*
*   + Production Company Name + Agency Name + Product Name*

This event response can only store data that come from the incoming flow; there is no other input to the process. We have already established that products may only be represented by one agency at a time, so the agency name is not essential if the system is told the product name. If you reduce INFORMATION ABOUT COPY to whatever comes from the incoming flow and remove the redundancy from your working definition, you get

Information About Copy = Commercial Copy Number + Product Name
*   + Production Company Name + Automated Cassette Recording*

Now attribute the elements to entities. The first three elements can be attributed to these entities: COMMERCIAL COPY, PRODUCT, and PRODUCTION COMPANY.

     A FILMING relationship between PRODUCTION COMPANY and COMMERCIAL COPY enables Piccadilly to follow up faulty automated cassette recordings. We know that the PUBLICIZING relationship between PRODUCT and COMMERCIAL COPY is certainly interesting (a previous event used it). The data model that emerges is shown in Figure 3.12.17.

     The entities from the event-response data model are now used to replace the physical files in the process model, giving the model in Figure 3.12.18.

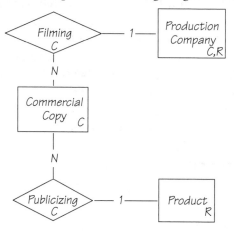

Figure 3.12.17: The data model for event 7.

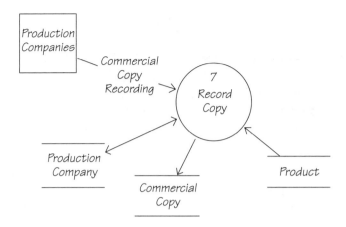

Figure 3.12.18: The essential model for event 7.

Note that the essential version ignores the reality of the ACR's being recorded on videotape. How does something tangible come to be recorded along with the intangible data? Remember that you are dealing with the essential view. In this view, there is no technology within your context, so the medium of the data is irrelevant. Commercial copy, the images of the actors and actresses pushing their products, is treated in the same way as the name of the production house and the number of the recording.

What do you gain by ignoring the technology? If you assume for the essential view that there is no technology, then when the time comes to implement the new system, you are not bound by the technological decisions of the past. The essential data model shows an entity COMMERCIAL COPY with attributes of an identifying number and the PHYSICAL COPY itself.

Wouldn't it be a better implementation if the copy were stored digitally in the same place as the information about the copy? If the cassettes were eliminated, there would be no need for a librarian to extract the cassette some hours ahead of transmission time. If a digital version of the copy could be broadcast, then Piccadilly could continue selling the airtime right up to a few seconds before the break.

Programme ratings can be affected by events outside the station's control. A sensational, late-breaking story can lift the ratings for the news beyond its expected audience. Advertisers would like to be able to take advantage of this phenomenon, and buy into a break minutes before it went to air.

The moral of the story is the purer the essential model, the better the implementation.

Event 8   Personnel hires a sales executive

In Figure 3.12.19, note that the process ASSIGN AGENCIES TO EXECUTIVE has another incoming flow (NEW AGENCY) and an outgoing flow (NEW AGENCY FORM) in the full version of the current physical model. You can ignore these flows because when modeling this event response, you are tracking the ripple effect caused by the flow NEW SALES EXECUTIVE. You have already included the other two flows in their appropriate place when you modeled event 9 *New agency wants to do business* (Chapter 1.9 *Modeling an Event Response* and its associated review chapter, 3.8). Remember that the current model reflects how the users currently carry out the business, and they feel that assigning executives is a process regardless of whether it is triggered by a new executive or a new agency.

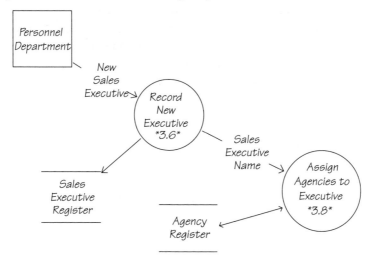

Figure 3.12.19: The current physical model for event 8.

Looking at the current event-response model, you probably suspect that there is some overlap between the two data stores. The dictionary shows

Agency Register = * Data store. File of agencies and responsible sales execu-
tives kept by the Sales Department. *
{Agency Name + Agency Address + Agency Phone Number
+ Sales Executive Name + Servicing Start Date}

Sales Executive Register = * Data store. File of sales executives' responsibili-
ties kept by the Sales Department. *
{Sales Executive Name
+ {Agency Name + Servicing Start Date + Servicing End Date}}

Suspicions confirmed! If you eliminate the overlap, you get the working definition:

Stored Data For This Event = * *This is a hypothetical data store, used only to explain this event response* *
                {*Sales Executive Name*
                + {*Agency Name* + *Servicing Start Date* + *Agency Address*
                + *Agency Phone Number* + *Servicing End Date*}}

Note the use of braces. The agency register shows one sales executive for one agency, but the executive register shows that one sales executive may have many agencies. An agency would have only one address of interest to the system, so all the agency information is collected after the agency name.

These data yield two entities, SALES EXECUTIVE and ADVERTISING AGENCY, and because they have appeared together in files, they must be related. (An earlier exercise discussed this relationship.) The data model appears in Figure 3.12.20.

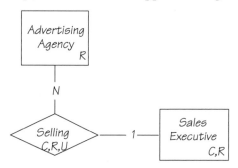

Figure 3.12.20: The data model for event 7.

The first part of the response is straightforward. The new sales executive's details are to be recorded by the system. The incoming data flow is this:

New Sales Executive = * *Data flow* *
                Sales Executive Name + *Sales Executive Address*
                + *Sales Executive Start Date*

All of these elements become attributes of the SALES EXECUTIVE entity. Thus, you can replace the SALES EXECUTIVE REGISTER with the SALES EXECUTIVE entity. The second part of the response is to make the executive work by making him responsible for selling to one or more agencies. To do that, the process must find appropriate agencies and establish the relationship between them and the executive. The mini specification for event 8 reads

Mini specification 8 Assign Agencies to Executive

Input: NEW SALES EXECUTIVE
With reference to the event-response data model

Create an instance of SALES EXECUTIVE
If there are any ADVERTISING AGENCY(s) without a current SELLING
relationship
       For up to 10 ADVERTISING AGENCY(s)
               Create a SELLING relationship with SALES EXECUTIVE
               SERVICING START DATE = today's date

This mini spec raises some questions: What if there are no unassigned agencies? Is there some way of re-allocating agencies already assigned? What happens if there are more than ten agencies that are not assigned? These will have to be resolved and the answers added to the mini specification. It is unlikely that the users' input will change the essential model. We show it in Figure 3.12.21.

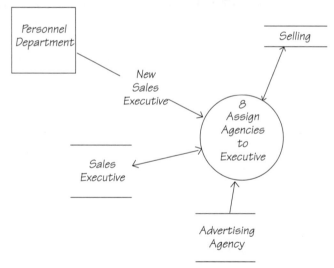

Figure 3.12.21: The essential process model for event 8. The physical files have been replaced with the essential entities. The relationship is shown because it contains a data attribute stored by this event response.

Event 10 Agency cancels a spot

To model event 10, let's start by looking at what happens in the present system. The current physical model is shown in Figure 3.12.22.

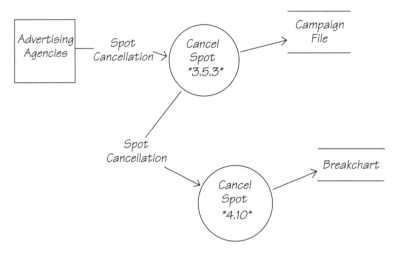

Figure 3.12.22: The current physical model for event 10.

The BREAKCHART and the CAMPAIGN FILE probably hold a lot of data that aren't needed by this event response. A quick glance at the data dictionary confirms this theory:

Breakchart = * Data store. Board containing hanging files plus breaksheets, used to record available time and sold time. *
{Breakchart Date
+ {Programme Name + Episode Start Time + Episode End Time}
+ {Break Start Time + Break End Time + Break Minimum Rate
+ {Spot Number + Product Name + Spot Duration + Spot Price
+ Rate Moveability + Spot Booking Agreement}}}
+ {Programming Rule}

Campaign File = * Data store. File kept in the Sales Department. *
{Campaign Requirements + Suggested Campaign}

Campaign Requirements = * Data flow. An agency's description of require-
ments for an advertising campaign. *
Agency Name + Product Name + Campaign Budget Total
+ Piccadilly Budget Amount + Target Audience
+ Target Rating Percentage + Campaign Duration
+ Campaign Start Date + Campaign End Date
+ {Required Spot Duration}

Suggested Campaign = * Data flow. Suggestions to an agency about the
makeup of an advertising campaign. *
Agency Name + Product Name + Campaign Number
+ Campaign Start Date + Campaign End Date
+ {Required Spot Duration + Spot Price + Rate Moveability
+ ([Breakchart Date + Break Start Time | {Breakchart Date}])}
+ Target Rating Percentage + Campaign Predicted Rating

We could make a data model from these definitions, but this time let's take a short-cut and use the data model from Chapter 3.6 (Figure 3.6.10). Note the COMMERCIAL SPOT entity. It is identified by a SPOT NUMBER, and that is central to the whole response. The data flow that triggers this event response looks like this:

Spot Cancellation = * Data flow *
Agency Name + Product Name + Campaign Number
+ Spot Number

A reasonable approach here is to say that the process can locate the spot to be canceled using the SPOT NUMBER. Once that is found, the task is to delete it and sever the relationships between it and the surrounding entities. Look at the data model in Figure 3.6.10 and ask, "Which relationships and which entities are affected?" The SPOT CANCELLATION data flow gives you some clues. It includes AGENCY NAME and CAMPAIGN NUMBER. These are the identifying attributes of ADVERTISING AGENCY and ADVERTISING CAMPAIGN, so it is reasonable to say that those entities are referenced, and the relationship between them and COMMERCIAL SPOT is deleted. The product name is also supplied in the data flow, but as there is no direct connection between COMMERCIAL SPOT and PRODUCT, you cannot delete the relationship. Why is PRODUCT NAME supplied by the agency? It is probably used as a double check by the spot manipulators, but we must ask the question in order to find out whether PRODUCT NAME is essential for this event.

Look further at the data model. Remember that this spot has not yet been transmitted (Piccadilly's rules sensibly do not allow agencies to cancel spots after transmission), so only the pre-transmission relationships are affected. For example, the spot is OCCUPYING a COMMERCIAL BREAK. That relationship is deleted by this event response. Similarly, once canceled, the spot does not have a PRICING AGREEMENT relationship with a SPOT RATE, a MOVEABILITY relationship with RATECARD SEGMENTs, a CAMPAIGN MAKE UP relationship with an ADVERTISING CAMPAIGN, a BUYING relationship with an ADVERTISING AGENCY, or an ALLOCATING relationship with a COMMERCIAL COPY. The SPOT MEASURING and BILLING relationships do not yet exist for this untransmitted COMMERCIAL SPOT.

Remember that the data model in Chapter 3.6 is only the first cut, and it doesn't necessarily show all the relationships and entities needed by this event response. You will see later how your event-response modeling will result in a proved and updated system data model. Look through the data dictionary definitions of the data stores to see if you find any other relationships or entities that might be concerned with this event response.

The data that we consider relevant to this response are shown in Figure 3.12.23. The essential process model develops after we figure out the data. Ours is in Figure 3.12.24.

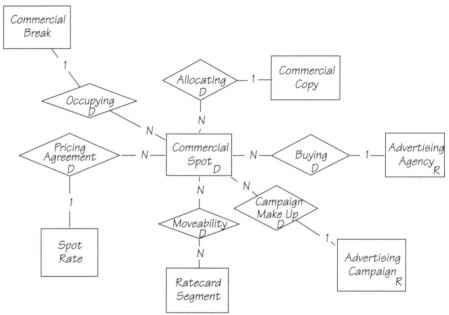

Figure 3.12.23: The data model for event 10, showing the relationships and entities that are affected by this response. Note the CRUD operators.

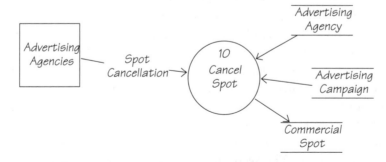

Figure 3.12.24: The essential process model for event 10. No relationship is shown because none has any data attributes. Note the flow going into COMMERCIAL SPOT. This is the correct notation for a deletion.

As all the hard work was done when you figured out the data model, let's make use of it. This is the mini specification we wrote:

Mini specification 10 Cancel Spot

Input: SPOT CANCELLATION
With reference to the event-response data model

The CRUD notation contained in the entities and relationships, along with the data dictionary definitions, is sufficient to specify this process.

Event 11 Agency wants to upgrade a spot

The current physical model for event 11 is shown in Figure 3.12.25.

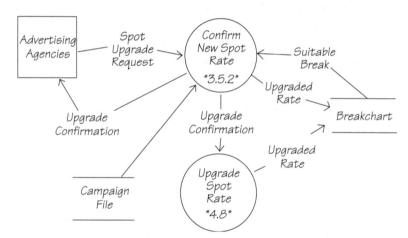

Figure 3.12.25: The current physical model for event 11. This response brings together processes from two departments.

Because the current response to this event takes place in two departments, there will probably be some redundancy in the processing. First, examine the data flows:

*Spot Upgrade Request = * Data flow. Increase in rate to avoid preemption. *
Agency Name + Product Name + Campaign Number
+ {Spot Number + Spot Duration + Rate Moveability}*

*Upgrade Confirmation = * Data flow. Agreement with an agency to upgrade*
*the rate of a spot. ***
Agency Name + Campaign Number + Product Name
+ Spot Number + Spot Duration + Spot Price
+ Rate Moveability

From these definitions, we can deduce that CONFIRM NEW SPOT RATE adds SPOT PRICE to the data in SPOT UPGRADE REQUEST to produce UPGRADE CONFIRMATION. The process references the CAMPAIGN FILE, presumably because the agency supplies the CAMPAIGN NUMBER. However, as the breakchart can be accessed using the SPOT NUMBER, there is no need to access the campaign file. The current view also shows a duplicated update of the breakchart. This may be due to incorrect modeling, or it may be because some executives upgrade the breakchart while others leave it to the Commercial Booking Department. Whatever the reason, essentially we only need to update the data once.

This event is now similar to event 10 because essentially it is concerned with a COMMERCIAL SPOT, which we know is identified by SPOT NUMBER. Look at the stored data surrounding COMMERCIAL SPOT and determine which relationships and entities are altered by this response. We derived the data model shown in Figure 3.12.26.

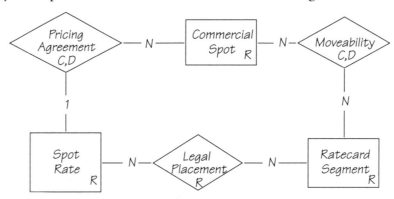

Figure 3.12.26: The data model for event 11. Note how the CRUD operators reveal the essential processing.

Once you have the data, the essential process model is straightforward to build. Ours is shown in Figure 3.12.27. Note that it needs only one bubble, and that none of the relationships is shown as they have no data attributes. The combination of the event-response process and data models tells you everything you need to know about this event. If you feel that by omitting the relationships from the essential process model you are not showing the complete picture, you may add them. However, we have found that the best way to work is to show relationships in the process model only when they contain data utilized by the event response. This still tells the full story, and avoids unnecessary clutter in your models.

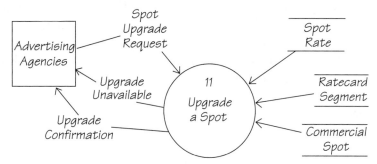

Figure 3.12.27: The essential event-response process model for event 11.

We shall reuse the mini specification from event 10.

Mini specification 11 Upgrade a Spot

Input: SPOT UPGRADE REQUEST
With reference to the event-response data model

The CRUD notation contained in the entities and relationships, along with the data dictionary definitions, is sufficient to specify this process.

Output: UPGRADE CONFIRMATION, UPGRADE UNAVAILABLE

Event 12 Agency chooses spots for a campaign

This event takes place sometime after the sales executive has selected a suggested campaign. The agency considers the executive's suggestions, and then picks the spots that it considers best for the campaign. The current response involves the Sales Department and the Commercial Booking Department. The processes are collected to make the response shown in Figure 3.12.28.

Because people in different departments want to hold private data, chances are this event response contains redundant processing and data. Looking at the processes, we can say that essentially the task is to get the selected spots onto the breakchart, and if that activity displaces any spots, to find new homes for the homeless. That there are only two essential processes is an assumption on our part.

However, starting with an assumption, and then proving it, is a reasonable way to proceed for this kind of event response.

We also assumed that the bubbles in the chain before SLOT SPOT are not essential. PREPARE SPOT STICKER is a prime example of an implementation process. Stickers are what Commercial Booking uses at the moment to identify the duration of a spot. It could be done in many other ways, and so is not essential. FINALIZE DEAL is the executive's double-checking the campaign availability that was originally offered and recording the selected spots in a file. While both these tasks are essential, we will find they are duplicated by other processes in this event response. The point is that it is not essential for the same task to be done more than once.

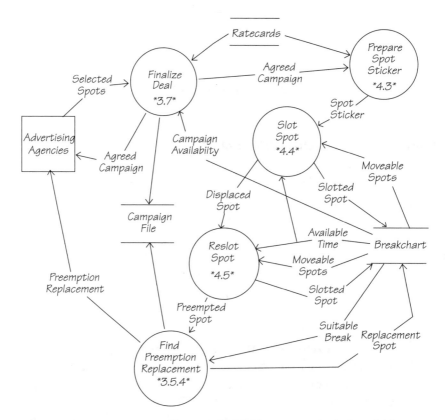

Figure 3.12.28: The current physical model for event 12, a fairly convoluted response to the event.

Now let's examine the task of getting the spot onto the breakchart. The breakchart is a current physical file, so we must find the essential equivalent. Its dictionary entry reads

Breakchart = * Data store. Board containing hanging files plus breaksheets,
used to record available time and sold time. *
{Breakchart Date
+ {Programme Name + Episode Start Time + Episode End Time}
+ {Break Start Time + Break End Time + Break Minimum Rate
+ {Spot Number + Product Name + Spot Duration + Spot Price
+ Rate Moveability + Spot Booking Agreement}}}
+ {Programming Rule}

The entry for the incoming flow is

Selected Spots = * Data flow. Spots selected by an agency as part of a cam-
paign. *
Agency Name + Campaign Number + Product Name
+ Campaign Start Date + Campaign End Date
+ {Spot Number + Spot Duration + Spot Price
+ Rate Moveability
+ ([Breakchart Date + Break Start Time | {Breakchart Date}])}

The process of slotting the spots is next. When the sales executive suggests the
campaign to the agency, he finds suitable breaks for the suggested spots. However,
suggested spots are not booked into breaks before the agency makes its selection.
The spot–slotting process is a matter of matching to the spot number; if the agency
hasn't changed the suggested price and moveability, and if there is availability in the
time period selected by the agency, the spot is slotted. If the agency changes the
suggested price or moveability for a spot, the existing relationships must be deleted
and new ones created (Figure 3.12.29).

Mini specification 12.1 Slot Selected Spots

Input: SELECTED SPOTS
With reference to the event-response data model

For each SPOT NUMBER on SELECTED SPOTS
Find the matching COMMERCIAL SPOT
* Identify the criteria for a suitable slot for the spot *
Refer to related SPOT RATE to ensure selection is appropriate for rate
Refer to related RATECARD SEGMENTs to find all legal segments for
the rate

If the existing PRICING AGREEMENT and/or MOVEABILITY
relationships are not consistent with those for SELECTED SPOTS
    Delete the existing relationships and create new ones

Refer to BREAK START TIME on SELECTED SPOTS to find
selected break
Refer to {BREAKCHART DATE} on SELECTED SPOTS to find
selected days
Refer to SPOT DURATION on SELECTED SPOTS to find duration of
required slot
Identify the COMMERCIAL BREAKs that satisfy the above criteria
* Select a suitable slot *
If there is a slot that does not have an OCCUPYING relationship
with another COMMERCIAL SPOT
    Then you have found a SUITABLE SLOT
Otherwise
    If there is a slot that does have an OCCUPYING relationship
    with another COMMERCIAL SPOT
        If the other COMMERCIAL SPOT has a lower SPOT RATE
        Delete the OCCUPYING relationship
        Issue DISPLACED SPOT
        You have found a SUITABLE SLOT
If you have found a SUITABLE SLOT
    Create an OCCUPYING relationship between COMMERCIAL SPOT and
    COMMERCIAL BREAK
    Create a BUYING relationship between COMMERCIAL SPOT and
    ADVERTISING AGENCY
    Add COMMERCIAL SPOT to AGREED CAMPAIGN
Otherwise
    Add UNAVAILABLE SLOT to AGREED CAMPAIGN

Output: DISPLACED SPOT, AGREED CAMPAIGN

* It would be better to partition this process into smaller processes, partly
because this mini specification is too long and complex and partly because
many parts of this process are repeated in other parts of the system. For
example, RESLOT DISPLACED SPOT uses much of the same logic contained in this
process. With this strategy, each smaller process would be specified once and
that specification would be pointed to by other processes in the system. *

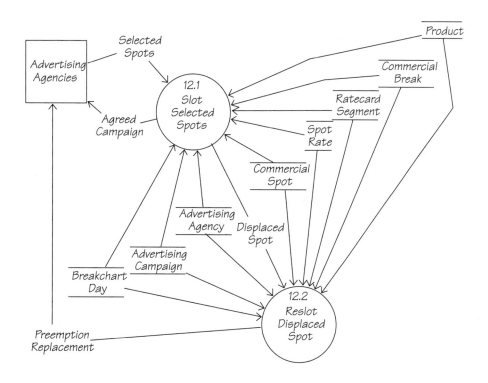

Figure 3.12.29: The essential event-response process model for event 12.

Mini specification 12.2 Reslot Displaced Spot

Input: DISPLACED SPOT
With reference to the event-response data model

✳ *This specification is virtually identical to* SLOT SELECTED SPOTS.
The difference is when there is not a suitable slot for DISPLACED SPOT,
then the SPOT RATE *is upgraded to the next highest rate and another attempt
is made to slot the spot. This continues until the spot is successfully slotted
or until there are no slots available at the highest rate.* ✳

If the spot has been preempted
     PREEMPTION REPLACEMENT *is issued to inform the agency*

Output: PREEMPTION REPLACEMENT

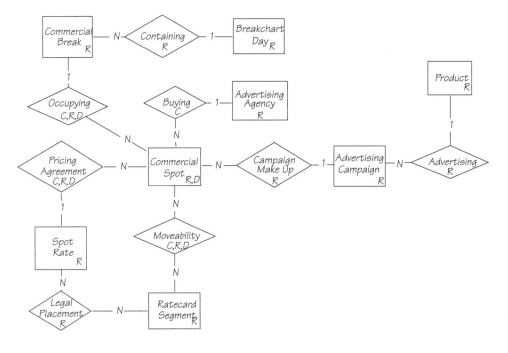

Figure 3.12.30: The data model for event 12.

Event 13 Spots are transmitted

Event 13 is an interesting temporal event. In Figure 3.12.31, the data flow BREAK TRANSMISSION SCHEDULE is actually a false data store. It is created each evening by the Commercial Booking Department and held by the programme transmission controller until the breaks are transmitted. If we ignored the geographical boundary between the two departments and the paper technology that they use, there would be no reason to make up this schedule. The transmission process could look directly at the essential stored data.

This event response is unusual in that it takes all day. The invoices are sent to the agencies at the end of each broadcasting day. Why is there no temporal event called *Time to produce invoices?* Because the production of invoices is part of the response to the transmission event; it takes place after the last spot has been broadcast.

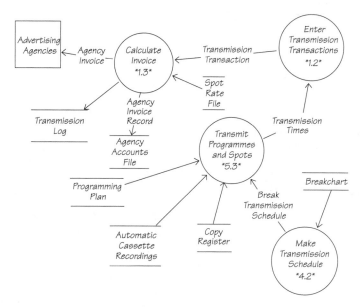

Figure 3.12.31: The current physical model for event 13.

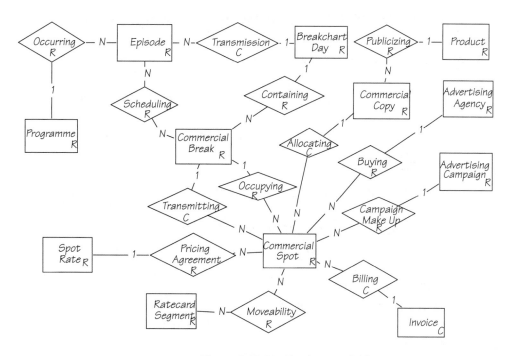

Figure 3.12.32: The data model for event 13.

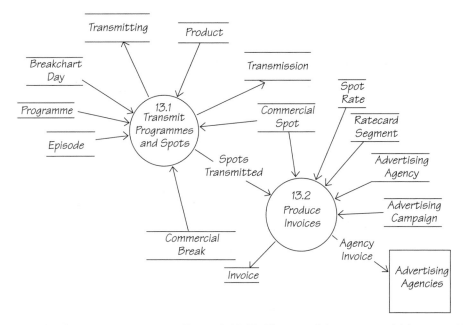

Figure 3.12.33: The essential process model for event 13.

Mini specification 13.1 Transmit Programmes and Spots

With reference to the event-response data model

Find the BREAKCHART DAY that has a BREAKCHART DATE corresponding to
today's date
Transmit EPISODEs and COMMERCIAL SPOTs according to the EPISODE START TIME
and the BREAK START TIME
For each EPISODE transmitted
> Create a TRANSMISSION relationship between the EPISODE and the
> BREAKCHART DAY
> PROGRAMME TRANSMISSION DATE and PROGRAMME
> TRANSMISSION TIME are the actual transmission date and time of the
> EPISODE
For each COMMERCIAL SPOT transmitted
> Find the PRODUCT associated with this spot
> Find the COMMERCIAL COPY whose COPY TRANSMISSION DATE
> matches today's date
> Create an ALLOCATION relationship between the COMMERCIAL SPOT
> and the COMMERCIAL COPY

> *Create a TRANSMITTING relationship between the*
> *COMMERCIAL SPOT and the COMMERCIAL BREAK*
> *SPOT TRANSMITTED TIME and SPOT TRANSMITTED DATE*
> *are the actual transmission time and date of the spot*
> *Add this COMMERCIAL SPOT to the SPOTS TRANSMITTED*
> *Issue SPOTS TRANSMITTED*

Output: SPOTS TRANSMITTED

Note how each spot is added to SPOTS TRANSMITTED when it is broadcast. SPOTS TRANSMITTED is input to process 13.2.

Mini specification 13.2 Produce Invoices

Input: SPOTS TRANSMITTED
With reference to the event-response data model

> *For each AGENCY NAME*
> > *For each CAMPAIGN NUMBER*
> > > *For each SPOT NUMBER on SPOTS TRANSMITTED*
> > > > *Find SPOT RATE with a PRICING AGREEMENT relationship*
> > > > *Add SPOT PRICE to INVOICE TOTAL*

INVOICE DATE = today's date
INVOICE NUMBER = next available number
Create an instance of INVOICE
Create an instance of BILLING relationship between the COMMERCIAL SPOT and the INVOICE
Issue AGENCY INVOICE

Output: AGENCY INVOICE

Event 14 Time to analyze revenue

Event 14 is a temporal event. The current response, shown in Figure 3.12.34, indicates that the essential process will be revealed by inspecting the data flow and determining what stored data are necessary to provide sufficient data for the flow.

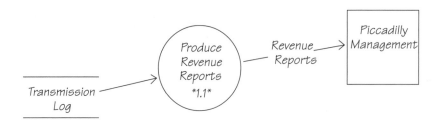

Figure 3.12.34: The current physical model for event 14.

*Revenue Reports = * Data flow. Computer reports used by management to*
*help set sales targets. ***
Breakchart Date + Daily Revenue
+ {Spot Number + Spot Rate + Product Name
+ Spot Transmitted Time + Spot Price}

Let's work out what stored data in the form of entities and relationships are needed. The data model in Chapter 3.6 will be of some help, but by now you know enough about the data in this system to figure out the entities needed.

Start with an easy one. SPOT NUMBER is the identifier of a COMMERCIAL SPOT. Even if you have forgotten the name or used a different one, you know that Piccadilly is interested in all the spots that it broadcasts. SPOT RATE is not an attribute of the spot (there are many spots with the same rate), so SPOT RATE must be an attribute of another entity related to the spot. For SPOT TRANSMITTED TIME, there are several spots in a break and they all take up time during the break. So the spot must relate to a break, or COMMERCIAL BREAK as we have called it.

We determined the rest of the necessary entities and relationships by going through the data flow definition. We then checked the data stores to make sure we hadn't missed anything:

*Transmission Log = * Data store ***
Spot Transmitted Date
+ {Spot Number + Spot Booking Agreement + Spot Price
+ Rate Moveability + Product Name + Spot Transmitted Time}

The transmission log contains data that are not used by this event. The data may either be used by another event response or simply be redundant data. You can ignore it. Our data model is in Figure 3.12.35.

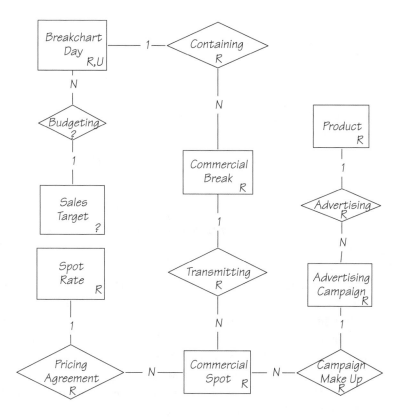

Figure 3.12.35: The data model for event 14, derived from the output data flow. The question mark indicates a question we had about an entity and relationship that are not accessed, but could be useful if they were. Perhaps the report doesn't use them now because of current technology limitations.

The essential process model is in Figure 3.12.36. We didn't write a mini specification for this event response because the data model, its CRUD operators, and this data dictionary entry are sufficient specification in themselves.

Daily Revenue = * Data element. Revenue made by Piccadilly for all the commercial spots transmitted on one breakchart day. Derivable: total of all Spot Prices for every Commercial Spot having a Transmitting relationship with the Commercial Breaks on today's date. *

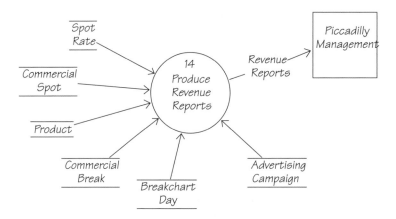

Figure 3.12.36: The essential process model for event 14.

Event 15 Time to analyze the breakchart

Event 15 is another temporal event. Its purpose is to issue preemption warnings to the agencies, with the goal of persuading them to raise the rate they are paying for a spot. Agencies want to spend the advertisers' money, as most agencies are paid a percentage of the airtime booking fee, but they also want to get the best value for their clients' money, so there will be more of it next year.

The current response (Figure 3.12.37) shows three processes. One, *4.7*, analyzes the airtime position; another, *3.5.1*, prepares the warning and sends it to the agency; and the third, *3.4*, sets the minimum rate for the breaks. Notice that

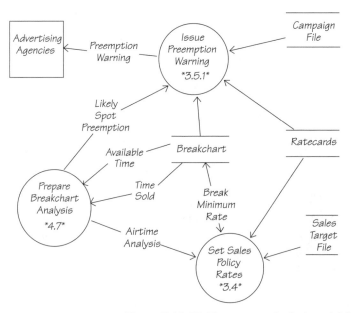

Figure 3.12.37: The current physical model for event 15.

we have only extracted part of process *3.4* from the original physical model. The process in the physical model responds partly to event 15 and partly to event 2. Look back at event 2 to see that the other parts of the original *3.4* are included in the response to this event.

Replacing the breakchart with its essential data equivalent is the next step. Our essential view is shown in Figure 3.12.38.

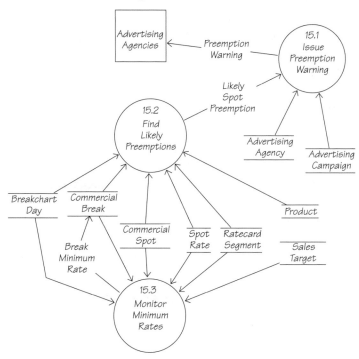

Figure 3.12.38: The essential process model for event 15.

How did we arrive at the essential data? Again, by looking at the definition of the files to be replaced, and rationalizing them into entities and relationships. The dictionary entry for the first file is

Breakchart = * Data store. Board containing hanging files plus breaksheets,
               used to record available time and sold time. *
               {Breakchart Date
               + {Programme Name + Episode Start Time + Episode End Time}
               + {Break Start Time + Break End Time + Break Minimum Rate
               + {Spot Number + Product Name + Spot Duration + Spot Price
               + Rate Moveability + Spot Booking Agreement}}}
               + {Programming Rule}

Since we've done this several times already in this chapter, we'll simply present the data model (Figure 3.12.39) and let you figure it out. Note that not all of the data in this model are used by all the processes. For example, the process that issues the warning to the agency is the only one that uses the agency and campaign data. In the essential model, the interfaces around each process contain only the data that are necessary for that process to do its work.

Mini specification 15.2 Find Likely Preemptions

With reference to the event-response data model

At 9 AM each day
For three BREAKCHART DAYs starting from today
        For each COMMERCIAL BREAK that has a CONTAINING relationship
            ✳ Determine if the break is full ✳
            Accumulate the SPOT DURATIONS of all COMMERCIAL SPOTS having an
            OCCUPYING relationship
                If the accumulation = BREAK DURATION
                    For each COMMERCIAL SPOT having an OCCUPYING relationship
                        Find RATE MOVEABILITY on the SPOT RATE that has a
                        PRICING AGREEMENT relationship
                        If RATE MOVEABILITY is not = Fixed
                            Issue a LIKELY SPOT PREEMPTION

Output: LIKELY SPOT PREEMPTION

You can specify the issuing process by referring to the data model in Chapter 3.6. You can use the technique of referring to the data model to specify most of the policy in processes 15.1 and 15.3. We will need to ask the users about the decision-making process for setting the BREAK MINIMUM RATE.

Notice how the process MONITOR MINIMUM RATES interfaces with the other two processes through the essential stored data. This indicates that we really have two event responses here. Recall that we thought it was only one event because it was presented as TIME TO ANALYZE THE BREAKCHART. Our modeling work has uncovered the truth. There are two very different reasons for analyzing the breakchart, and both are separate temporal events. At this point, we should break this event into two: TIME TO FIND LIKELY PREEMPTIONS and TIME TO MONITOR TARGETED REVENUE.

The event that monitors the targeted revenue is unusual because there are no flows connecting the processing to a terminator. The policy says that at a preset time before transmission, the unsold time is analyzed. If there is an excess of availability, the break minimum rate is adjusted using the sales target as a guideline to try to make the time more attractive to the agencies. This also raises a question about policy: Should there be a flow to the agencies telling them that some cheaper time has become available?

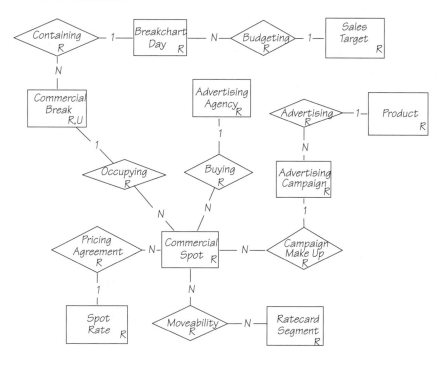

Figure 3.12.39: The data model for event 15. Note how the CRUD operators reveal the essential processing.

Event 16 Time to finalize new programme schedule

We made the temporal event response for event 16 into an essential model in the same way as for event 14. One difference, though, is that the data are more complex. The current model is in Figure 3.12.40.

The output data are

Programme Transmission Schedule = * Data flow. Piccadilly's planned transmission for the next quarter. *
{Programme Transmission Date
+ {Episode Number + Episode Start Time + Episode End Time
+ Programme Name + Programme Description
+ Programme Type + Predicted Rating}
+ {Break Start Time + Break End Time}}

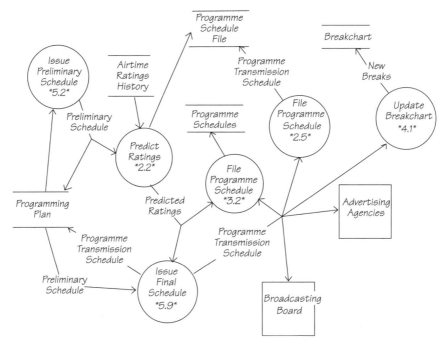

Figure 3.12.40: The current physical model for event 16.

We worked through these data elements and came up with the data model shown in Figure 3.12.41.

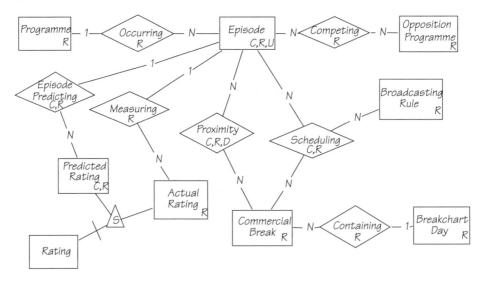

Figure 3.12.41: The data model for event 16.

There is much overlap in the data stored by the physical event response. For example, the PROGRAMME TRANSMISSION SCHEDULE is stored three times. The PRELIMINARY SCHEDULE is stored and later retrieved by the same event response. When the data are reduced to the essentials, the need to have multiple bubbles for this event response disappears. We fitted all the essentials into the one process shown in Figure 3.12.42.

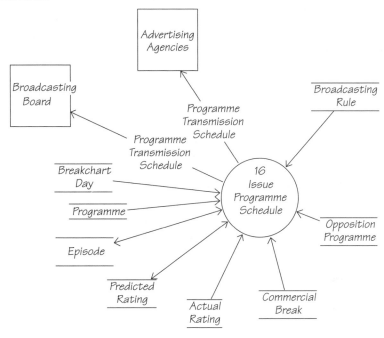

Figure 3.12.42: The essential process model for event 16, using the entities from the data model.

The mini specification for this event is more complex than we expected:

Mini specification 16 Issue Programme Schedule

With reference to the event-response data model

Taking the BROADCASTING RULES into account

On the first day of the quarter
Prepare a PROGRAMME TRANSMISSION SCHEDULE starting on the first day of the next quarter

For each BREAKCHART DAY within the next quarter
>For each EPISODE having an EPISODE SCHEDULED DATE within the next quarter
>>Add defined details to the PROGRAMME TRANSMISSION SCHEDULE
>>For each COMMERCIAL BREAK having a BREAK SCHEDULING relationship
>>>Add defined details to the PROGRAMME TRANSMISSION SCHEDULE
>If there is any unoccupied time during the BREAKCHART DAY
>>Search for a PROGRAMME with a PROGRAMME DURATION to fit into the unoccupied time
>>Create an instance of EPISODE
>>Create a PROXIMITY relationship with the affected COMMERCIAL BREAKS
>>>If the chosen PROGRAMME does not fit exactly into the unoccupied time
>>>>For each of the affected EPISODEs
>>>>>Update the EPISODE START TIME and EPISODE END TIME
>>>>>Delete the PROXIMITY relationship with the COMMERCIAL BREAKS that no longer fall during, directly before, or directly after the EPISODE
>>>If the chosen EPISODE does not have an EPISODE PREDICTING relationship with a PREDICTED RATING
>>>>For each AUDIENCE TYPE
>>>>>Take an average of the ACTUAL RATINGs for the other EPISODEs that are related to PROGRAMMEs of the same PROGRAMME TYPE of the same type
>>>>>Create PREDICTED RATING
>>>>>Create PREDICTING relationship with EPISODE
Issue PROGRAMME TRANSMISSION SCHEDULE

Output: PROGRAMME TRANSMISSION SCHEDULE

After we wrote this mini specification, we decided that the process is too complex and would be better if it were split into three: one process responsible for creating a preliminary schedule, another for predicting ratings, and the third for issuing the final programme transmission schedule.

Event 17   Another channel sets a schedule

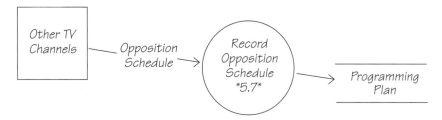

Figure 3.12.43: The current physical model for event 17.

This is a reasonably straightforward event that has the same pattern as many other external events. The purpose of this event response is to keep track of which opposition programmes occupy what time slots.

The data that this event response stores can only come from the incoming data flow, so let's start there.

Opposition Schedule = * Data flow *
          Television Company Name
          + {Opposition Transmission Date
          + Opposition Transmission Time + Opposition Programme Name
          + (Opposition Predicted Rating)}

The definition gives up its first entity right away. TELEVISION COMPANY NAME must be the identifier of an entity. Let's call it OPPOSITION COMPANY. The braces indicate a number of programmes are attached to this company, so that reveals the next entity. Let's call it OPPOSITION PROGRAMME and give it a relationship to OPPOSITION COMPANY.

Note that each opposition company's programme has a time and date (and sometimes a rating) attached to it. This enables the system to locate the appropriate Piccadilly programme and record the opposition company's programme against it. As we saw earlier, some of the programmes shown by Piccadilly are in the form of a series. We used the entity EPISODE to model each showing of a programme. The opposition company's programme is competing in the same time slot as one of Piccadilly's, so a relationship between the OPPOSITION PROGRAMME and EPISODE is appropriate. The data model appears in Figure 3.12.44, and the essential process model in Figure 3.12.45.

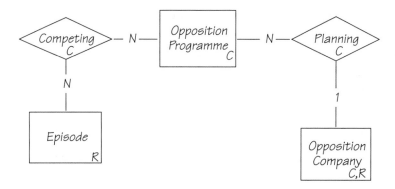

Figure 3.12.44: The data model for event 17. Note the CRUD operators for OPPOSITION COMPANY. If the correct instance of this entity does not exist, the response will create one.

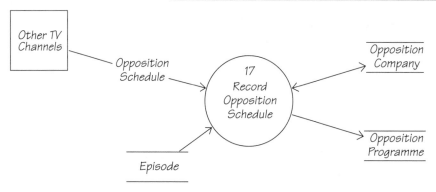

Figure 3.12.45: The essential process model for event 17.

Mini specification 17 Record Opposition Schedule

Input: OPPOSITION SCHEDULE
With reference to the event-response data model

Find the matching OPPOSITION COMPANY *that matches the* TELEVISION COMPANY
NAME *or create one if it does not exist*
For each OPPOSITION PROGRAMME NAME
    Locate the EPISODE *whose* EPISODE SCHEDULE DATE *and* EPISODE START
    TIME *match the dates and times in the* OPPOSITION SCHEDULE
    Create an instance of OPPOSITION PROGRAMME
    Create the relationships PLANNING *and* COMPETING

Event 18 Broadcasting Board makes rules

The Broadcasting Board issues programming rules to all television stations. The Board's intention is to set standards for television programming and for the showing of commercials within these programmes. Piccadilly's current response is shown in Figure 3.12.46.

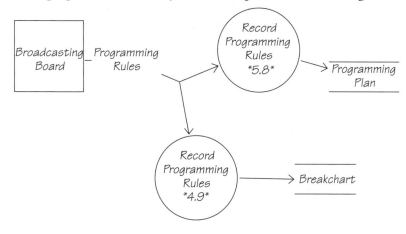

Figure 3.12.46: The current physical model for event 18.

Note that the current system records the rules in two places. This happens because both the Commercial Booking and the Programme Transmission departments need to access the rules. An obvious step to make this into an essential model is to eliminate the duplication.

When the rules arrive, the system's response is to record the rules, and record the current date as the RULE EFFECTIVE DATE and do nothing else. These rules are used when new programme schedules are being put together and when spots are being added to the breakchart. However, those actions are part of other event responses. Therefore, the only possible action for the essential model for this particular event is to record the rules. The essential model is shown in Figure 3.12.47.

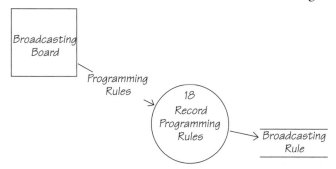

Figure 3.12.47: The essential process model for event 18.

We have not drawn a data model for this event response, as there is only one entity type, BROADCASTING RULE. Nor have we written a mini specification. The process is specified by a combination of the event-response process model, data dictionary, and data model. Now look at the system data model to see how the BROADCASTING RULE entity is used. The system data model, brought up-to-date with the discoveries you have made by doing essential modeling, appears in Figure 3.12.48.

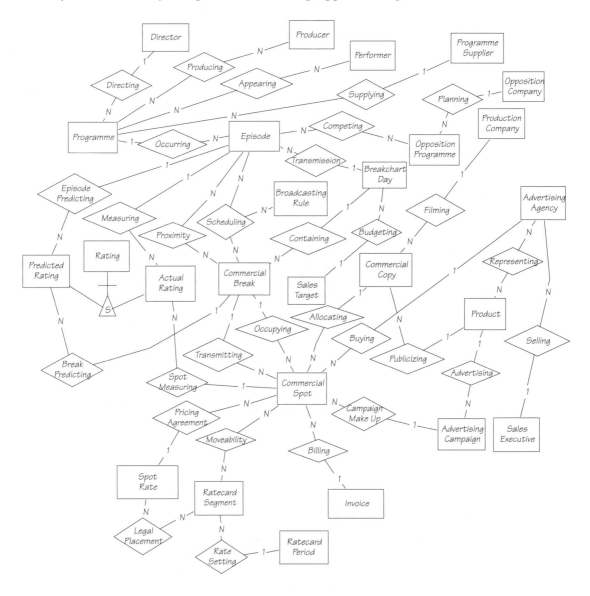

Figure 3.12.48: The complete essential data model for Piccadilly.

The Combined Data Model

We derived the system data model in Figure 3.12.48 by combining all the event-response data models, and matching the entities from the various event-response data models. When the same entity appears in several models, combine them by joining at the entity.

Study this model. It represents the final version of the essential data.

✚ Ski Patrol

This is not a true ✚ Ski Patrol. If you are having problems, the ✚ Ski Patrol in Chapter 1.13 *More Events* has remedial advice. If you made it to this point, we want to congratulate you. It takes a truly heroic effort to work your way through all the events.

Along the way, we have tried to show you various approaches to event-response modeling, and we hope you've become thoroughly familiar with them.

Take a break now to review what you know about event-response modeling. Also think about any areas that are less than clear. You may wish to go back over some of the models, compare them to what we've said in the Textbook chapters, and clarify your understanding of event responses. You will be using them a lot in your future.

Trail Guide

You have now finished gathering all of the essential requirements from the current system. We suggest you celebrate this milestone in an appropriate manner. Now you are going to look at a different type of requirement.

● Easiest and ■ More Difficult: Go to Chapter 2.13 *Modeling New Requirements* to see how the next stage of the Piccadilly Project unfolds.

◆ Most Difficult: Proceed to Chapter 1.14 *Some New Requirements.*

✳ Promenade: Chapter 2.13 *Modeling New Requirements* is the appropriate destination for you.

3.13 REVIEW: SOME NEW REQUIREMENTS

Before You Reached Here ...
You have modeled the new requirements described in Chapter 1.14 *Some New Requirements*. These new requirements have resulted in changes to your data model, context diagram, mini specifications, event list, and data dictionary.

First Model the Requirement

According to our strategy, you first build stand–alone event-response models of the new requirements, and then integrate them with any existing models later. Your analysis of Sales Director Blake Hall's requirements results in three event-response models.

The first of the three new events, *Agency requires phone-in service,* records an agency's request to use the phone-in facility for a particular campaign. Figure 3.13.1 shows the system's response to this event.

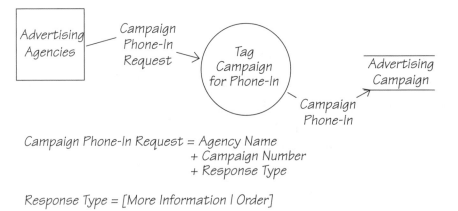

Campaign Phone-In Request = Agency Name
+ Campaign Number
+ Response Type

Response Type = [More Information | Order]

Figure 3.13.1: The event-response process model for the event **Agency requires phone-in service.**

RESPONSE TYPE is the kind of service requested by the agency. Our data dictionary entry shows that the agency can request that the telephone service take orders for the product or receive requests for more information. The users must confirm if both services will ever be needed.

Since the telephone answering service does not apply to all campaigns, the people who answer the telephones may be part-time workers. If so, the system's response may have to include a notification to whoever manages the telephone system to enable him to assemble enough staff for the expected telephone calls.

There is only one entity so we did not build a data model.

The next new event, *Viewer responds to commercial,* happens when the viewer makes a telephone call. Remember that Blake Hall said, "Viewers of a commercial can telephone immediately with orders or requests for more information about the product." The system's response to this event is to record the response details and relate them to the particular commercial spot.

Let's start this event-response model by thinking about the data. The incoming data flow is defined like this:

Viewer Response = Product Name + Approximate Viewing Time
+ Response Type + Viewer Name + Viewer Address

In this case, building an event-response data model is particularly helpful so that you can consider how all the data will be accessed and recorded. Our data model for this event is shown in Figure 3.13.2.

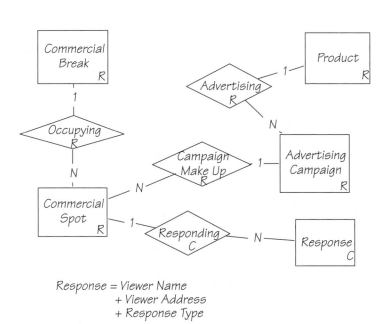

Response = Viewer Name
+ Viewer Address
+ Response Type

Figure 3.13.2:
The data model for the event **Viewer responds to commercial.**

Note that the response relates to the commercial spot. Blake Hall said that Piccadilly wants to be able to tell the agencies which spots attracted the responses. Because the data model portrays the business policy, it helps you to raise questions.

Once you know the data, the process model for this event is easier to build. Ours is in Figure 3.13.3.

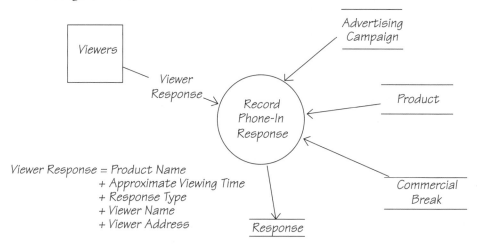

Figure 3.13.3: The process model for the event **Viewer responds to commercial.**

The process model raises some questions. For instance, what happens if the viewer's approximate viewing time does not correspond to a commercial spot for the product in question? Suppose there are two commercial spots equally close to the approximate viewing time. To which one should the response relate? Alternatively, could it relate to both spots? Our dictionary entry for VIEWER RESPONSE shows only the viewer's name and address. Does the agency require Piccadilly to accept complete order information including credit card details?

Only the users can answer these questions. Your task is to make sure you ask *all* the right questions.

The last new event is temporal. It's concerned with producing viewer response reports that are sent to the relevant agencies every morning. You will have to confirm the data in the VIEWER RESPONSE REPORT with the users, as it determines the processing for this event.

Once you model the new requirements, review them with the users, and clear them with the project manager, your next task is to integrate the new requirements.

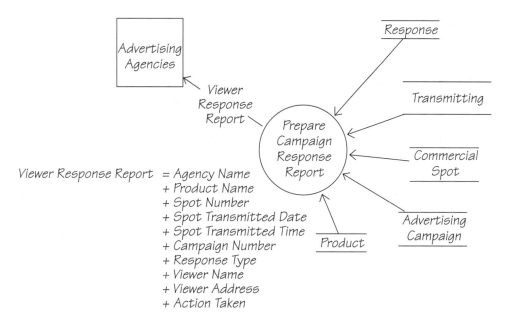

Viewer Response Report = Agency Name
+ Product Name
+ Spot Number
+ Spot Transmitted Date
+ Spot Transmitted Time
+ Campaign Number
+ Response Type
+ Viewer Name
+ Viewer Address
+ Action Taken

Figure 3.13.4: *The process model for the temporal event* **Time to report viewer response to agency.**

Integrating the Requirements with Your Existing Models

Some of the new requirements to the Project are added as brand-new, stand-alone events, while other new requirements will be additions to already existing events. Let's consider each event-response model in turn.

The agency will probably request the campaign phone-in service when it gives Piccadilly the campaign requirements. This means that you can integrate the new event *Agency requires phone-in service* with event 1 *Agency wants to run a campaign.* To add the new requirements to the event, you need to add the field RESPONSE TYPE to the flow CAMPAIGN REQUIREMENTS and to the entity ADVERTISING CAMPAIGN.

Your second new requirement, *Viewer responds to commercial,* is a new event. To integrate the new event into your Project, follow these steps:

- Add the terminator VIEWERS and the flow VIEWER RESPONSE to your context diagram.
- Define VIEWER RESPONSE, RESPONSE TYPE, VIEWER NAME, VIEWER ADDRESS, and ACTION TAKEN in your data dictionary.
- Add the entity RESPONSE and the relationship RESPONDING to your data model (per the event-response data model), and define them in the data dictionary.
- Write a mini specification for RECORD PHONE-IN RESPONSE.
- Add the event to your event list.

The last new requirement, *Time to report viewer response to agency,* is a new temporal event response, which requires you to

- Add a data flow called VIEWER RESPONSE REPORT to your context diagram, going from the context to the terminator ADVERTISING AGENCIES.
- Define VIEWER RESPONSE REPORT in the data dictionary.
- Write a mini specification for PREPARE CAMPAIGN RESPONSE REPORT.
- Add the event to your event list.

When you have done all that, you have successfully incorporated three changes into your Project.

We have left the changes until late in the Project, but more often than not in the real world, the new requirements are mentioned early on in the analysis. Our strategy of building stand-alone models and then integrating them means that you can add new requirements on demand.

✚ Ski Patrol

You may have taken a shortcut and modeled the first event, *Agency requires phone-in service,* as an alteration to event 1 *Agency wants to run a campaign.* We suggest you not do this because the new requirement is often an event in its own right. Also, by submerging it into an existing event response, you may overlook some vital questions. For example, you wouldn't think to raise the issue about notifying the staff for the telephones if you had simply altered the incoming flow for event 1.

You can identify the requirements the same way as you previously did: Draw a context diagram from the interview with Blake Hall, and use the boundary data flows to identify events. Modeling new requirements is the same as modeling existing event responses. You have already modeled eighteen or so of them, so you are relatively experienced in this area.

If you are having problems and you skipped over it, Chapter 2.13 *Modeling New Requirements* would be beneficial for you to review.

Trail Guide
● Easiest, ■ More Difficult, and ◆ Most Difficult: Go to Chapter 1.15 *CRUD Check* to verify the integrity of your essential models.

✳ Promenade: The others are off to do some hard work. That's not why you are reading this book, so Chapter 2.14 *New Physical Viewpoint* is a good place to go next.

REVIEW: 3.14
CRUD CHECK

Before You Reached Here ...
You have completed the CRUD check problem described in Chapter 1.15 *CRUD Check*.

Using the CRUD Table

Your CRUD table shows which event responses access each of the entities and relationships in your data model. Ours is presented in Figure 3.14.1.

Entity/Relationship	Create	Reference	Update	Delete
Actual Rating	3	6,16		
Advertising	1	12,14,15		
Advertising Agency	9	1,8,10,12,13,15		
Advertising Campaign	1	10,12,13,14,15		
Allocating	13			10
Appearing	6	6		
Billing	13			
Breakchart Day		1,2,3,5,12,13, 14,15,16	14	
Break Predicting				
Broadcasting Rule	18	6,16		
Budgeting	2	15		
Buying	12	13,15		10
Campaign Make Up	1	12,13,14,15		10
Commercial Break		1,2,3,6,12,13, 14,15,16	5,15	
Commercial Copy	4,7	5,13	4,5	
Commercial Spot	1	1,3,11,12,13,14,15		10,12

Entity/Relationship	Create	Reference	Update	Delete
Competing	6,17	16		
Containing		1,2,3,12,13,14 15,16		
Directing	6	6		
Director	6	6		
Episode	6,16	1,2,3,13,16,17	16	
Episode Predicting	16	1,16		
Filming	7			
Invoice	13			
Legal Placement	2	1,2,11,12		
Measuring	3	6,16		
Moveability	1,11,12	12,13,15		10,11,12
Occupying	1,12	1,3,12,13,15		10,12
Occurring	6	3,13,16		
Opposition Company	17	6,17		
Opposition Programme	17	6,16		
Performer	6	6		
Planning	17	6		
Predicted Rating	16	1,16		
Pricing Agreement	1,11,12	12,13,14,15		10,11,12
Producer	6	6		
Producing	6	6		
Product	1	1,4,5,7,12,13, 14,15		
Production Company	7	7		
Programme	6	3,6,13,16		
Programme Supplier	6	6		
Proximity	16	1,2,3,16		16
Publicizing	4,7	13		
Ratecard Period	2	1,2		
Ratecard Segment	2	1,2,11,12,13,15		
Rate Setting	2	1,2		
Rating	3			
Representing	1	1		
Sales Executive	8	9		

Entity/Relationship	Create	Reference	Update	Delete
Sales Target	2	15		
Scheduling	6,16	13,16		
Selling	8,9	8,9	8	
Spot Measuring	3			
Spot Rate	2	1,2,11,12,13,14, 15		
Supplying	6	6		
Transmission	13			
Transmitting	13	14		

Figure 3.14.1: The CRUD table for the Piccadilly system. It cross-references the entities and relationships with the event responses.

First, compare your CRUD table against this one. Yours should be substantially the same. However, the purpose of the exercise is to find missing events and thus confirm the completeness of the essential model. Our table has raised some questions about the model's completeness, and we want to discuss them. Before we do, let's see how the table works.

Find the entry for the entity ADVERTISING AGENCY. The table indicates that event 9 creates an instance of this entity. Your event list shows that event 9 is *New agency wants to do business.* The event-response process model shows that a new instance of the entity ADVERTISING AGENCY is created when the NEW AGENCY data flow arrives from one of the advertising agencies.

The CRUD table also shows that events 1, 8, 10, 12, 13, and 15 all have reason to reference instances of the ADVERTISING AGENCY entity. You can find out the reason by looking at the corresponding event-response models for each of these events. This means that the system both creates and references the entity.

Note that the table does not show an event that updates an ADVERTISING AGENCY, so it raises a question about the need to change any of the attributes of this entity. Let's use the data dictionary definition for ADVERTISING AGENCY to help answer the question:

Advertising Agency = Agency Name + Agency Address + Agency Phone Number

It seems reasonable to have some provision for an agency to change its address or phone number or even its name. Thus, we see that the CRUD table has exposed an event that has not yet been included in our analysis.

You'll need to update the specification to reflect the missing requirement:

1.   Add the data flow AGENCY DETAILS CHANGE to the context diagram.

2.   Define AGENCY DETAILS CHANGE in the data dictionary:

Agency Details Change = Agency Name
*   + (New Agency Name)*
*   + (New Agency Address)*
*   + (New Agency Phone)*

Strictly speaking, this definition allows a flow that changes nothing. However, when faced with the alternative lengthy definition that lists each of the possible combinations of the three fields, we find this is a better, less error-prone alternative. Also, the risk of misunderstanding is very low because even if someone did input a data flow that changes nothing, it will not harm the system's data.

1.   Add event 19 *Agency changes details* to the event list.

2.   Build event-response process and data models for this event.

3.   Add an entry to the CRUD table to show that ADVERTISING AGENCY is now updated by the new event 19.

According to the CRUD table, there is no event that deletes instances of the ADVERTISING AGENCY entity. Is there ever any circumstance under which Piccadilly stops doing business with an agency and deletes all reference to the agency? If the answer is yes, then you will need to add another event to your list, just as you did for the agency details change. If the answer is no, then put NOT APPLICABLE in the appropriate column in the table.

Note that not all data are deleted. This is an essential model, and deleting data to make room in the database is not a consideration. Data are only deleted when the business policy requires that you must expunge all reference to the data.

Work through the rest of the entities to ensure that all the events are captured.

Don't Forget the Relationships

You may also use the CRUD table to check the relationships in the same way. For instance, find the entry in the table for the BILLING relationship. This relationship between INVOICE and COMMERCIAL SPOT is created by event 13 *Spots are transmitted.* However, the relationship is never referenced, updated, or deleted.

This definitely seems odd. Surely, Piccadilly would need to reference the relationship when an agency pays an invoice. Now look at the context diagram. Shouldn't we have a data flow called something like AGENCY PAYMENT? If our system is responsible for sending out invoices, should it also be responsible for accepting payments? Or are payments part of another system? If the users decide that payments are part of this system, you'll need to add another event to the list, model the event, and update the appropriate data definitions.

CRUD Checks Your Context

The few examples mentioned here affect the context of the system. Not only does the CRUD check discover any missing boundary data flows, but it also verifies the integrity of the existing ones. Once you have resolved the questions raised by this check, then you can be certain that your context is complete and that your analysis effort will capture all the requirements for the new system.

The CRUD table above reveals some flaws in our context. For example, the ALLOCATING relationship between COMMERCIAL COPY and COMMERCIAL SPOT is created and deleted, but never referenced. So it appears that the Piccadilly system remembers which commercial was played in which break, but never uses that information. Either there is no need to remember it, or we have missed a requirement. Perhaps the information about which copy was transmitted should appear on the invoice. This is something you should take up with the users.

From the table, we can see that BREAKCHART DAY is never created. We went back through the events to find out why. Event 16 *Time to finalize new programme schedule* is the most likely candidate to create this entity. Look at the mini specification for the process ISSUE PROGRAMME SCHEDULE. A few lines down, it says,

> *For each* BREAKCHART DAY *within the next quarter*

Yet this process is the one that sets up the breakchart and its programmes for the coming quarter. We should add a line before this to say,

> *Create an instance of* BREAKCHART DAY *for every day of the quarter*

and then

> *For each* BREAKCHART DAY

We then update the CRUD table to show that event 16 creates the entity BREAKCHART DAY, and we update the event-response data model.

Another entity, COMMERCIAL BREAK, is never created. The problem again lies with the mini specification for event 16. When the schedule is being finalized, the broadcasting rules are used to determine where the breaks in each programme are put. It is this process that establishes the breaks along with the CONTAINING relationship. The specification has failed to fully describe the process.

You were not expected to know how breaks are created. None of the text has mentioned it. However, the CRUD check revealed that something was wrong. Your task as an analyst is to find out what went wrong.

The FILMING relationship is created, but not referenced. The system remembers which production companies produced which commercial copy, but it doesn't seem interested in this information. It may be remembered for the benefit of another system, or this system may be missing an event. Back to the users you go. While there, you can also ask about the SUPPLYING relationship between PROGRAMME SUPPLIER and PROGRAMME. For perhaps the same reasons, it is not referenced either. Whatever the reasons, the errors can be corrected before we start to implement the system.

We have not examined the entities and relationships to see if they are updated or deleted. Nor do we intend to. For the purposes of this book, the check you ran is sufficient for you to understand the process. We feel that the time invested in examining each of the attributes to determine if it can be updated or deleted would be wasted *in the context of this book*. No such dispensation applies for your real projects.

What Have You Achieved?

The answer is quite a lot. Once you have fully completed the CRUD check (see below), you can be certain that your specification is complete. The errors that we revealed above were genuine. We've done our best to build a specification without gaps. We missed, but now we can correct them. The important lesson is that a rigorous specification is possible.

How Much Detail Is Enough?

We mentioned that although we didn't expect you to go into detail in this exercise, your final CRUD table should cross-reference each data element with the appropriate events.

There is one last detail that you need to add to your table. When an event-response process model is made up of more than one process, it is useful to put the process number, rather than the event number, in the CRUD table. For instance, the process model for event 1 *Agency wants to run a campaign* contains three processes. One of those processes, 1.2 CREATE SUGGESTED CAMPAIGN, creates an instance of ADVERTISING CAMPAIGN. There are two reasons for this mind-bending detail. First, as input to the task of file design, it provides precise information about exactly which processes access which data. Second, if you are planning to implement a system in an object-oriented environment (more about this in Chapter 2.15 *Object-Oriented Viewpoint),* this detail provides input to the task of object partitioning.

The correlation of the data to the individual event responses is largely a clerical task. As such, it is a perfect candidate for automation and for use of a CASE tool. Now that you understand the technique, let the CASE tool do it for you. If you don't have a CASE tool or if your tool doesn't have this facility, you'll have to do it manually. It is not such a huge task. We used a spreadsheet for the CRUD table in this book. We have also seen CRUD tables written on whiteboards, in notebooks, and most things in between.

✛ Ski Patrol

The CRUD table is mostly a clerical task, as we said, and should not give you too many problems, especially if you do it progressively. Every time you finish modeling an event response, fill in the CRUD table cells concerned with that event. If you had difficulty relating the event response to the stored data, try doing it another way. Look up the data dictionary definition for each data flow that triggers an event response. Each of the attributes in this flow is probably stored. Look through the definitions of the entities and relationships of the stored data. Whenever a data flow element matches a stored attribute, that indicates the response triggered by this data flow is the one to create, reference, or update the entity or relationship. Check the event-response process model and the mini specs to determine which CRUD action is appropriate.

When the elements of the outgoing data flows match the stored attributes, it usually indicates a reference action. Again, look to the event-response process models and mini specs to confirm.

There is a section on CRUD checking in Chapter 2.11 *Event-Response Models* if you wish to review this topic. Pay careful attention to the discussion of the event-response data model.

Also, study the Piccadilly event-response models in Chapter 3.12, and note their use of stored data. Keep the CRUD table beside you, and ensure that you can correlate each of the table's accesses with the event-response models.

Trail Guide

● Easiest and ■ More Difficult: Go to Chapter 2.14 *New Physical Viewpoint* to learn how to turn the essential requirements into a new implementation.

◆ Most Difficult: A word of warning: You are about to enter an area of systems analysis that has undergone many changes. If you feel confident with your analytical skills, plunge onward to Chapter 1.16 *Strategy: Toward Implementation.* Otherwise, it's no loss of face to follow the others to Chapter 2.14 *New Physical Viewpoint.*

❈ Promenade: Proceed to Chapter 2.14 *New Physical Viewpoint* to see how all this becomes an operational system in the real, physical world.

SECTION 4

Textbook Solutions

SOLUTIONS: ANALYSIS MODELS 4.1

Exercise 1: Woolly Mammoths

The objective of this model is to show the data and the functions of the system. By combining the two processes into one bubble, some important functionality would have been hidden. In this case, we know that the functions are not related because had they been combined, the resulting bubble would have been very difficult to name. Similarly, you could say that the processes do different work, and that the tool makers and the hunters are probably different people. As each of the processes needs to be explained to the people doing the work, showing the functions separately makes the explanations simpler.

If all that is not enough, the case for making them separate bubbles is that the two processes are most likely done at different times. If they are active independently of each other, then your model will portray the system better by showing them separately. Later in the book, when we visit event-response models, you will see how important it is to separate activities that happen at different times.

Exercise 2: Other Uses for the Model

The model had another purpose before it was used to explain the hunting system to the elders. It helped the cavemen analysts themselves to understand the system. Like the cavemen, you cannot complete such a model without understanding the system, but building the model helps you to understand it.

This model also specifies the system in the same way that an architect's working drawings specify the building to be constructed. Perhaps if you wanted to implement this hunting system in another tribe, the model would be a good starting point. You'd also need some supporting documentation, explaining how each of the bubbles worked. The necessary logic would not be immediately obvious to a new and less wise tribe. For instance, you would need to explain how to analyze brontosaurus habits, and how to make tools. You will read about these explanations, or mini specifications, in Chapter 2.12 *Mini Specifications*.

Exercise 3: The System Remembers

The piece of system memory, WOOLLY MAMMOTHS FILE, correctly called a *data store,* provides information for the process SELECT AMBUSH LOCATIONS. You can reasonably expect that anything in the store relates to where the mammoths are traveling. The store is fed by a data flow called MAMMOTH SIGHTINGS coming from the hunting process. You can expect the store to contain such items as WHERE SIGHTED or TRAIL NAME, TIME OF SIGHTING, TO OR FROM THE WATERING HOLE, NUMBER OF MAMMOTHS SIGHTED, and so on. The exact contents of the store will be revealed by a detailed investigation, as you'll see in Chapter 2.9 *Data Dictionary*.

Trail Guide

This guide duplicates the one at the end of Chapter 2.1 *Analysis Models*. (All of the Textbook Solutions chapters follow this convention.)

● Easiest: Go to Chapter 2.2 *Data Flow Diagrams* to learn how to use this type of model.

■ More Difficult: You didn't have to come here. However, if you want to know more about the data flow model as we use it, jump to the ● Easiest Trail and turn to Chapter 2.2 *Data Flow Diagrams*. If you are already familiar with data flow diagrams, proceed to Chapter 1.2 *Start with the Context,* and resume reading.

◆ Most Difficult: This is not on your selected trail. We suggest that you return to Chapter 1.1 *Your Project Starts Here* for information about using the Trail Guides, and select a destination from there.

❋ Promenade: Your destination is Chapter 2.2 *Data Flow Diagrams*. You already saw this model in the cavemen's hunting system.

SOLUTIONS: DATA FLOW DIAGRAMS 4.2

Before You Reached Here ...

You have attempted one or more of the exercises in Chapter 2.2 *Data Flow Diagrams*. Even if you haven't, read through this chapter for the commentary provided. It'll be worth it.

Naturally, if you have done the exercises, you'll want to compare your answers with ours. Any glaring errors should be reported immediately. Write a short description on the back of large denomination banknotes and send them to the authors.

Exercise 1: Nelson Buzzcott's Employment Agency

a) What is the most probable data content of MATCHED APPLICATION?

The best way to answer this question is to ask your users. If they are not available, imagine that you are the process receiving the data flow. Now, what data do you need to do your job?

You have to interact with CLIENTS and APPLICANTS, so you can reasonably expect the incoming data flow to deliver an identifier and telephone number for each of these. If the interview is arranged by mail (unlikely), you'd want a postal address. You'd also like to know what job you are talking about, so the flow must contain an identifier and description of the job.

We haven't yet discussed the topic of the data dictionary, but here is the definition in data dictionary format:

Matched Application = Client Identifier + Client Telephone + Job Identifier
+ Job Description + Applicant Identifier + Applicant Telephone

The reason that we asked you this question is that the data drive the processes. If you do not know the content of a flow, you cannot understand the process that receives that flow. We will talk much more about data in coming chapters because the role of data is so critical to the system.

b) What kind of information would you find in JOB REGISTER?

Typically, we'd find a number of jobs, each one having some kind of identifier, a description of the job, the salary being offered, an identifier, and telephone (and possibly address) of the client.

Our answer is partly from speculation, and partly from deductions made by looking at the model. Most of the data in MATCHED APPLICATION must come from the JOB REGISTER: If we know one, we will know the other. The next step is to take our preliminary definition to the users and ask them to confirm or amend it. Naturally, if the users have a job register in use at the moment, we would also want to look at that.

c) Which is processed first: process 1 FIND SUITABLE APPLICANTS, or process 2 REGISTER APPLICANT?

This was something of a trick question, as there is no "first." The data flow diagram does not show any sequence for the processes. It does show the data produced by each process and the dependencies between processes, and from this you could work out a sequence if you wanted. But it really doesn't matter. A process is active whenever it has some data to process. Both of the bubbles mentioned are triggered by data flows from the outside world, so it is impossible to predict when the flow will arrive. For all you know (or care), processes 1 and 2 could be active at the same time. Probably the best answer to this question is it doesn't matter because the data flow diagram models asynchronous processes.

d) Why isn't CLIENTS shown as a bubble?

CLIENTS isn't a process that we're studying. CLIENTS is a shorthand notation for the clients' systems. These systems have processes, but we are not interested in them. When we defined the context, we decided that the internal workings of the clients' systems were beyond the scope of our study.

e) What happens to the data flow UNFILLED JOB? Why?

UNFILLED JOB carries the details of a new job for which there are no applicants on file. This file is recorded in the data store JOB REGISTER, where it awaits a match to a new applicant. Its data content is the same as one of the entries in JOB REGISTER.

f) How do we know that a job has been filled?

The current model doesn't show any way of recording that a job has been filled, so the model is incomplete. You must now raise this question with your users and find out how they record filled jobs. Note here the success (not failure) of the model. By demonstrating an inconsistency, it has helped you to raise a necessary question. By raising it, you are nearing your goal of finding *all* the requirements for the system.

Exercise 2: The FastBuck Book Company

You were asked to list the errors you found in the model of the FastBuck Book Company (Figure 2.2.15).

Process 1 PROCESS ORDER doesn't appear to do anything. This is a very poor name for a bubble, and its vagueness suggests that the analyst is rather unsure about just exactly what is happening. The solution to this problem is to write a coherent description of the bubble. If you cannot, then you should de-bubble it.

There is no rejection flow from this bubble, but recall we were told that it rejects orders without payment.

Process 2 GENERATE BOOK has no output. All functions must produce something. In this case, a BOOK would be the most likely output.

Both the data stores being accessed by this function are read-only. There is no source of input. This suggests an everlasting fountain of data, which needs no maintenance and has no data added to it. Read-only or write-only data stores must be investigated further (as should the dealings of the FastBuck Book Company, but that's a job for someone else).

SEND PHONY INVOICE TO REPEAT CUSTOMER has no way of knowing that the incoming order is to a repeat customer. To know this, the system needs access to a data store of previous orders.

The data flow PHONY INVOICE should flow to the CUSTOMER terminator. We redrew this model (Figure 4.2.1) to show what we think Benedict should have produced.

In this model, note that we have added an ORDERS FILE so that the process CHECK ORDER can tell whether an incoming order is a repeat order from a customer. The process also stores all orders.

Both the data stores in the previous model were read-only. Process 5 puts material into the SUPPLY OF COVERS store. We are guessing that the book covers come from a printer. Since we are even less sure where the generic books come from and since the people who run this company are very tight-lipped on this one, we have left it as a question mark.

The last paragraph was a joke, but the point of it is that when you are unsure of something, the best approach is always to put down your best guess. Put a large question mark beside it to indicate that you are guessing, and to remind yourself to ask the users.

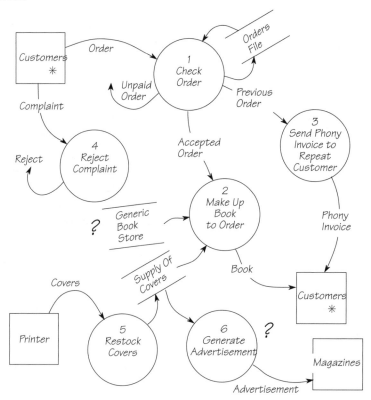

Figure 4.2.1: Data flow model, second edition, for the FastBuck Book Company.

The question mark beside process 6 is there because we don't know if the company wants us to study this part of the operation. It also looks suspicious because the way we have drawn it, the company appears only to advertise "books" it has in stock. We think this company is not above advertising books that it does not have.

Exercise 3: The Government Research Paper Clearing House

There is no right answer to this exercise, for there can be no right context diagram until you talk to the users and all the other people who have an interest in the system. We cannot transport you to the Clearing House, but we can give you some alternatives, and some observations.

The first alternative, shown in Figure 4.2.2, represents the most rational approach. It includes all the functionality mentioned in the problem statement.

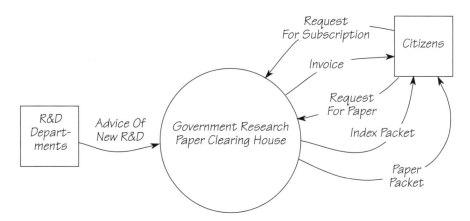

Figure 4.2.2: Context diagram, first edition, for the Government Research Paper Clearing House.

First, some technical observations: All of the data flows in your diagram should be named, and the names recognizable to the system users. The name of each terminator should reflect its role. From a mechanical point of view, context diagrams are reasonably simple. However, before proceeding, take time to ensure that all the notation on your diagram is correct.

The ADVICE OF NEW R&D flow was not specifically mentioned, but the Clearing House must have some way of knowing what papers are available for requests. The other flows in the diagram are referenced in the problem statement.

You were given details of the processing done by the Clearing House, but for the moment you hide these details inside the context bubble. Later you can break them out; the task here is to define the scope of the problem.

You may well have come up with a different solution to that given in Figure 4.2.2. For example, you were told that the Accounting Office handles the invoicing. Although you may feel it wise to study the way that the office produces the invoices, doing so implies that the Clearing House also must look after the receipt of payments, as well as the dunning of customers who fail to pay on time. In this case, you'll have to add data flows such as PAYMENT and DUNNING LETTER to the context diagram. Alternatively, the Accounting Office may have no desire for you to study this part of its activities, in which case you must show the ACCOUNTING OFFICE as a terminator. As all the accounting activity will take place inside this terminator, there can be no INVOICE flow from the context to the CITIZENS terminator, but there must be a flow to advise the ACCOUNTING OFFICE of the charges. Figure 4.2.3 shows this arrangement.

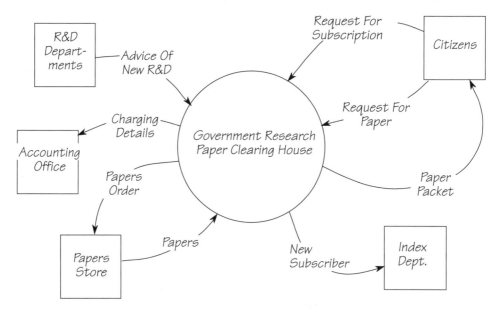

Figure 4.2.3: An alternative context diagram. The Accounting Office and Index Department are now outside the scope of the study, along with the Papers Store.

The Index Department may also be outside the scope of your study. The two women who run it are adamant that you are not to study their activities, and that their efficiency awards mean that the job is being done as well as possible, and that there is no reason to try to improve it. This decision does mean that the subscribers' list is maintained in two places; the Clearing House has to keep an up-to-date list and advise the Index Department of new subscribers to update its list. Now you can remove the indexes from your diagram, because the Index Department sends the index packet directly to the citizen. Note that it is incorrect to show a data flow between terminators.

As we said, there is no one right answer, but either of these diagrams gives you a starting point to raise questions with the users. Take a moment to reconcile the data flows in our diagram with yours.

When you are satisfied with your grasp of the drawing conventions, and with your answer versus the alternatives presented here, it's time for some more work.

Trail Guide

This Trail Guide duplicates the one in Chapter 2.2 *Data Flow Diagrams.*

● Easiest: Now that you know about data flow diagrams, go to Chapter 1.2 *Start with the Context,* where you will build a context diagram for the Piccadilly Project.

■ More Difficult: If you made it here, perhaps you should be following the ● Easiest Trail. If you have not already built the Piccadilly context diagram in Chapter 1.2 *Start with the Context,* that is your destination. Otherwise, go to Chapter 2.3 *A Variety of Viewpoints.*

◆ Most Difficult: Any chapter number beginning with a "4" is not part of your trail. Try picking up your trail in Chapter 1.2 *Start with the Context.*

❋ Promenade: The purpose of data flow diagrams is to model the processes, and the flows of data between the processes. The data in a system are just as important as the processes, so the next logical step for you is to learn about the data. Jump to Chapter 2.9 *Data Dictionary.* Don't worry about the chapters you pass over to get there. Eventually, you will return to learn about the models in those chapters.

4.3 SOLUTIONS: DATA MODELS

Before You Reached Here ...
You have attempted the data model problem given in Chapter 2.5 *Data Models*. Even if you haven't, you will benefit from reading this discussion and relating it to the problem statement.

Exercise: The Barbican Data Model

The data model that we derived from the user's statement and from the sample documents is shown in Figure 4.3.1.

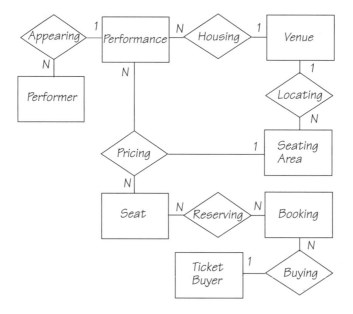

Figure 4.3.1: Data model for the Barbican Centre exercise.

A partial data dictionary to accompany the data model is this:

Booking = Booking Date + Booking Price Paid + Booking Discount Amount

*Discount Qualification = * Data element **
        ["School Child" | "Senior Citizen" | "Coach Party" |
        "Youth & Music Member" | "Student" | "Unemployed"]

Performance = Performance Date + Performance Description
        + Performance Start Time

Performer = Performer Name

*Pricing = * Relationship **
        Seat Price + Venue Discount Price + Party Size
        + Party Discounted Price

Seat = Seat Number

Seating Area = Seating Area Identity + Seating Area Name

Ticket Buyer = Ticket Buyer Identification + Discount Qualification
        + Mailing List Subscriber + Children's Cinema Club Member
        + Party Size

Venue = Venue Name + Venue Discount Conditions

*Venue Name = * Data element **
        ["Hall" | "Theatre" | "Pit" | "Cinema" | "Music Hall" | "Lecture
        Recital Room"]
        ** Are Hall and Music Hall the same? **

To provide a manageable problem, we gave you a context that was limited to the part of the Barbican's business related to booking seats. You can ignore any other data. For example, although the name of a performance's sponsor may be of great concern to the fund-raising part of the business, it doesn't appear to affect the context that concerns you. For this reason, we didn't include SPONSOR NAME as one of the attributes of PERFORMANCE.

Let's discuss this data model by considering the processing circumstances under which the entities and relationships are created and referenced. When the Barbican people schedule a PERFORMANCE, a HOUSING relationship is created between the PERFORMANCE and the VENUE they have chosen to house the performance. At the same time, they know the performers who will be appearing, and so instances of PERFORMER and APPEARING are established.

Each venue has a number of SEATING AREAs, each having a number of SEATs in them. The pricing for a seat is dependent on both the seating area and the performance. Therefore, we have chosen to have a trinary (three-way) relationship between PERFORMANCE, SEATING AREA, and SEAT. It is this PRICING relationship that holds the attributes SEAT PRICE, VENUE DISCOUNT PRICE, PARTY SIZE, and PARTY DISCOUNTED PRICE. In other words, this is the pricing for a seat in a designated seating area for a given performance.

However, the price of the seat may be different from the price actually paid if the ticket buyer qualifies for a discount. When someone wants to book a seat, the BOOKING PRICE PAID and the BOOKING DISCOUNT AMOUNT are calculated using the ticket buyer's DISCOUNT QUALIFICATION, the SEAT PRICE, the VENUE DISCOUNT CONDITIONS, the VENUE DISCOUNT PRICE, and the PARTY SIZE. A BOOKING entity is created, a RESERVING relationship keeps track of which seats have been booked, and a BUYING relationship keeps track of who made the booking.

Let's look a little more closely at the pricing policy. The VENUE DISCOUNT PRICE is invoked if the selected VENUE has one and if the TICKET BUYER has the appropriate qualification. For example, the user's statement says, "In the Pit ... school children and senior citizens can get a £5 ticket for matinées only. Cinema prices are £3.50 for adults. Senior citizens and children pay £2.50." Party discounts are available for some performances. The qualifying party size and discounted price for a party are attributes of PRICING. Other discounts are available, but are calculated differently. For example, reductions are offered for coach parties, to members of Youth & Music, and to students. On the assumption that the reduction is calculated as a percentage of the SEAT PRICE, we did not store the reduced prices as attributes. However, this is an assumption, and you must check carefully with the user.

The TICKET BUYER entity has an attribute CHILDREN'S CINEMA CLUB MEMBER. This is necessary, as the Cinema has screenings that are restricted to members of this club. The MAILING LIST SUBSCRIBER is a necessary attribute as subscribers are allowed to book earlier than others.

When you do some more process modeling, you will verify everything in the data model against your process models. In other words, even though the data model at this stage may contain assumptions, you will confirm or reject each of them by the end of the analysis effort. The reason for building this model is to understand the data well enough to be able to raise relevant questions with the users.

Your model should be substantially the same as ours, although there will be differences if you made different assumptions about the data.

Trail Guide

This guide duplicates the one in Chapter 2.5 *Data Models*.

● Easiest: Now that you know how to build data models, it's time to do one for Piccadilly. The task is in Chapter 1.3 *What About the Business Data?*

■ More Difficult and ◆ Most Difficult: Although this was not officially part of your trail, we're glad you made it through the chapter. Your next task is to build a data model for the Piccadilly Project. It's waiting for you in Chapter 1.3 *What About the Business Data?*

❋ Promenade: By now you have seen different types of system models, and read about viewpoints that are useful when building system models. Your trail to this point may have been a little disjointed, but there is method to this madness. We wanted to present the views and models in a way that we felt would be most comfortable for you. We also wanted you to have the correct preparation for the next important topic: the essential viewpoint of the system's process. This is in Chapter 2.10 *Essential Viewpoint*.

4.4 SOLUTIONS: MORE ON DATA FLOW DIAGRAMS

Before You Reached Here ...
You have done the exercises in Chapter 2.6 *More on Data Flow Diagrams*. This chapter contains sample solutions and discussions.

Exercise 1: Any Defects?

The errors, in no particular order, found in Figure 2.6.15 are these:

1. All but one of the data flows have a single-character name; these should be replaced with clearly functional names.

2. Bubble 3 has no output. There is no reason for a process to have input but no output.

3. Bubble 4 has no input. Apart from a random number generator, or the production of pure fiction, processes must have some kind of data input to transform into their output.

4. The data flow SOMETHING TO DO WITH INVOICES is badly named. It appears that the analyst has not studied the system sufficiently to fully understand what this data flow represents. Also, the terminator that is the source of the flow is unnamed.

5. There is a data flow Q between the terminators X and Y. If there is a real need to know about this flow, then one or the other terminators should be a process inside the context of study.

6. The STORE RRR is write-only, and there is no business reason for such a store. Either a process should read this store, or it should be a terminating data store.

7. The data flow called J is stored in STORE SSS. However, the only data retrieved from it are called M. Either some of the stored data are not being used, or not enough data are stored. You need definitions of the flows and store to determine precisely what is wrong.

8. The data flow out of bubble 5 has no name.

Exercise 2: Can You Improve This?

The model shown in Figure 2.6.16 has quite a few problems. The major problem is that it breaks the Rule of Data Conservation on several occasions. For example, bubble 1 REGISTER PASSENGER has two inputs: PASSENGER NAME and SEAT PREFERENCE. The process cannot possibly derive BAG WEIGHT and FLIGHT NUMBER from these inputs. Similarly, bubble 3 magically generates output data with insufficient input. Added to that, the flow SEAT PREFERENCE does not appear to be used by bubble 1. Our first act to improve the model is to give the system the correct inputs.

Some other problems with the model are

1. FLIGHT RECORD is a write-only data store. It may be created for use by another system, in which case it's better to annotate the store with a terminator symbol, like so: Flight Record

2. PASSENGER IS REGISTERED appears to be a control flow. Its name does not indicate that it carries any data, and so it should be removed from the model.

3. The name of bubble 3, PREFERENCES, indicates that it is not a well-defined function. The output data flows suggest confused functionality within the bubble.

We have reworked the model to correct its errors. The result is shown in Figure 4.4.1. Note that most airlines require advance notice for special meals. The model shows SPECIAL MEAL REQUEST flowing from the write-only data store FLIGHT RECORD. Therefore, the meal request must be entered at the time the reservation is made, or at least some time before the passenger arrives to check in for the flight. This implies that there is some other system to maintain some of the data in the FLIGHT RECORD data store.

The SEATING PLAN store is also created outside this system. This store reflects the seating layout of the airplane assigned to the flight. Airlines change the seats in their planes, and the type of plane assigned to a flight may differ from day to day. Both of these factors are outside the scope of a check-in system, so we chose to show data stores that are created by another system.

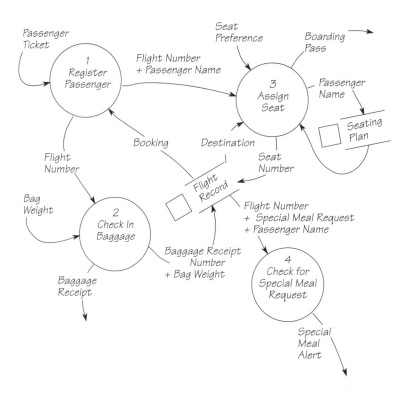

Figure 4.4.1: An alternative airline check-in model.

We introduced information into our model that was based on our interpretation of the requirements. Whatever improvements you choose to make, your model must produce the same output as the original. After all, you cannot change the system's requirements to make the model more convenient.

> When your system is concerned with the flow of physical material, you will include that physical material in a system model; otherwise, you will omit it in favor of the material's attributes.

We have shown the PASSENGER TICKET as an input. The reason for this is that part of the process of registering a passenger is checking that the passenger has a valid ticket. So here, the existence of a ticket, as well as its data, is important. However, in other circumstances, you omit physical articles from the model. For example, the passenger's bags in and of themselves are not interesting, but the system needs to know the weight of each bag. You'll notice in many of the other models in this book that there is no reference to inanimate objects, only to their properties. Why is this? Systems analysis is usually the prelude to designing an information system, and most of the time you are more interested in studying the information used by the system than the material that comes in contact with the system.

Exercise 3: The Clearing House Revisited

The exercise asked you to draw a data flow diagram for the Government Research Paper Clearing House operation by decomposing the context diagram shown in Figure 2.6.17. The lower-level diagram is called Diagram 0. We always do this by first drawing all the boundary data flows. All the input and output data flows that appear in the context diagram must also be part of any breakdown.

Once the flows around the edge of the system are in place, you follow each one to find the functionality. We came up with the diagram shown in Figure 4.4.2.

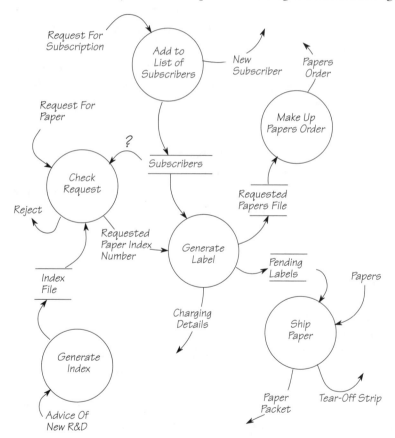

Figure 4.4.2: Diagram 0 for the Government Research Paper Clearing House system. Note that the two data flows REJECT and TEAR-OFF STRIP were not shown in the context diagram.

Your model may differ because you partitioned the system differently. There is no fixed rule that dictates the correct partitioning, so different analysts on the first attempt usually come up with different models. However, as you spend time working with the users, you'll eventually arrive at a partitioning that satisfies all concerned. The important objective for this exercise is that you use the data flow notation correctly, and show all the boundary data flows.

We do not show terminators in this model. They are in the context diagram, and to repeat them here would be redundant. If you want to know where a given data flow comes from, look at the context diagram. However, some analysts prefer to show terminators at all levels. While there is nothing technically incorrect with doing that, we feel the extra overhead of maintenance outweighs the slight gain in readability of the model. It is better to get used to working with more than one level of model simultaneously rather than creating an incomprehensible monster by trying to fit everything you know on the one model. Some CASE tools maintain the repetition of terminators on lower-level diagrams, in which case you have no choice. The best CASE tools give you a choice. We suggest that you show terminators in some lower-level models and leave them off others, in order for you to decide for yourself which you prefer. For the remainder of this book, we will show terminators at the highest level only, normally the context diagram, and thereafter omit them.

The data flow into the CHECK REQUEST bubble carries a question mark. The wording of the problem is vague about exactly what is being identified: Does it simply make sure that the paper exists, or does it also check that the request is coming from a registered subscriber? The problem statement doesn't say what happens to rejected requests for papers. If they are merely discarded, then REJECT is a trivial reject and is not shown on the context diagram. However if rejected requests are returned to the subscriber, you should update the context diagram to show the flow to the subscriber.

You weren't told if the charge for all papers is the same. This model assumes the same fee, but this must be checked with the users. The TEAR-OFF STRIP is shown as a rejection arrow. As it is simply discarded, there is no need to show it on the context diagram.

REQUESTED PAPERS FILE and PENDING LABELS are data stores. The policy of the system states that the order to the PAPERS STORE (a terminator) is sent only once a day. Yet, the Clearing House processes many individual requests during the day. This means that the requests must be stored (in the REQUESTED PAPERS FILE) until it is time to send the order. For a similar reason, the PENDING LABELS store is necessary because the labels have to wait until the store delivers the papers each morning.

Look through the model and reconcile it with your own. The important lesson is your use of the data flow notation, and your growing familiarity with the idea of building a model of the system that you are studying.

Exercise 4: La Cave du Morey Saint-Denis

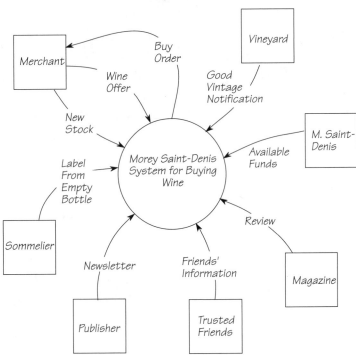

Figure 4.4.3: Context diagram of the Morey Saint-Denis System for Buying Wine.

Again, your partitioning for the Morey Saint-Denis system may be different from ours, but it is important that you portray the same functionality and use the equivalent data. For example, the bubbles DETERMINE IF NEEDED and DETERMINE IF SUITABLE may be combined into one process. Your data stores may be different provided you correctly recognized the stored data needed by the system.

The data flows UNSTOCKED WINE/LOW STOCK and SUITABLE WINE + SUGGESTED QUANTITY must both contain the price, as well as the name of the wine. This price comes from the WINE OFFER.

Whenever he buys, Monsieur Saint-Denis updates the CELLAR LOG with the order. This prevents him from reordering the same wine should he be offered it again before the arrival of his order.

UPDATE GOOD PERFORMERS is definitely a judgmental process. The shading on the sides of the bubble helps you to explain to Morey Saint-Denis that this process consists mainly of his own opinions. On the other hand, DETERMINE IF SUITABLE can be formalized into a set of rules, and so is not shaded.

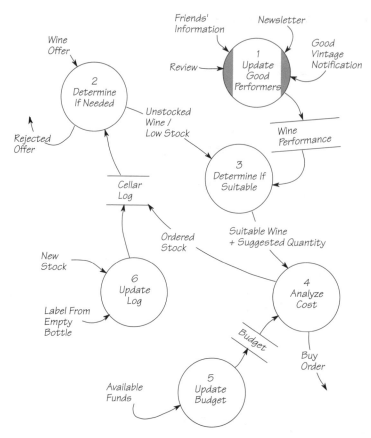

Figure 4.4.4: Lower-level diagram of the Morey Saint-Denis system.

Feeling good about your models? We hope so. At this stage, we want you to be comfortable with the modeling conventions, and very happy with the idea of building models of systems. You will have many opportunities to draw data flow diagrams as you work your way through this book, but we want you to concentrate on the system, not the diagramming conventions.

Trail Guide

This guide points to the same destinations as in Chapter 2.6 *More on Data Flow Diagrams.*

● Easiest: We will continue our study of data flow models. If you are building a model of a large system, you'll need to build a leveled model. You'll find out how in Chapter 2.7 *Leveled Data Flow Diagrams.*

■ More Difficult: This chapter is not on your trail. If you are reading it, that suggests you will rediscover your trail in either Chapter 2.8 *Current Physical Viewpoint;* or, if you have already been there, see Chapter 1.4 *The Piccadilly Organization.*

◆ Most Difficult: This was not intended reading for you. Chapter 1.4 *The Piccadilly Organization* is the best destination for you.

✳ Promenade: It's hard to know why you have strayed this far off your path. If you found this chapter interesting, you'll most likely want to go to Chapter 2.7 *Leveled Data Flow Diagrams.* If you want to get back to your correct trail, go to Chapter 2.8 *Current Physical Viewpoint.*

4.5 SOLUTIONS: LEVELED DATA FLOW DIAGRAMS

Before You Reached Here ...
● Easiest: You have worked on the data flow exercises in Chapter 2.7 *Leveled Data Flow Diagrams.* This chapter provides sample answers and a discussion.

Exercise 1: Find the Leveling Problems

Here are the problems with the parent and child diagrams:

1. The child diagram should be called Diagram 3, not Diagram 1.

2. G is an output in the child, but an input in the parent diagram. It is either incorrectly drawn in one of the diagrams, or the analyst is confused about the use of this piece of data.

3. B is shown as an input to the child Diagram 3, but the parent diagram shows it as an input to bubble 2.

4. Bubble 3.5 is missing. A missing bubble is not necessarily an error because the only significance of the number is as a tag for referencing the bubble, but it is confusing to a user. It may also indicate that the diagram has been incorrectly drawn.

5. There is an unnamed, unbalanced data flow out of process 3.6. However, if you remove it, that bubble has no output. This could indicate a problem at the parent level.

Exercise 2: Balancing Data Stores
The data store problems in the three diagrams are

1. The data store XXX is not shown as an output in Diagram 1, although both the parent Diagram 0 and the grandchild Diagram 1.1 show it.

2. YYY is an output from bubble 1.1 in Diagram 1, but an input to bubble 1.1.2 in the child Diagram 1.1.

3. This is a sneaky one. You were asked to check the data stores, but you can claim bonus points if you spotted that data flow E is missing from Diagram 1.1.

4. Note that data store ZZZ is shown correctly. It was hidden until the leveling reached down to Diagram 1.1, the highest level where it is used by more than one bubble.

Exercise 3: Draw the Parent Bubble
The system can be leveled upward to form the diagram in Figure 4.5.1.

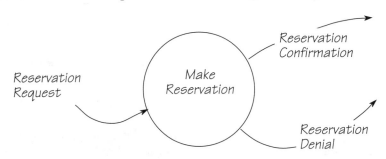

Figure 4.5.1: Context diagram for the hotel room exercise.

The input data flow is fragmented in the lower-level diagram because it is used by several bubbles. However, any customer wishing to make a reservation at this hotel would supply all the information at the time of booking. So it is reasonable to group it together in the higher-level model. The RESERVATION REQUEST would contain the ARRIVAL DATE, ARRIVAL TIME, LENGTH OF STAY, ROOM PREFERENCES, the customer's NAME & ADDRESS, and CREDIT CARD DETAILS. The RESERVATION CONFIRMATION would confirm the customer's NAME & ADDRESS, the ARRIVAL DATE, LENGTH OF STAY, SELECTED ROOM NUMBER if applicable, and a GUARANTEED STATUS. The RESERVATION DENIAL is simply a message saying, "Sorry, no room available."

Exercise 4: Repartition the Model

The model in this exercise is poorly leveled. It has only three bubbles at each of its two levels. Additionally, process 1 has far too many interfaces for it to be functionally partitioned. By replacing bubble 1 with its child level, we made the expanded diagram shown in Figure 4.5.2. In this case, the expanded diagram is a better model. It does not contain too many bubbles, so we stopped there.

Keep in mind that leveling allows you to decompose, recompose, and generally manipulate your bubbles to give you whatever view of the system you need.

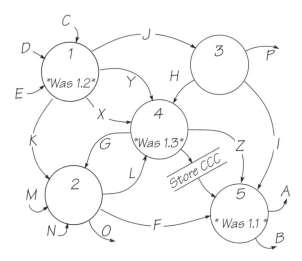

Figure 4.5.2: Revised Diagram 0 for the exercise.

Trail Guide

● Easiest: You now have all the information you need about data flow models. But when you build a model, it needs a viewpoint. One that you will find useful as you continue to explore the Piccadilly system is the current physical viewpoint. We discuss this in your next chapter, 2.8 *Current Physical Viewpoint.*

■ More Difficult: This chapter is not part of your trail; your next chapter is 2.8 *Current Physical Viewpoint.*

◆ Most Difficult: This was not intended reading for you, and Chapter 1.4 *The Piccadilly Organization* is the best destination for you.

❊ Promenade: We didn't intend you to be here, but hope that you found leveling interesting. Turn now to Chapter 2.8 *Current Physical Viewpoint.*

SOLUTIONS: DATA DICTIONARY 4.6

Before You Reached Here ...
You have written some data dictionary entries from Chapter 2.9 *Data Dictionary*. With the exception of the first solution, we have omitted the definitions of the data elements, not because we're lazy, but because it will save you wading through scores of elementary definitions.

1. *Title Page = Book Title*
    + {Author Name}
    + Foreword Writer Name
    + Publisher Name + Publisher Address

    *Author Name = * Data element **

    *Book Title = * Data element **

    It would also be correct to say that BOOK TITLE is made up of a main title and a subtitle.

    *City Name = * Data element **

    *Foreword Writer Name = * Data element **

    Publisher Address = Street Number + Street Name + City Name
    + State + Zip Code

    *Publisher Name = * Data element **

    *State = * Data element **

    *Street Name = * Data element **

Street Number = * Data element *

Zip Code = * Data element *

2.   Selling Price = [Standard Price | Special Price | Discounted Price]

The selling price can be one of three prices. Notice that this defines SELLING PRICE, not how each of the prices is calculated. One approach to defining the calculations for data elements is to use the mini specification. However, if you have a calculation that is referred to in many specifications, then you can define the calculation in the dictionary entry for the calculable data element. For example, to specify the calculation for DISCOUNTED PRICE:

Discounted Price = * Data element. Calculable as follows: (Standard Price – (Standard Price multiplied by Discount Rate)). *

3.   Personal Identity = [Name + Address + Date Of Birth | Social Security Number]

Note what this definition is saying. *Either* you can have a name, address, and date of birth, *or* you can have a social security number.

4.   Client Identifier = [Client Name | Client Acronym | Client Name + Client Acronym]

Again, note what it says: either one, the other, or both.

5.   This was a trick of the type that analysts have to deal with constantly. Because the English language has no precedence for and's and or's, the description could be

Invoice Line = Product Description + Quantity Sold + Selling Price

with the selling price defined as

Selling Price = [Undiscounted Price | Discounted Price | Special Price
+ Handling Charge]

Or the invoice line could be

Invoice Line = Product Description + Quantity Sold + Selling Price
+ Handling Charge

Selling Price = [Undiscounted Price | Discounted Price | Special Price]

The definitions in the data dictionary must have one and only one meaning.
From the description you were given, though, you couldn't possibly make a
definitive entry. In this case, you can only establish the correct meaning with
the users' assistance.

6. *Application = Name + Address*
+ (Telephone Number)

The telephone number is not elementary, so you need another entry:

Telephone Number = Area Code + Local Number
+ (Extension)

7. *Brewery Delivery Note = Pub Name*
+ {Container Count + Ale Container Name
+ Ale Container Capacity}

If only one size of container is delivered to each pub, remove the braces. An
additional entry is needed:

Ale Container Name = ["Pin" | "Firkin" | " Hogshead"
| "Barrel" | "Puncheon"]

Ale Container Capacity = [4 | 8 | 16 | 32 | 64]
∗ Units: gallons ∗

An alternative way to describe the data is

Brewery Delivery Note = Pub Name
+ {Pin Count}
+ {Firkin Count}
+ {Hogshead Count}
+ {Barrel Count}
+ {Puncheon Count}

Note that either definition of BREWERY DELIVERY NOTE allows the pub to receive, for example, a number of firkins and a number of barrels. In fact, they could get some of each size container. Because of the unusual nature of the names, a comment stating the capacity of each container name would be enlightening.

 ✳ *Each ale container name has a fixed capacity in gallons:* PIN 4, FIRKIN 8, HOGSHEAD 16, BARREL 32, PUNCHEON 64. ✳

We leave it to you, dear reader, to find your own method of researching the capacities of beer containers.

8. *Client File = {Client Acronym + Client Name*
     + {Job Ident + Job Type + Start Date}
     + Contact Name}

An equivalent answer is

Client File = {Client Record}

and then

Client Record = Client Acronym + Client Name
     + {Job Record}
     + Contact Name

with

Job Record = Job Ident + Job Type + Start Date

The advantage of the second solution is that each entry is simpler. The disadvantage is that you have to look up more entries. Selecting an alternative depends on the number of items in an entry. Up to about ten items is okay; beyond that the entry gets a bit unwieldy.

9. *Actor File = {Actor Header*
        + {Part Application Record}
        + {Part History Record}}

The remainder of the definitions look like this:

*Actor Header = * Green card **
        Actor Name + Actor Address + Actor Date Of Birth

*Part Application Record = * White card **
        Part Classification + Producer Identity + Start Date
        + Salary Offered + Role Description

*Part History Record = * Yellow card **
        Part Classification + Date Started + Date Ended
        + Salary + Producer Identity
        + {Review}

Since the color of the card indicates its data content, the color is not part of the definition. However, the comment serves to help the users verify the entry. Note the use of DATE STARTED in the PART HISTORY RECORD, and START DATE in the PART APPLICATION RECORD. Different names are used because these elements have different meanings. DATE STARTED is an actual date, whereas START DATE is a proposal. Your finished data dictionary will need to contain a comment specifying the meaning of each of these data elements. For example,

*Date Started = * Data element. The date that an actor started playing*
        *a part. **

You could lessen potential confusion and remove the need for some of the data dictionary comments by improving the names of the data elements so they aren't ambiguous. You could rename the three data elements in question as follows:

Date Part Started
Date Part Ended
Proposed Start Date

Trail Guide

This guide duplicates the one in Chapter 2.9 *Data Dictionary.*

● Easiest: The next step is to return to the Piccadilly Project to begin its data dictionary. Go to Chapter 1.5 *Building the Data Dictionary.*

■ More Difficult and ◆ Most Difficult: You didn't have to come here. Your next task for the Piccadilly Project is in Chapter 1.5 *Building the Data Dictionary.*

❋ Promenade: You've now seen the models that systems analysts use when they try to understand and specify systems. But there are more questions: What is the point of modeling the current system if that is to be replaced by something new? How can an analyst build a model of a new system without first understanding the true policy of the current one? To solve this dilemma, we introduce the concept of *viewpoints.* When applied to system modeling, using a certain viewpoint means that any unwanted information is filtered out and … Wait. Turn to Chapter 2.3 *A Variety of Viewpoints,* and all will be explained.

SOLUTIONS: EVENT-RESPONSE MODELS 4.7

Before You Reached Here ...
You have built some event-response models in Chapter 2.11 *Event-Response Models.*

Exercise 1: Dentist Performs Service

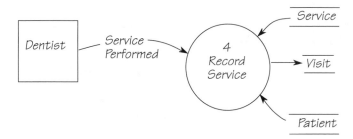

Figure 4.7.1: The essential event-response process model
for event 4 **Dentist performs service.**

The essential activity for the system is to record the services that have been performed. The service takes place outside the system; both the patients and the dentists are terminators in the context diagram. Note that this event–response establishes the VISIT and relates it to the SERVICE.

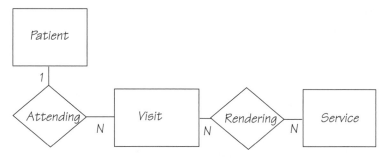

Figure 4.7.2: The essential event-response data model for
event 4.

Note that the VISIT entity is different from the APPOINTMENT entity. A visit is a record of a patient being treated by a dentist, whereas an appointment is a reservation of the dentist's time. Not all appointments are kept, and appointments can be canceled, but visits can't. Patients can also make walk-in emergency visits without appointments. This event may establish the relationship between visit and appointment, or the receptionist may establish it when the patient arrives at the office. In that case, there may be another event, *Patient arrives at office*. We must go back to the users to ask them.

Exercise 2: Time to Produce Appointment Schedule

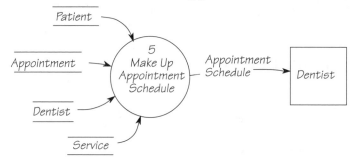

Figure 4.7.3: The essential event-response for event 5 **Time to produce appointment schedule.**

The data dictionary entry looks like this:

Appointment Schedule = Appointment Date + Dentist Name
+ {Appointment Time + Patient Name
+ {Service Identification}}

The receptionist makes up a schedule for each dentist. This is why there is only one dentist's name on the schedule. The {SERVICE IDENTIFICATION} is on the schedule so that the dentist has some warning of any major operations.

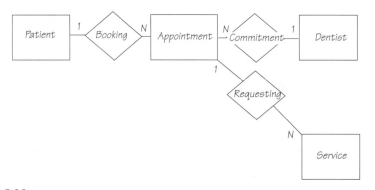

Figure 4.7.4: The essential event-response data model for event 5.

Exercise 3: Sid Edison's Radio Repairs

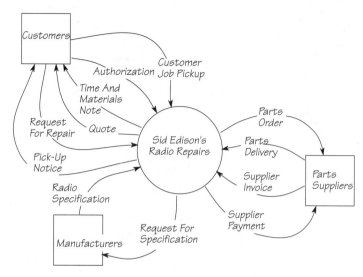

Figure 4.7.5: *Context diagram for Sid Edison's Radio Repairs system.*

Event Name	Associated Data Flows
1. Customer needs radio to be repaired	REQUEST FOR REPAIR (IN) QUOTE (OUT) TIME AND MATERIALS NOTE (OUT)
2. Customer authorizes repair	AUTHORIZATION (IN) PICK-UP NOTICE (OUT) REQUEST FOR SPECIFICATION (OUT) PARTS ORDER (OUT)
3. Manufacturer sends specification	RADIO SPECIFICATION (IN) PICK-UP NOTICE (OUT) PARTS ORDER (OUT)
4. Supplier delivers parts	PARTS DELIVERY (IN) PICK-UP NOTICE (OUT)
5. Supplier sends invoice	SUPPLIER INVOICE (IN)
6. Time to pay supplier	SUPPLIER PAYMENT (OUT)
7. Customer collects radio	CUSTOMER JOB PICKUP (IN)

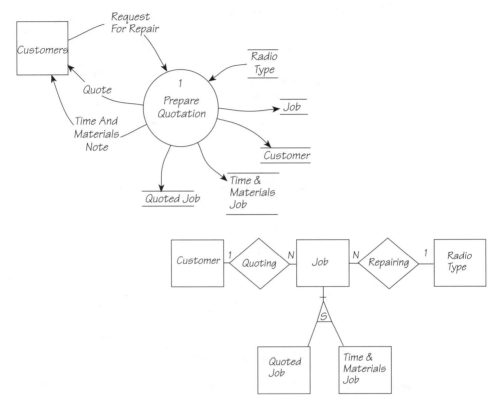

Figure 4.7.6: Event-response process and data models for event 1 **Customer needs radio to be repaired.**

The event-response model for event 1 is shown in Figure 4.7.6. This response happens when a customer brings in a radio to be repaired. First, Sid inspects the radio to assess the problem. Depending on what he finds, he either gives the customer a quote for the job, or agrees with the customer that the repairs are to be done on a time-and-materials basis. In the latter case, he gives the customer a TIME AND MATERIALS NOTE to outline his charges.

The response to event 2 *Customer authorizes repair* can have several outcomes. If Sid has both the specification and the parts, he can repair the radio and send the pick-up notice to the customer. On the other hand, if he needs a specification, the processing for the *Customer authorizes repair* event response stops, and the system waits for the event *Manufacturer sends specification.* Similarly, if he needs parts, the event response stops, and the system waits for *Supplier delivers parts.* PICK-UP NOTICE is also an output of the response to these other events.

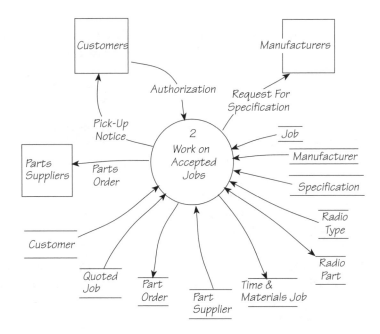

Figure 4.7.7: Event-response process model, showing one essential activity for event 2 **Customer authorizes repair.**

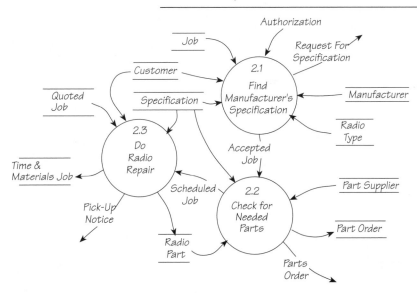

Figure 4.7.8: Lower-level model for event 2 **Customer authorizes repair.** The one bubble in the previous model is too complex for one mini specification, so the bubble is partitioned to produce this model with three primitive processes.

Note that in the response to event 2 (Figure 4.7.9), there are two relationships between JOB and CUSTOMER. The QUOTING relationship is established when Sid gives the customer a note explaining the charges involved in fixing the radio. The AUTHORIZING relationship is established when the customer decides to have the radio repaired.

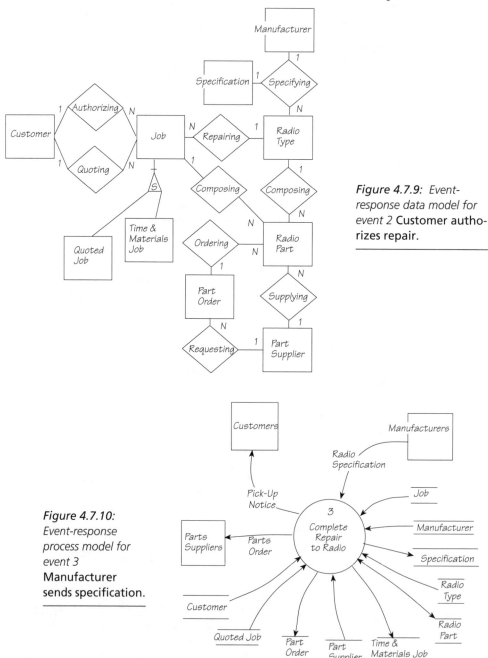

Figure 4.7.9: Event-response data model for event 2 **Customer authorizes repair.**

Figure 4.7.10: Event-response process model for event 3 **Manufacturer sends specification.**

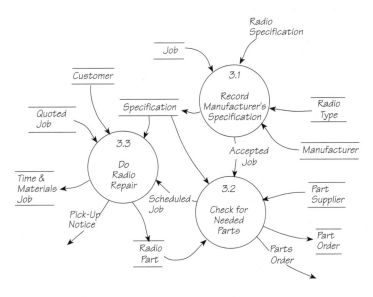

Figure 4.7.11: *Leveled process model for event 3* **Manufacturer sends specification.** *Notice that the processes* DO RADIO REPAIR *and* CHECK FOR NEEDED PARTS *are identical with those in the model for event 2. Although you draw the processes in both models, you'll write the mini specifications only once.*

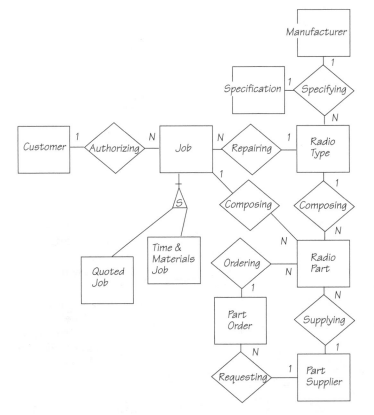

Figure 4.7.12: *Event-response data model for event 3* **Manufacturer sends specification.** *This data model is almost identical with that for event 2.*

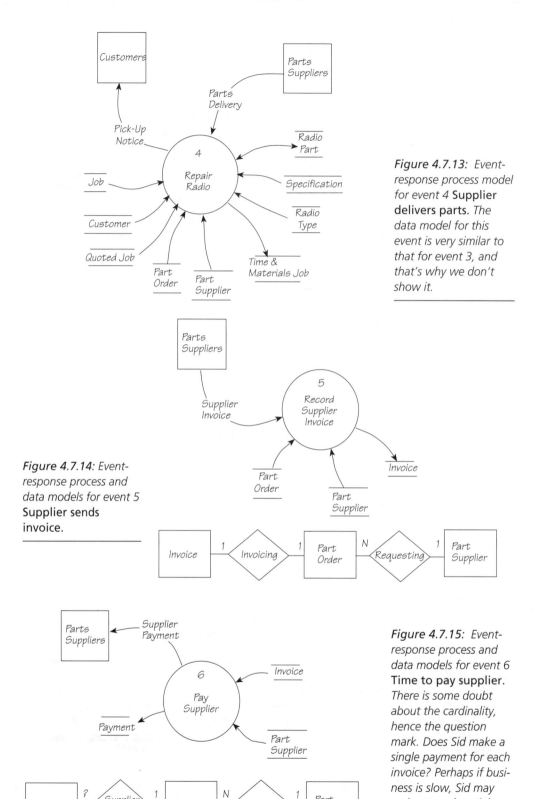

Figure 4.7.13: Event-response process model for event 4 **Supplier delivers parts.** The data model for this event is very similar to that for event 3, and that's why we don't show it.

Figure 4.7.14: Event-response process and data models for event 5 **Supplier sends invoice.**

Figure 4.7.15: Event-response process and data models for event 6 **Time to pay supplier.** There is some doubt about the cardinality, hence the question mark. Does Sid make a single payment for each invoice? Perhaps if business is slow, Sid may make several partial payments.

The response to event 7 is shown in Figure 4.7.16. The triggering data flow CUS-TOMER JOB PICKUP, carries the customer's identity, the job identifier, and the cash payment for the job. Note the 1:1 cardinality of the CUSTOMER PAYMENT to the JOB. This is because Sid demands that his customers pay in full before they remove their radios from his shop. The 1:N CUSTOMER to CUSTOMER PAYMENT is because some customers may have several radios repaired over a period of time.

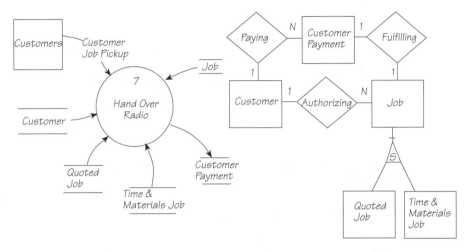

*Figure 4.7.16: Event-response process and data models for event 7 **Customer collects radio**.*

Note that there is no event *Repair is completed*. If you disregard the implementation of the system (Sid, who is only human, takes lunch breaks and works at normal human speed), any repair would be carried out instantaneously. In other words, the radio is repaired in response to the authorization, the delivery of the parts, or the arrival of the specification.

Custodial Activities

As Sid sends data flows to suppliers and manufacturers, he has to keep information about them, so there are additional events whose responses will be mainly custodial activities:

8. *Create new manufacturer*
9. *Update manufacturer details*
10. *Delete manufacturer*
11. *Create new supplier*
12. *Change supplier details*
13. *Delete supplier*

There are no data flows in the context diagram that suggest how Sid gets this information, so you will have to question him further to find the missing boundary data flows.

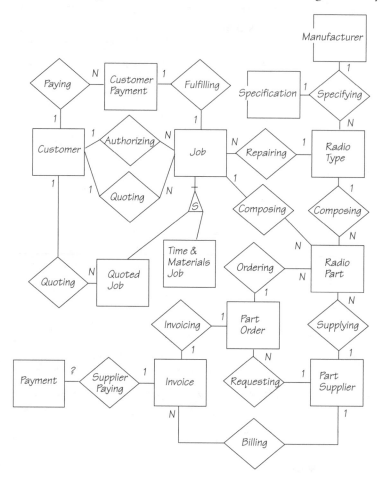

Figure 4.7.17: This system data model for Sid Edison's Radio Repairs is an amalgamation of all the event-response data models.

Sid doesn't indicate what information he keeps about his customers. He may keep none at all once the radio is collected. If he does store information about customers, then you will need

14. *Update customer*
15. *Delete customer*

We suggest that you ask Sid if you need

16. *Update supplier invoice details*
17. *Delete supplier invoice* (although we think that this last one is questionable)

The system stores specifications. This probably means you need

18. *Update specification* (He may not do this, as the manufacturer may send a new specification to replace an old one.)
19. *Delete specification* (The manufacturer would have to send notice of a deletion.)

Note that there are more events concerned with custodial activities than events for fundamental processing. This is a normal pattern for most systems. Note, too, that the custodial events indicate there are missing data flows that will have to be added to the context diagram. This is also normal for most systems.

Trail Guide
● ■ ◆ ✳ All trails: Return to Chapter 2.11 *Event-Response Models.*

4.8 SOLUTIONS: MINI SPECIFICATIONS

Before You Reached Here ...

You have written some mini specifications for the exercises in Chapter 2.12 *Mini Specifications*.

Exercise 1: Hopper's Choppers

There were so many decisions to make that we thought a decision table was the best way to express the policy.

Capacity?	Y	Y	Y	Y	N	N	N	N	N	N	N	N	N
Emergency?	Y	N	N	N	Y	Y	N	N	N	N	Y	N	N
Full fare?	-	Y	N	N	-	-	Y	Y	N	N	-	Y	N
Cheap fare?	-	-	Y	N	-	-	-	-	Y	N	-	-	Y
Employee?	Y	-	-	Y	Y	Y	-	-	-	Y	Y	-	-
Provisional bookings?	-	-	-	-	Y	N	Y	N	-	-	-	N	-
Employee bookings?	-	-	-	-	N	N	N	N	N	-	Y	Y	Y
Issue conf. booking	✓	✓			✓	✓	✓				✓	✓	
Issue prov. booking			✓	✓									✓
Bump newest prov.					✓		✓						
Bump newest conf.						✓							
Bump newest employee											✓	✓	✓
Record booking	✓	✓	✓	✓	✓	✓	✓				✓	✓	✓
Deny booking								✓	✓	✓			

The questions "Provisional bookings?" and "Employee bookings?" are contractions of "Are there any provisional or employee bookings already established?" The dash is used to show indifference. Given the other answers in the column, the answer to this question doesn't matter.

The RECORD BOOKING action results in the data model in Figure 4.8.1.

Figure 4.8.1: *The data model that results from the RECORD BOOKING action.*

Seats are allocated on a first-come, first-served basis. Larry didn't mention whether customers could ask for particular seats. If Grace's data dictionary is correct, then Larry must allocate seats starting with seat number 1 and continue until there are no more. However, we should check this with Larry.

Exercise 2: Terry's Ski Tuning Service

This one looks better written in structured language:

For each SERVICE TICKET
    For each SERVICE PERFORMED
        Find the corresponding entry in the RATES TABLE
        Accumulate the SERVICE RATE *into* TOTAL CHARGE
    If TECHNICIAN *is* "Terry" *or* "Joel"
        Increment the TOTAL CHARGE *by* 15%
    *If (*DAY OF WEEK *is* "Tuesday" *or* "Wednesday"*) and (*TIME OF DAY *is* "Night"*)*
        Discount TOTAL CHARGE *by* 20%
    If TIME OF DAY *is* "Day"
        Add $20 to TOTAL CHARGE
Issue CUSTOMER BILL

Sometimes, it is only when you come to write the mini specification that you discover the policy is not as clear as it could be. For instance, our specification takes the view that the $20 is neither discountable, nor subject to a premium. It is added after the other calculations. We also interpret the policy to say that all services are discounted on Tuesday or Wednesday nights, as is Terry's and Joel's loading. We will have to check these assumptions with Terry.

This mini specification demonstrates one of the problems of systems analysis: It never seems to end. However, it is far better to ask Terry now than to implement an incorrect system.

When you are satisfied with your mini specifications, it's time to write some for Piccadilly.

Trail Guide

● Easiest: Now it's time to return to the Piccadilly Project to specify some of the processes that you have identified. Your destination is Chapter 1.11 *Writing Mini Specifications*.

■ More Difficult and ◆ Most Difficult: This chapter was not part of your trail, but there are no penalties for reading it. Your Piccadilly Project work resumes in Chapter 1.11 *Writing Mini Specifications*.

❄ Promenade: You now know how to specify the requirements for a system, but there are other requirements that you'll meet in Chapter 2.13 *Modeling New Requirements*.

BIBLIOGRAPHY

The following works are either central to the ideas presented in this book or simply interesting reading for systems analysts.

Abdel-Hamid, Tarek, and Stuart Madnick. *Software Project Dynamics: An Integrated Approach.* Englewood Cliffs, N.J.: Prentice-Hall, 1991.
> The authors have investigated the process of building software, modeled the process, and used the model to predict the implications of managerial policies. An outstanding book.

Alexander, Christopher, Sara Ishikawa, and Murray Silverstein. *A Pattern Language.* New York: Oxford University Press, 1977.
> This fascinating book discusses how to make use of patterns when constructing buildings. By following the patterns, a builder will construct a livable and beautiful house. A similar approach to software construction would undoubtedly lead to more usable and maintainable systems.

Arango, Guillermo, Eric Schoen, and Robert Pettengill. "Design as Evolution and Reuse." *Proceedings of the Second International Workshop on Software Reusability.* Luca, Italy: IEEE Computer Society, March 1993.
> This paper describes practical experience in the evolution and reuse of product knowledge across wide geographical boundaries.

Boehm, Barry W. *Software Engineering Economics.* Englewood Cliffs, N.J.: Prentice-Hall, 1981.
> This is a monumental work in the field of measuring the software process. The author includes work on the COCOMO estimation and measuring model.

_____ , **ed.** *Software Risk Management.* Washington, D.C.: IEEE Computer Society Press, 1989.

> In today's world, the majority of software projects either fail completely, or fall far short of original expectations. Often, this is due to the development team failing to evaluate the risks. Read this collection and lessen your own risks.

Böhm, C., and G. Jacopini. "Flow Diagrams, Turing Machines and Languages with Only Two Formation Rules." *Communications of the ACM,* Vol. 9, No. 5 (May 1966), pp. 366-71.

> In this landmark paper, the authors proved that any program could be constructed using only selection and repetition as the constructs.

Booch, Grady. *Object-Oriented Design with Applications.* Redwood City, Calif.: Benjamin/Cummings Publishing Company, 1991.

> A very good book on using the object paradigm for design. Booch's notation is somewhat complex, but the sample applications and the discussions of several object-oriented languages make this very worthwhile.

Chen, Peter, ed. *Entity-Relationship Approach to Information Modeling and Analysis.* Amsterdam: North-Holland Publishing Company, 1983.

Delskov, Lis, and Therese Lange. *Strukturet Analyse — Integreret Systemanalyse.* Copenhagen: Teknisk Forlag, 1991.

> This is in Danish, so it has a somewhat restricted readership. The charming authors present an intelligent integration of event-response and data modeling.

DeMarco, Tom. *Structured Analysis and System Specification.* Englewood Cliffs, N.J.: Prentice-Hall, 1978.

> Although the analysis technique DeMarco describes has evolved since 1978, this book is an entertaining and informative read on basic structured analysis and how it got started.

_____ . *Controlling Software Projects.* Englewood Cliffs, N.J.: Prentice-Hall, 1982.

> Presents a practical method for making estimates based on the analysis model. As the analysis model is a plan for the construction of the system, this method makes a lot of sense.

_____ , **and Timothy Lister.** *Peopleware: Productive Projects and Teams.* New York: Dorset House Publishing, 1987.

> A series of fascinating anecdotes and lessons that reveal more about project management than almost all conventional books on the subject.

_____ , **eds.** *Software State-of-the-Art: Selected Papers.* New York: Dorset House Publishing, 1990.

> From the Editors' Preface: "Do yourself a huge favor: Read these thirty-one papers and throw out all the stacks of unread material dating back as far as 1980. You'll feel better immediately."

Dickinson, Brian. *Developing Quality Systems,* 2nd ed. New York: McGraw-Hill, 1989.

> The author uses data flow diagrams to describe his method for doing systems analysis, design, and development.

Dijkstra, Edsger. *Selected Writings on Computing: A Personal Perspective.* New York: Springer-Verlag, 1982.

> This noted pioneer's writings are more of a memoir than a text. The advice he gives, however, is lovely, some focused on the human side, the rest on the technology.

Flavin, Matt. *Fundamental Concepts of Information Modeling.* Englewood Cliffs, N.J.: Prentice-Hall, 1981.

> The definitive book on information modeling. It pays back the effort necessary to understand its concepts.

Freedman, Daniel P., and Gerald M. Weinberg. *Handbook of Walkthroughs, Inspections, and Technical Reviews,* 3rd ed. New York: Dorset House Publishing, 1990.

> We consider walkthroughs and reviews to be an important part of systems development. This is the most accessible book on the subject.

Gause, Donald C., and Gerald M. Weinberg. *Exploring Requirements: Quality Before Design.* New York: Dorset House Publishing, 1989.

> This book takes a very close look at the importance of systems analysis. Highly recommended reading.

Gilb, Tom. *Principles of Software Engineering Management.* Reading, Mass.: Addison-Wesley, 1988.

> An especially good treatment of how to define measurable project goals.

Hatley, Derek J., and Imtiaz A. Pirbhai. *Strategies for Real-Time System Specification*. New York: Dorset House Publishing, 1987.

> Real-time systems have few differences from commercial systems. However, the differences may be important to you. The Hatley-Pirbhai approach is difficult to understand at first, but worth the effort.

Heckel, Paul. *The Elements of Friendly Software Design*. New York: Warner Books, 1984.

> One of those paperbacks that you can read in two lunchtimes, but Heckel's thoughts will stay with you much longer.

Jacobson, Ivar, Magnus Christerson, Patrik Jonsson, and Gunnar Overgaard. *Object-Oriented Software Engineering: A Use-Case Driven Approach.* Wokingham, England: Addison-Wesley, 1992.

> An object-oriented approach that builds on established concepts.

Kidder, Tracy. *The Soul of a New Machine*. Boston: Little, Brown and Company, 1981.

> This Pulitzer Prize-winning book brilliantly follows the development of a new computer. Powerful reading.

McMenamin, Stephen M., and John F. Palmer. *Essential Systems Analysis*. Englewood Cliffs, N.J.: Prentice-Hall, 1984.

> This outstanding book introduces event partitioning. We are indebted to McMenamin and Palmer for this work, and it is definitely recommended reading.

Meyer, Bertrand. *Object-Oriented Software Construction*. Hemel Hempstead, England: Prentice-Hall, 1988.

> Meyer's book is a good explanation of the object paradigm. One must fault his (understandable) bias towards his own Eiffel language and (incorrect) attacks on structured design.

Miller, George A. "The Magical Number Seven, Plus or Minus Two: Some Limits on Our Capacity for Processing Information." *The Psychological Review,* Vol. 63, No. 2 (March 1956), pp. 81-97.

> An oft-quoted paper dealing with conceptual limits.

Myers, Glenford J. *Composite/Structured Design.* New York: Van Nostrand Reinhold, 1978.
> Myers is one of the originators of structured design, and his books on the subject are readable and informative.

Page-Jones, Meilir. *The Practical Guide to Structured Systems Design,* 2nd ed. Englewood Cliffs, N.J.: Prentice-Hall, 1988.
> The most readable book on the subject of software design. Page-Jones links his subject to event-partitioned systems analysis.

Robertson, Suzanne, and Kenneth Strunch. "Reusing the Products of Analysis." *Proceedings of the Second International Conference on Software Reusability.* Luca, Italy: IEEE Computer Society, March 1993.
> An account of how project teams built and reused analysis models.

Rock-Evans, Rosemary. *Analysis Within the Systems Life Cycle,* Vols. 1-4. Maidenhead, England: Pergamon Infotech, 1987.
> An exhaustive treatment of systems from the data analysis viewpoint.

Ross, Ronald G. *Entity Modeling: Techniques and Application.* Boston, Mass.: Database Research Group, 1988.
> Entity-relationship modeling is important, and not always easy to understand. We think that different authors throw different lights on the subject, and that one of those lights might be the one that illuminates it for you.

Rumbaugh, James, Michael Blaha, William Premerlani, Frederick Eddy, and William Lorensen. *Object-Oriented Modeling and Design.* Englewood Cliffs, N.J.: Prentice-Hall, 1991.
> This book connects the ideas of structured analysis and essential systems analysis to the development of object-oriented systems.

Russell, Peter. *The Brain Book.* London: Routledge and Kegan Paul, Ltd., 1980.
> Describes mind maps, a useful tool for taking notes and capturing ideas.

Shlaer, Sally, and Stephen J. Mellor. *Object-Oriented Systems Analysis.* Englewood Cliffs, N.J.: Prentice-Hall, 1988.

_____ . *Object Lifecycles: Modeling the World in States.* Englewood Cliffs, N.J.: Prentice-Hall, 1992.

> These two books by Shlaer and Mellor provide excellent coverage of data modeling. The first introduces the topic in a highly readable way, and the second continues the good work by showing how to model the states of an entity's life history.

Stevens, Wayne P. *Using Structured Design.* New York: John Wiley & Sons, 1981.

> A good, practical book on structured design.

Tsichritzis, Dionysios C., and Frederick H. Lochovsky. *Data Models.* Englewood Cliffs, N.J.: Prentice-Hall, 1982.

> A complete reference work covering different types of data models.

Tufte, Edward. *Envisioning Information.* Cheshire, Conn.: Graphics Press, 1983.

> This and the next book will give you new insights into how data can be meaningfully displayed.

_____ . *The Visual Display of Quantitative Information.* Cheshire, Conn.: Graphics Press, 1983.

> This is a wonderful book. Tufte explains how to avoid the junky graphs and charts that are being endlessly churned out of desktop publishing systems. He further explains his concept of "data-ink": how to get the most out of your graphics. It was the best book we read that year. Buy a copy and give it to yourself for your next birthday.

Ward, Paul T. *Systems Development Without Pain: A User's Guide to Modeling Organizational Patterns.* Englewood Cliffs, N.J.: Prentice-Hall, 1984.

> A good introduction to structured analysis for users and junior team members alike.

_____ , **and Stephen J. Mellor.** *Structured Development for Real-Time Systems,* Vols. 1-3. Englewood Cliffs, N.J.: Prentice-Hall, 1986.

> Presents a method for specifying real-time systems.

Weinberg, Gerald M. *Quality Software Management, Vol. 1: Systems Thinking.* New York: Dorset House Publishing, 1992.

_____ . *Quality Software Management, Vol. 2: First-Order Measurement.* New York: Dorset House Publishing, 1993.

> In these two books of a multi-volume set, Weinberg discusses management of projects with the emphasis on the two most important issues: people and honesty.

_____ . *Rethinking Systems Analysis & Design.* New York: Dorset House Publishing, 1988.

> Weinberg has the ability to see things that nobody else can. His different approach here is worthwhile reading.

Yourdon, Edward, ed. *Classics in Software Engineering.* Englewood Cliffs, N.J.: Prentice-Hall, 1979.

_____ , **ed.** *Writings of the Revolution.* Englewood Cliffs, N.J.: Prentice-Hall, 1982.

> The collections of papers in these two books trace the history of software engineering.

GLOSSARY

Alias The existence of two names in a data dictionary for the same item.

Allocated event-response model An event-response model showing the devices that will be used to implement the response.

Analysis model An abstracted representation of the system being studied. The model may include a set of data flow diagrams, event-response models, data models, data dictionary, and mini specifications.

Association The relationship between entities.

Asynchronous model A model in which the components act independently or in parallel. A data flow diagram is an asynchronous model, as it represents the system as a network of independent processes.

Attribute A data element that describes an entity or a relationship. Each attribute applies to every occurrence of its entity or relationship.

Balancing rule The requirement that the inputs and outputs of a parent diagram match the inputs and outputs of its child diagrams. For data stores, the rule is to show data stores at the highest level where they are used by more than one bubble and at every relevant lower level.

Behavioral model A model that specifies the control and/or behavior of a system. Commonly used to specify the behavior of interfaces between processors. Examples are transaction synchronization models, prototypes, and any type of state transition or synchronization model.

Boundary data flow A data flow that enters or leaves the system context. The term "boundary" is used because it crosses the perimeter of the system.

BASE (brain-aided software engineering) A term coined by our colleague Tim Lister to emphasize the importance of understanding the thinking behind software development before a CASE tool can be successfully used. The focus of this book is on BASE.

Bubble A popular name for process in a data flow diagram.

Business policy The set of rules and data that exist because of the essential purpose of the system; they are independent of the technology used to carry out the rules.

Cardinality The number of entities of each type participating in a relationship.

CASE (computer-aided software engineering) A software tool that automates the diagramming, storage, balancing, and accessing of the system specification.

Child diagram A diagram showing the decomposition of a higher-level process into its component processes, data flows, and data stores.

Context diagram The highest-level diagram of a leveled set of data flow diagrams. It shows the system being studied as a single bubble connected to the outside world by its boundary data flows. This diagram, or more precisely the boundary data flows, defines the domain of the analysis study.

Context of study The system as delimited by the boundary data flows in the context diagram. Sometimes this is shortened to "context."

CRUD check A way of verifying that every data element within the context is being created, retrieved, updated, or deleted by at least one process; a way of checking that every data element needed by every process within the context of study has been defined. This powerful check leads to the discovery of missing and redundant events, processes, and data, and is an effective assessment of the completeness of the analysis.

Current physical model A model that focuses on how a system is currently implemented. It is a replica of the users' existing operation; its purpose is to serve as a starting point for the analyst to understand unfamiliar subject matter and to communicate with the users.

Custodial activity An activity whose sole purpose is to maintain the system's stored data and keep it current. For example, changing a customer's address is custodial, and is done only so the business can continue to carry out its fundamental activity of sending that customer's invoices to the correct address.

Custodial processing Processing related to maintaining the system's essential data.

Data conservation See **Rule of Data Conservation**.

Data dictionary The collection of definitions of every piece of data used within a context of study; the part of the analysis model that provides definitions of the data flows, data elements, stores, entities, and relationships.

Data element A primitive item of data; one that has a value within the context of study and is not further decomposed.

Data element grouping A name that refers to a collection of data elements and that is defined in the data dictionary. The name might be included as part of many data flows or stores.

Data flow A path that carries packets of information of known composition; a roadway for data. Every data flow's composition is recorded in the data dictionary.

Data flow diagram A model of the system that shows the system's processes, the data that flow between them (hence the name), and the data stores used by the processes. The data flow diagram shows the system as a network of processes, and is thought to be the most easily recognized of all the analysis models.

Data model A statement of the business policy showing the essential entities and the relationships between them; it ignores the processing part of the system and focuses only on the information that is essential to the system. Also known as an entity-relationship diagram, Chen diagram, or information model.

Data object See **entity**.

Data store A time-delayed repository of information.

Decision table A tabular tool for specifying the action that will result from each combination of a set of conditions.

Decision tree A graphic tool for specifying the action that will result from each combination of a set of conditions.

Design The process of defining how a system will be implemented; the objective is to use the available technology to implement the essential requirements so that the implemented system looks as much like the essential system, and hence the problem, as possible.

Detailed analysis A comprehensive inspection and modeling effort to record the details of the system.

Detailed design The activity of deciding how the chosen technology can best be used to make the essential requirements work in the real world. Detailed design includes deciding the software components to make up a computer program, and determining how the people are to interact with the automated system.

Discrete element An element with a strictly limited number of possible values.

Domain of study The part of the business being analyzed. This is synonymous with context of study.

Entity A rational collection of data elements that describes something from the real world of importance to the business. Every entity must have a unique and definable role in the business and must have at least one attribute to describe it. Another name for data object.

Essence The part of the system that is implementation independent; the underlying business policy.

Essential Relating to the "perfect" view of the system, which concerns only the system requirements and excludes anything having to do with how the system is designed or implemented. Sometimes called "logical."

Essential activity model An event-response model in which all the processes have been consolidated into a single bubble. For the least complex event responses, this model is the same as the event-response model.

Essential analysis A major extension to structured analysis that focuses on the essential or logical view of analysis; the approach was introduced by the team of McMenamin and Palmer (see the Bibliography).

Essential policy The logic behind the system's current activities.

Essential requirements The requirements that describe the business being done by the system with no bias toward implementation. They represent the fundamental business of the enterprise. Also known as logical or business policy requirements.

Essential requirements model Usually made up of event-response process models and data models. This model shows only the essential, or logical, requirements.

Essential viewpoint An abstract view of the system, showing only the requirements specific to the subject matter within the context of study and excluding anything that exists because of how the system is designed and implemented.

Event A happening that stimulates a system to respond in a unique fashion; each event produces a unique data flow and has a unique name. Events are of two types: An external event takes place outside the context of study; and a temporal event is stimulated by the arrival of a predetermined time.

Event list A practical tool for inventorying all the events to which the system responds. It contains the event name, along with its associated input and output, for every event that is the concern of a context of study.

Event partitioning A technique for breaking a system into the responses to each of its events. The result is a collection of logical, minimally connected pieces.

Event response The system's preplanned response to an event, specified using a combination of event-response data and process models. It is a collection of all the actions that respond to the event regardless of where they happen in the current implementation.

Event-response data model A data model containing the entities, relationships, and attributes used, or stored by, one event response.

Event–response model A model of the system's reaction to an event containing the event–response process model, event–response data model, data dictionary definitions, and mini specifications.

Event–response process model A data flow diagram showing the processes, flows, and stores involved in one event response.

Expanded diagram A diagram containing details that are normally shown in a set of low-level diagrams.

External design The design of the interfaces between the different processors in a system environment.

External event See **event**.

False data store A data store that is part of the current system for some reason connected to the implementation.

Foreign key One or more attributes that are included in one entity for the purpose of identifying another.

Functional partitioning A way of modeling the system so that similar processes are grouped together in such a way that the interfaces between them are minimal. This is often similar to logical partitioning.

Functional primitive A process that will not be further decomposed, because it is small enough to be specified in a one-page specification.

Function model See **process model**.

Fundamental activity This is an activity that is part of the reason for the system's existence, such as selling goods, monitoring traffic, and so on.

Fundamental processing Processing related to one of the system's primary business activities.

Head-sized piece The portion of a system that fits comfortably inside an analyst's head and is thus readily understood; it is a functional primitive and is described by a mini specification.

Implementation The activity of making the essential requirements work in the real world.

Implementation engine See **processor**.

Implementation model The highest-level blueprint for the construction of the system. Part of the new physical model, it summarizes the interprocessor interfaces for all the system events; it serves as a project management tool.

Instance One occurrence of something that has many occurrences, such as entities or objects. For example, April 4 is one instance of the BREAKCHART DAY entity.

Interface The connection between two components of a system. Analysis is concerned with the data content of an interface, and design is concerned with the behavior and appearance of the interface.

Internal design The design of the processes and interfaces within a processor in the system environment.

Key field The unique identifier of an entity or data store.

Leveling A technique for decomposing a system and modeling it at various levels of detail.

Logical See **essential**.

Logical partitioning A procedure for showing the system broken down into units that relate to the essential requirements of the system, and not simply to mirror the current situation.

Maxi specification A specification written for a process that is not a functional primitive.

Mini specification An analysis tool, named for its manageable size, for describing the policy to be carried out by a functional primitive, which can usually be described in a page or less. Also called a process specification, P-spec, function spec, or transformation spec; describes a single, discrete functional component of the system.

Model A miniature representation of chosen aspects of a system.

Narrow interface An interface in which only a small amount of data connects one process to another; functional processes have narrow interfaces.

New physical model A model used to illustrate, negotiate, and define the implementation of the new system. The model shows computers, humans, and machines; it also illustrates the physical details of interfaces. Also referred to as the new implementation environment or the preliminary design model.

New requirement A requirement that is outside the declared context of study.

Object An encapsulation of data and the processes that operate on the data.

Object-oriented design The craft of partitioning the system into objects, organizing the objects into class hierarchies, and devising messages that communicate between the objects. See the Bibliography for references on this subject.

Parent diagram A high-level summary of several detailed models.

Participation Participation rules are attached to relationships. They specify whether every instance of a connected entity has mandatory or optional participation in that relationship.

Partitioning Breaking a system into manageable and, finally, specifiable pieces.

Partitioning theme A way of breaking down the system. Current physical models typically use the existing processors and data flows as their theme. Essential models use events and logical data as their partitioning theme.

Perfect requirements Another name for essential requirements.

Perfect technology A concept proposed by McMenamin and Palmer as a way of seeing past the current implementation.

Physical A term used to refer to models that portray the implementation features of a system. Such features are "physical" because you can touch the devices, as opposed to the "essential" processes and data, which are conceptual in nature.

Preliminary design The process of defining the implementation interfaces. Doing enough of the external and internal design to reveal the design tasks and the interfaces between them.

Problem space See **context of study**.

Procedural Concerned with the sequence of actions used to carry out a task.

Process A system component that transforms input data into output data according to defined business policy.

Process model A model showing the processes carried out by a system and the data interfaces between those processes; same as a data flow model.

Processor Any person, job, department, organization, mechanical tool, or computer capable of implementing an essential requirement; also referred to as an implementation engine.

Process specification See **mini specification**.

Prototype A simulation, usually automated, of the computer system to be implemented.

Real requirements See **essential requirements**.

Relationship The association of two or more entities; through this association, it expresses the business policy of the data model.

Repeating group A collection of data items that occur more than once.

Repetition A structured programming construct that states that an action will be repeated whenever a condition is true.

Risk management A policy of determining the greatest potential failure associated with a project.

Rule of Data Conservation The rule that each system component must receive data that are both necessary and sufficient to produce its output.

Selection A structured programming construct whereby the control flow is diverted according to the results of a choice. This is implemented in most programming languages, and in structured language for mini specifications, as an IF statement.

Sequence This means the statements of a mini specification are read from top to bottom with each statement following the previous one.

Spiral development A systems development strategy that breaks a project into a number of subprojects whose size can range from a subsystem to an event.

The systems development tasks within and between pieces overlap depending on the constraints of the particular project.

Structured analysis A system of analysis, popularized by Tom DeMarco and others, whereby the system is analyzed by building data flow models of it. This book proposes an analysis technique that is a progression beyond structured analysis.

Structured design A set of tools, concepts, and strategies involving hierarchical partitioning in a top-down manner, using coupling and cohesion analysis to refine the design.

Structured language A subset of natural language, used for writing mini specifications and restricted to a few verbs that manipulate the data items in the data dictionary. The language follows the same rules for combining statements as does structured programming.

Structured programming The convention, first proposed by the Italians Böhm and Jacopini, that computer programs are written using only selection and repetition to join the statements. This is commonly, and somewhat incorrectly, known as "goto-less" programming.

Subtype An entity that has its own unique characteristics and also shares the characteristics of its supertype entity.

Supertype A generalized entity; its business role and its attributes are common to all the subtypes.

Synchronization model See **behavioral model**.

System data model A data model showing all the entities and relationships within a context of study.

System environment The processing, data carrying, and data storage technology that is available for the implementation of a system.

System environment model A formal specification of the technology that is available to implement the system.

Systems analysis The craft of understanding and specifying systems by building models of them.

System scope See **context of study**.

Temporal event See **event**.

Terminating data store A data store that is owned by another system and so acts as a terminator.

Terminator A person, place, or system that is connected to the context via one or more boundary data flows. A terminator is considered outside the context of study; the systems analyst is only interested in the terminator as a provider or receiver of the boundary data flows.

Top-down approach A technique that first produces an overview before the analyst models the details.

Transaction synchronization model A model that shows the interaction between two processors during a defined time period. Typically used for defining the processing, screens, and control sequence across a human-machine boundary.

Trivial reject A data flow that shows rejected data leaving the system.

User knowledge Information held by the user or users about the business purpose and rules of the system.

Viewpoint A way of focusing on a system that highlights the characteristics germane to a particular viewpoint. The four viewpoints of use to systems analysts are current physical, essential, data, and new physical.

Waterfall model A systems development methodology that dictates the completion of one activity before beginning the next.

Working model A model that demonstrates that each process in the data flow diagram can manufacture its outputs from its inputs, and each entity and relationship in the data model can supply or store the data needed by all the processes.

INDEX

593

Reusability:
  of analysis components, 101
  of design components, 88, 315, 321–22
  object-oriented approach and, 338–44
  templates and, 92, 321, 329
Reviews, project, 99–101
Risk management, 330, 590
Rule of Data Conservation, 116, 124, 175, 176, 188, 278, 455, 545, 590
Rules of thumb:
  for estimating length of analysis, 98–99
  for event-response processing, 431–37, 451
  for finding entities, 156–58, 353–54
  for finding relationships, 151–52, 159–60, 355–56
  for grouping processes, 389
  for naming events, 49–52
  for naming relationships, 152
  for repeating group, 231
  for size of context of study, 124
Rumbaugh, J., 96, 344, 579

S

Sales Department, 24–25, 30, 40–44, 58–59, 79–80, 342, 372, 387–91, 434, 439, 441, 466, 493
  Diagram 0 of, 369, 391, 397
  Diagram 3 of, 45, 388, 389, 400
  lower-level data flow diagrams of, 390, 391, 401
Sales policy, 40–43, 387, 464–66
Sales target, 16, 24–30, 40, 464–67
  in data dictionary, 424
Scope, project: *See* Project scope
Selection construct, 106, 225, 284

Sequence construct, 282, 325
Shlaer, S., 96, 344, 579, 580
Single processor, 96, 466
Specification, 108–10, 114, 115, 125, 174–75, 193–95, 204, 211, 279–82
  See also Mini specification
  completeness of, 75, 279, 367, 524–27
  of data storage and retrieval, 161–63, 291–94
  techniques, 278–94
  of technology, 307–13
Spiral development strategy, 99
Spot: *See* Commercial spot
Stevens, W., 96, 580
Stored data, 7, 20, 37, 93, 137, 142–43, 145–73, 214, 308, 390–92
  common usage of, 275–77, 390
  CRUD check and, 72, 525–27
  in data dictionary, 228–31
  in event-response model, 62, 73, 386, 443–45
  modeling, 115–16, 145–63, 255–59, 390–93
  private, 63
  specifying, 161–63, 291–94
Structure chart, 96
Structured analyst, 338
Structured design, 106
Structured language, 282–87, 447–48
  for author/book enquiry system, 291–94
  for data storage and retrieval, 291–92
  for Terry's Ski Tuning Service, 573
Structured programming, 106, 282
Subtype, 160–61, 591
  of RATING, 360–61, 471
Supertype, 160–61, 471, 591

The following interview originally appeared in Volume 3, Number 4 of The Dorset House Quarterly, *a free newsletter for readers of Dorset House books. Call (800) 342-6657 or (212) 620-4053 to request a subscription.*

What were the circumstances that led you to write *Complete Systems Analysis*?

We had clocked many years of systems analysis experience, most of it as freelance consultants. One thing that we noticed was that each time we left a project, our experience walked out the door with us. We tried several experiments at training in-house staff to replace us, but managers who were paying a lot for our services (it wasn't really that much) wanted us to do "real work" rather than induct their people. This book is our attempt to pass on real project experience. It's also a teaching book. We love to teach, and this is one way that we can do it.

How did Piccadilly Television become the subject of the book's main project?

We worked for a television company in London. It was a fascinating project, and we didn't want that experience just to pass into memory. So we wrote a book about it. The company gave us permission to do it, although we changed some aspects to protect any confidential information. The name Piccadilly is a little joke. The real company is called Central Television. Central is also the name of one of the London tube lines, so we named our case study after another line: Piccadilly. Various aspects of the tube crop up in the book, and some mysteries in

the book can be solved with a map of the London Underground.

What does *CSA* offer people who have already learned analysis techniques on the job?

Complete Systems Analysis reflects ideas that have developed over at least fifteen years. New ideas like event-response data and process models, integrated analysis, and viewpoints are presented, along with earlier practices like top-down functional decomposition and data flow analysis techniques. The book is written so that analysts with previous experience can add to it rather than having to relate to a completely new set of terminology.

How did you structure the book to adapt to the diversity of readers' systems analysis backgrounds?

The first section of the book presents one project, the Piccadilly Television case study. The reader has to work through it. We kept the project separate from the textbook section so that someone with a little experience could work straight through the project, without being diverted by the textbook. After the project is finished, the textbook, because it is separate, serves as a reference. We wanted to mimic the way people work on projects: They sit at their desks or, better still, the users' desks and build models of systems. If they get stuck, they refer to manuals, textbooks, other people, and so on, for help. The book follows this structure.

(continued)

609

(continued)

How did you come up with the idea of using ski trails as an organizing motif?

We were searching for a way to give people paths through complicated subject matter. Tim Lister, one of our partners in the Atlantic Systems Guild, pointed out that such a device already existed: ski trails. In the years that we have been skiing, we have spent many hours staring at signposts giving the names of the trails, and about the same amount of time trying to reconcile ski trail maps with the terrain. It somehow seemed natural, when we needed a device to guide people around this book, to use ski trails.

The motif appealed to us for another reason. Not every reader comes with the same level of experience, or has the same requirements from this book. Ski trails are graded to beginner, intermediate, and expert levels, so once again that seemed the perfect analogy. We have the three types of trails running through the book, and the reader may follow any one he or she wishes. We've also added a fourth path — a promenade for managers that covers the fundamentals without assigning any modeling work.

The skiing analogy led to one other thing: the ski patrol. In the book, the ski patrol arrives at the end of a workshop, discusses what may have gone wrong, and offers advice on how to proceed. We loved writing the ski patrol pieces. It was really tough (and embarrassing) to go back over all the mistakes we made in our careers, and offer advice to anyone about to make the same mistakes. It was teaching at its best to write a piece on how to recover from a mistake and get back on the trail again. This underpins our unshakable belief that learning anything, particularly systems analysis, requires one to experiment, make mistakes, and then — most importantly — understand why the mistake happened and how to avoid it in the future. We try to get our readers to make mistakes (there are a few deliberately set traps in the book) so that they understand the process better.

In what ways does *Complete Systems Analysis* add to the theory of systems analysis?

The word "complete" says it. We have avoided an exclusive treatment of small (but nevertheless interesting) parts of the subject in favor of demonstrating how systems analysis is really done. In other words, the book is not about a theoretical aspect of one facet of the subject; it involves the whole process of systems analysis.

Thank you, James and Suzanne!

Reviews

"a masterful job. . . . a thoroughly detailed case study."
—**Ed Yourdon**, *Guerrilla Programmer*

"the Robertsons' theory is heavily integrated with practical exercises. . . . you will appreciate the tremendous effort the Robs have put into making this . . . a true learning tool."
—**Warren Keuffel**, *Software Development*

"The authors make years of practical experience available to the readers, providing valuable guidance to the analyst. . . . This is one well-written book. It succeeds in making a difficult subject easily understandable."
—**Erik Hansen**, Kommunedata

TRAIL GUIDES

TRAIL GUIDES

◆ Most Difficult Trail

❄ Promenade Trail